Geometrical Theory of Analytic Functions

Geometrical Theory of Analytic Functions

Editor

Georgia Irina Oros

MDPI • Basel • Beijing • Wuhan • Barcelona • Belgrade • Manchester • Tokyo • Cluj • Tianjin

Editor
Georgia Irina Oros
University of Oradea
Romania

Editorial Office
MDPI
St. Alban-Anlage 66
4052 Basel, Switzerland

This is a reprint of articles from the Special Issue published online in the open access journal *Mathematics* (ISSN 2227-7390) (available at: https://www.mdpi.com/journal/mathematics/special_issues/Geometrical_theory_analytic_functions).

For citation purposes, cite each article independently as indicated on the article page online and as indicated below:

LastName, A.A.; LastName, B.B.; LastName, C.C. Article Title. *Journal Name* **Year**, *Volume Number*, Page Range.

ISBN 978-3-0365-2498-6 (Hbk)
ISBN 978-3-0365-2499-3 (PDF)

© 2022 by the authors. Articles in this book are Open Access and distributed under the Creative Commons Attribution (CC BY) license, which allows users to download, copy and build upon published articles, as long as the author and publisher are properly credited, which ensures maximum dissemination and a wider impact of our publications.

The book as a whole is distributed by MDPI under the terms and conditions of the Creative Commons license CC BY-NC-ND.

Contents

About the Editor vii

Preface to "Geometrical Theory of Analytic Functions" ix

Mugur Acu and Gheorghe Oros
Starlikeness Condition for a New Differential-Integral Operator
Reprinted from: *Mathematics* **2020**, *8*, 694, doi:10.3390/math8050694 1

Hiba Al-Janaby, Firs Ghanim and Maslina Darus
On The Third-Order Complex Differential Inequalities of ξ-Generalized-Hurwitz–Lerch
Zeta Functions
Reprinted from: *Mathematics* **2020**, *8*, 845, doi:10.3390/math8050845 11

Ibtisam Aldawish, Tariq Al-Hawary and B. A. Frasin
Subclasses of Bi-Univalent Functions Defined by Frasin Differential Operator
Reprinted from: *Mathematics* **2020**, *8*, 783, doi:10.3390/math8050783 33

Abdullah Alotaibi, Muhammad Arif, Mohammed A. Alghamdi and Shehzad Hussain
Starlikness Associated with Cosine Hyperbolic Function
Reprinted from: *Mathematics* **2020**, *8*, 1118, doi:10.3390/math8071118 45

José A. Antonino and Sanford S. Miller
Systems of Simultaneous Differential Inclusions Implying Function Containment
Reprinted from: *Mathematics* **2021**, *9*, 1252, doi:10.3390/math9111252 61

S. Melike Aydoğan and Zeliha Karahüseyin
Coefficient Estimates for Bi-Univalent Functions in Connection with Symmetric Conjugate
Points Related to Horadam Polynomial
Reprinted from: *Mathematics* **2020**, *8*, 1888, doi:10.3390/math8111888 71

Rizwan Salim Badar and Khalida Inayat Noor
q-Generalized Linear Operator on Bounded Functions of Complex Order
Reprinted from: *Mathematics* **2020**, *8*, 1149, doi:10.3390/math8071149 79

Ali Ebadian, Nak Eun Cho, Ebrahim Analouei Adegani and Sibel Yalçın
New Criteria for Meromorphic Starlikeness and Close-to-Convexity
Reprinted from: *Mathematics* **2020**, *8*, 847, doi:10.3390/math8050847 93

Sheza M. El-Deeb, Teodor Bulboacă and Bassant M. El-Matary
Maclaurin Coefficient Estimates of Bi-Univalent Functions Connected with the q-Derivative
Reprinted from: *Mathematics* **2020**, *8*, 418, doi:10.3390/math8030418 107

Firas Ghanim, Khalifa Al-Shaqsi, Maslina Darus and Hiba Fawzi Al-Janaby
Subordination Properties of Meromorphic Kummer Function Correlated with Hurwitz–Lerch
Zeta-Function
Reprinted from: *Mathematics* **2021**, *9*, 192, doi:10.3390/math9020192 121

Rabha W. Ibrahim, Rafida M. Elobaid and Suzan J. Obaiys
On the Connection Problem for Painlevé Differential Equation in View of Geometric
Function Theory
Reprinted from: *Mathematics* **2020**, *8*, 1198, doi:10.3390/math8071198 131

Rabha W. Ibrahim, Rafida M. Elobaid and Suzan J. Obaiys
Symmetric Conformable Fractional Derivative of Complex Variables
Reprinted from: *Mathematics* **2020**, *8*, 363, doi:10.3390/math8030363 . . . . **143**

Leah K. Mork, Keith Sullivan and Darin J. Ulness
Taming the Natural Boundary of Centered Polygonal Lacunary Functions—Restriction to the Symmetry Angle Space
Reprinted from: *Mathematics* **2020**, *8*, 568, doi:10.3390/math8040568 . . . . **157**

Georgia Irina Oros
Best Subordinant for Differential Superordinations of Harmonic Complex-Valued Functions
Reprinted from: *Mathematics* **2020**, *8*, 2041, doi:10.3390/math8112041 . . . . **175**

Ágnes Orsolya Páll-Szabó and Georgia Irina Oros
Coefficient Related Studies for New Classes of Bi-Univalent Functions
Reprinted from: *Mathematics* **2020**, *8*, 1110, doi:10.3390/math8071110 . . . . **183**

About the Editor

Georgia Irina Oros is associate Professor at Faculty of Informatics and Sciences, Department of Mathematics and Computer Science since 2013, she is teaching at University of Oradea since 2004. She obtained PhD in 2006 in Geometric Function Theory at Babeș-Bolyai University, Cluj-Napoca, Romania under the supervision of Prof. dr. Grigore Ștefan Sălăgean. Habilitation Thesis defended in 2018 at Babeș-Bolyai University, Cluj-Napoca, Romania. Over 100 her papers published in the field of Complex Analysis, Geometric Function Theory.

Preface to "Geometrical Theory of Analytic Functions"

Devoted to geometric function theory, this book brings together 15 research papers accepted for publication in the Special Issue for *Mathematics*, entitled "Geometrical Theory of Analytic Functions". Scholars studying complex-valued functions of one variable in all aspects starting with special classes of univalent functions or operator-related results using the theory of differential subordination and superordination have submitted their lastest findings.

I want to thank all the authors who have decided to submit their works and have contributed to the success of this Special Issue.

I also want to express my gratitude to Ms. Patty Hu, the Special Issue Managing Editor of Mathematics, for her continued support.

I believe the subjects of disscusion taken into consideration in the 15 papers published as part of this project are variate enough for researchers in geometric function theory and related topics to find something of interest. We hope for the success of this Special Issue to be validated by many citations of the published papers!

Georgia Irina Oros
Editor

Article

Starlikeness Condition for a New Differential-Integral Operator

Mugur Acu [1,*,†] **and Gheorghe Oros** [2,†]

[1] Department of Mathematics and Informatics, Lucian Blaga University of Sibiu, Str. Dr. I. Rațiu, No. 5-7, RO-550012 Sibiu, Romania
[2] Department of Mathematics, University of Oradea, Str. Universității, No.1, 410087 Oradea, Romania; gh_oros@yahoo.com
* Correspondence: mugur.acu@ulbsibiu.ro
† These authors contributed equally to this work.

Received: 28 March 2020; Accepted: 27 April 2020; Published: 2 May 2020

Abstract: A new differential-integral operator of the form $I^n f(z) = (1-\lambda) S^n f(z) + \lambda L^n f(z)$, $z \in U$, $f \in A$, $0 \leq \lambda \leq 1$, $n \in \mathbb{N}$ is introduced in this paper, where S^n is the Sălăgean differential operator and L^n is the Alexander integral operator. Using this operator, a new integral operator is defined as:
$F(z) = \left[\frac{\beta+\gamma}{z^\gamma} \int_0^z I^n f(z) \cdot t^{\beta+\gamma-2} dt\right]^{\frac{1}{\beta}}$, where $I^n f(z)$ is the differential-integral operator given above. Using a differential subordination, we prove that the integral operator $F(z)$ is starlike.

Keywords: differential subordination; analytic function; univalent function; convex function; starlike function; dominant; best dominant

1. Introduction and Preliminaries

The introduction and study of operators has been a topic that emerged at the very beginning of the theory of functions of a complex variable. The first operators were introduced during the first years of the twentieth century by mathematicians like J.W. Alexander, R. Libera, S. Bernardi, P. T. Mocanu and many more. The Alexander integral operator is such an example, defined by J. W. Alexander in 1915 [1]. This paper is cited in nearly 500 papers. The use of operators has facilitated the introduction of special classes of univalent functions and studying properties of the functions in those classes, such as convexity, starlikeness, coefficient estimates, and distortion properties. The Sălăgean differential operator was introduced in 1983 [2] and is cited by over 1300 papers. It has been used in obtaining new classes of functions and proving many interesting results related to them. The operator we introduce in this paper gives a new perspective in the theory related to operators by combining the integral Alexander operator and the differential Sălăgean operator. The results were obtained also using the means of the theory of differential subordinations introduced by Professors Miller and Mocanu in two papers in 1978 and 1980 and condensed in the monograph published by them in 2000 [3]. This theory has remarkable applications allowing easier proofs of already known results and facilitating the emergence of new ones. The idea of combining integral and differential operators is illustrated in the very recent paper [4] where a differential-integral operator was defined and using the method of the subordination chains, differential subordinations in their special Briot-Bouquet form were studied obtaining their best dominant and, as a consequence, criteria containing sufficient conditions for univalence were formulated. Similar work containing subordination results related to a class of univalent functions obtained by the use of an operator introduced by using a differential operator and an integral one can be seen in [5].

We use the well-known notations:

- $\mathcal{H}(U)$ is the class of functions analytic in the unit disc $U = \{z \in \mathbb{C} : |z| < 1\}$,
- For $a \in \mathbb{C}$, $n \in \mathbb{N}$, $\mathcal{H}_{[a,n]} = \{f \in \mathcal{H}(U) : f(z) = a + b_n z^n + \ldots, z \in U\}$,
- $\mathcal{A}_n = \{f \in \mathcal{H}(U) : f(z) = z + a_{n+1} z^{n+1} + \ldots, z \in U\}$, with $\mathcal{A}_1 = \mathcal{A}$,
- $S^* = \left\{ f \in \mathcal{A}, \operatorname{Re} \dfrac{zf'(z)}{f(z)} > 0, z \in U \right\}$ is the class of starlike functions in U,
- $K = \left\{ f \in \mathcal{A}, \operatorname{Re} \dfrac{zf''(z)}{f'(z)} + 1 > 0, z \in U \right\}$ is the class of normalized convex functions in U.

The definitions of subordination, solution of the differential subordination and best dominant of the solutions of the differential subordination are recalled next as they can be found in the monograph published by Professors Miller and Mocanu in 2000 [3], which gives the core of the theory of differential subordination:

If f and g are analytic in U, then we say that f is subordinate to g, written $f \prec g$ or $f(z) \prec g(z)$, if there is a function w analytic in U with $w(0) = 0$, $|w(z)| < 1$ for all $z \in U$ such that $f(z) = g(w(z))$, for $z \in U$. If g is univalent, then $f \prec g$ if and only if $f(0) = g(0)$ and $f(U) \subset g(U)$.

Let $\psi : \mathbb{C}^3 \times U \to \mathbb{C}$ and h be a univalent function in U. If p is a analytic function in U which satisfies the following (second-order) differential subordination:

$$\psi(p(z), zp'(z), z^2 p''(z); z) \prec h(z),$$

then p is called a solution of the differential subordination. The univalent function q is called a dominant of the solutions of the differential subordination or more simply a dominant, if $p \prec q$, for all p satisfying the differential subordination. A dominant $\tilde{q}$ that satisfies $\tilde{q} \prec q$ for every dominant q is said to be the best dominant.

A well-known lemma from the theory of differential subordinations that is used in proving the new results is shown as follows:

Lemma 1. *[3] Let g be univalent in U and let θ and ϕ be analytic in a domain D containing $g(U)$, with $\phi(w) \neq 0$, when $w \in g(U)$. Set*

$$Q(z) = zq'(z) \cdot \phi[q(z)], \quad h(z) = \theta[q(z)] + Q(z),$$

and suppose that

(i) Q is starlike and

(ii) $\operatorname{Re} \dfrac{zh'(z)}{Q(z)} = \operatorname{Re} \left[\dfrac{\theta'[q(z)]}{\phi[q(z)]} + \dfrac{zQ'(z)}{Q(z)} \right] > 0$, $z \in U$.

If p is analytic in U, with $p(0) = q(0)$, $p'(0) = \ldots = p^{(n-1)}(0) = 0$, $p(U) \subset D$ and

$$\theta[p(z)] + zp'(z)\phi[p(z)] \prec \theta[q(z)] + zq'(z) \cdot \phi[q(z)] = h(z),$$

then $p(z) \prec q(z)$, and q is the best dominant.

In order to define the new differential-integral operator, we need the following definitions:

Definition 1. [2] For $f \in A$, $n \in \mathbb{N} = \mathbb{N}^* \cup \{0\}$, let S^n be the differential operator given by $S^n : A \to A$ with

$$S^0 f(z) = f(z)$$
$$\vdots$$
$$S^{n+1} f(z) = z[S^n f(z)]', \quad z \in U.$$

Remark 1. If $f \in A$, $f(z) = z + \sum_{j=2}^{\infty} a_j z^j$, then

$$S^n f(z) = z + \sum_{j=2}^{\infty} j^n \cdot a_j z^j. \tag{1}$$

Definition 2. [6] For $f \in A$, $n \in \mathbb{N} = \mathbb{N}^* \cup \{0\}$, let L^n be the integral operator given by $L^n : A \to A$ with

$$L^0 f(z) = f(z)$$
$$L^1 f(z) = \int_0^z \frac{L^0 f(t)}{t} dt$$
$$\vdots$$
$$L^n f(z) = \int_0^z \frac{L^{n-1} f(t)}{t} dt, \quad z \in U.$$

Remark 2. (a) For $n = 1$, $L^1 f(z) = \int_0^z \frac{f(t)}{t} dt$ becomes Alexander integral operator [1].

(b) For $f \in A$, $f(z) = z + \sum_{j=2}^{\infty} a_j z^j$, we obtain:

$$L^n f(z) = z + \sum_{j=2}^{\infty} \frac{1}{j^n} \cdot a_j z^j. \tag{2}$$

2. Main Results

Using Definition 1 and Definition 2, we introduce a new operator, as follows:

Definition 3. Let $0 \leq \lambda \leq 1$, $n \in \mathbb{N} = \mathbb{N}^* \cup \{0\}$. Denote by I^n the differential-integral operator $I^n : A \to A$ given by

$$I^n f(z) = (1 - \lambda) S^n f(z) + \lambda L^n f(z), \quad z \in U, \tag{3}$$

where S^n is Sălăgean differential operator, and L^n is Alexander integral operator.

Remark 3. (a) For $\lambda = 0$, $I^n f(z) = S^n f(z)$, the differential-integral operator is equivalent to Sălăgean differential operator.

(b) For $\lambda = 1$, $I^n f(z) = L^n f(z)$, the differential-integral operator becomes Alexander integral operator.

(c) For $f \in A$, $f(z) = z + \sum_{j=2}^{\infty} a_j z^j$ we obtain

$$I^n f(z) = z + \sum_{j=2}^{\infty} \left[(1-\lambda)j^n + \lambda \cdot \frac{1}{j^n} \right] a_j z^j, \ z \in U. \tag{4}$$

Using the differential-integral operator introduced in Definition 3, we define a new integral operator, which can be seen as generalization of some well-known integral operators.

Definition 4. *Let $\gamma \geq 0$, $0 < \beta \leq 1$, $n \in \mathbb{N} = \mathbb{N}^* \cup \{0\}$, and $f \in A$, $I^n f \in A$, where I^n is given by Equation (3). The integral operator $F : A \to \mathcal{H}_{[a,n]}$ is defined as:*

$$F(z) = \left[\frac{\beta + \gamma}{z^\gamma} \int_0^z I^n f(t) \cdot t^{\beta + \gamma - 2} dt \right]^{\frac{1}{\beta}}. \tag{5}$$

Remark 4. *(a) For $n = 0$, $\beta = 1$, $\gamma > 0$, we have*

$$F(z) = \frac{1+\gamma}{z^\gamma} \int_0^z f(t) \cdot t^{\gamma - 1} dt,$$

which is the Bernardi integral operator [7].
(b) For $n = 0$, $\beta = 1$, $\gamma = 1$, we have

$$F(z) = \frac{2}{z} \int_0^z f(t) dt,$$

which is the Libera integral operator [8].
(c) For $n = 0$, $\beta = 1$, $\gamma = 0$, we have

$$F(z) = \int_0^z \frac{f(t)}{t} dt,$$

which is Alexander integral operator [1].
(d) For $\beta = 1$, $n \in \mathbb{N}^*$, $\gamma > 0$, we have

$$F(z) = \frac{1+\gamma}{z^\gamma} \int_0^z I^n f(t) \cdot t^{\gamma - 1} dt$$

which was studied in [6].
(e) For $n = 0$, we have

$$F(z) = \left[\frac{\beta + \gamma}{z^\gamma} \int_0^z f(t) \cdot t^{\beta + \gamma - 2} dt \right]^{\frac{1}{\beta}},$$

$\beta > 0$, $\beta + \gamma > 0$ and $\beta \geq 2\gamma(1 - \beta)$ *studied in [9] where the authors have proved that $F \in S^*$.*

Using a differential subordination, we prove that the operator given by Equation (5) is starlike.

Theorem 1. *Let $0 < \beta \leq 1$, $\gamma \geq 0$, $n \in \mathbb{N} = \mathbb{N}^* \cup \{0\}$, and let*

$$h(z) = \frac{1+z}{1-z} + \frac{2z}{(1-z)[1+\beta+\gamma+(1-\beta-\gamma)z]}, \ z \in U.$$

If $f \in A$ and
$$\frac{1}{\beta} \cdot \frac{z(I^n f(z))'}{I^n f(z)} + \frac{\beta-1}{\beta} \prec h(z), \quad z \in U, \tag{6}$$
then
$$F(z) = \left[\frac{\beta+\gamma}{z^\gamma} \int_0^z I^n f(z) \cdot t^{\beta+\gamma-2} dt\right]^{\frac{1}{\beta}}$$
is starlike, i.e., $F \in S^*$, where $I^n f$ is given by Equation (3).

Proof. From Equation (5) we have
$$F^\beta(z) = \frac{\beta+\gamma}{z^\gamma} \int_0^z I^n f(z) \cdot t^{\beta+\gamma-2} dt,$$
and
$$z^\gamma F^\beta(z) = (\beta+\gamma) \int_0^z I^n f(z) \cdot t^{\beta+\gamma-2} dt. \tag{7}$$
Differentiating Equation (7), we obtain
$$F^\beta(z) \left[\gamma + \beta z \cdot \frac{F'(z)}{F(z)}\right] = (\beta+\gamma) I^n f(z) \cdot z^{\beta-1}. \tag{8}$$
We let
$$p(z) = z \cdot \frac{F'(z)}{F(z)}, \quad z \in U. \tag{9}$$
Using Equations (9) in (8), we have
$$F^\beta(z)[\gamma + \beta p(z)] = (\beta+\gamma) I^n f(z) \cdot z^{\beta-1}. \tag{10}$$
Differentiating (10), we get
$$\beta \cdot \frac{zF'(z)}{F(z)} + \beta \cdot \frac{zp'(z)}{\gamma + \beta p(z)} = \frac{z(I^n f(z))'}{I^n f(z)} + \beta - 1. \tag{11}$$
Using (9) in (11), we have
$$p(z) + \frac{zp'(z)}{\gamma + \beta p(z)} = \frac{1}{\beta} \cdot \frac{z(I^n f(z))'}{I^n f(z)} + \frac{\beta-1}{\beta}. \tag{12}$$
Using Relation (12), the differential subordination of Equation (6) becomes:
$$p(z) + \frac{zp'(z)}{p(z)} \prec h(z) = \frac{1+z}{1-z} + \frac{2z}{(1-z)[1+\beta+\gamma+(1-\beta-\gamma)z]}. \tag{13}$$
In order to prove the theorem, we shall use Lemma 1.
If we let $\theta : D \subset \mathbb{C} \to \mathbb{C}$ and $\phi : D \subset \mathbb{C} \to \mathbb{C}$ be analytic,
$$\theta(w) = q, \; \phi(w) = \frac{1}{w+\beta+\gamma} \text{ in a domain } D.$$

For $w = q(z) = \dfrac{1+z}{1-z}$, we obtain

$$\phi[q(z)] = \dfrac{1}{\dfrac{1+z}{1-z} + \beta + \gamma} = \dfrac{1-z}{1+\beta+\gamma+z(1-\beta-\gamma)} \tag{14}$$

$$Q(z) = z \cdot q'(z) \cdot \phi[q(z)] = \dfrac{2z}{(1-z)[1+\beta+\gamma+(1-\beta-\gamma)z]} \tag{15}$$

and

$$h(z) = \theta[q(z)] + Q(z) = \dfrac{1+z}{1-z} + \dfrac{2z}{(1-z)[1+\beta+\gamma+(1-\beta-\gamma)z]}. \tag{16}$$

Next we show that conditions in Lemma 1 are satisfied. We prove that the function Q is starlike. Differentiating Equation (15), we have

$$\dfrac{zQ'(z)}{Q(z)} = \dfrac{z}{1-z} + \dfrac{1+\beta+\gamma}{1+\beta+\gamma+(1-\beta-\gamma)z}.$$

We take

$$\operatorname{Re} \dfrac{zQ'(z)}{Q(z)} = \operatorname{Re} \dfrac{z}{1-z} + (1+\beta+\gamma)\operatorname{Re} \dfrac{1}{1+\beta+\gamma+(1-\beta-\gamma)z}$$

$$= -\dfrac{1}{2} + \dfrac{(1+\beta+\gamma)^2 + (1+\beta+\gamma)(1-\beta-\gamma)\cos\alpha}{[1+\beta+\gamma+(1-\beta-\gamma)\cos\alpha]^2 + (1-\beta-\gamma)^2 \sin^2\alpha}$$

$$= \dfrac{2(\beta+\gamma)}{[1+\beta+\gamma+(1-\beta-\gamma)\cos\alpha]^2 + (1-\beta-\gamma)^2 \sin^2\alpha} > 0.$$

We have shown that $\operatorname{Re} \dfrac{zQ'(z)}{Q(z)} > 0$, $z \in U$, i.e., $Q \in S^*$, hence (i) from Lemma 1 is satisfied.

We evaluate now:

$$\operatorname{Re} \phi[q(z)] = \operatorname{Re} \dfrac{1-z}{1+\beta+\gamma+(1-\beta-\gamma)z}$$

$$= \dfrac{2\beta + 2\gamma(1-\cos\alpha)}{[1+\beta+\gamma+(1-\beta-\gamma)\cos\alpha]^2 + (1-\beta-\gamma)^2 \sin^2\alpha} > 0, \ 0 < \beta \leq 1, \ \gamma \geq 0.$$

Since Q is starlike and $\operatorname{Re} \phi[q(z)] > 0$, we have

$$\operatorname{Re} \dfrac{zh'(z)}{Q(z)} > 0, \ z \in U.$$

Next we prove that $p(0) = q(0)$, $p \in \mathcal{H}_{[1,1]}$ and p is analytic in U, where

$$p(z) = \dfrac{zF'(z)}{F(z)}, \ z \in U.$$

From Equation (4), we have

$$I^n f(z) = z + \sum_{j=2}^{\infty} \left[(1-\lambda)j^n + \lambda \cdot \dfrac{1}{j^n}\right] a_j z^j = z + \sum_{j=2}^{\infty} b_j z^j,$$

where
$$b_j = \left[(1-\lambda)j^n + \lambda \cdot \frac{1}{j^n}\right] a_j.$$

From Equation (5), we can write
$$F(z) = \left[\frac{\beta+\gamma}{z^\gamma} \int_0^z \left(t + \sum_{j=2}^\infty b_j z^j\right) t^{\beta+\gamma-2} dt\right]^{\frac{1}{\beta}}$$
$$= \left[z^\beta + \sum_{j=2}^\infty c_j z^{j+\beta-1}\right]^{\frac{1}{\beta}}$$

and we obtain
$$F^\beta(z) = z^\beta + \sum_{j=2}^\infty c_j z^{j+\beta-1}, \ z \in U. \tag{17}$$

Differentiating Equation (17), we have
$$\beta F^{\beta-1}(z) \cdot F'(z) = \beta z^{\beta-1} + \sum_{j=2}^\infty c_j(j+\beta-1) \cdot z^{j+\beta-2}.$$

Further, we deduce
$$p(z) = \frac{zF'(z)}{F(z)} = \frac{z^\beta + \sum_{j=2}^\infty d_j z^{j+\beta-1}}{z^\beta + \sum_{j=2}^\infty c_j z^{j+\beta-1}} = 1 + p_1 z + p_2 z^2 + \ldots. \tag{18}$$

For $z = 0$, we obtain $p(0) = 1$ and $p \in \mathcal{H}_{[1,1]}$, hence F it is analytic in U.
Since $q(z) = \frac{1+z}{1-z}$, we have $q(0) = 1$, $p(0) = q(0) = 1$ and
$$\theta[p(z)] + zp'(z) \cdot \phi[p(z)] \prec \theta[q(z)] + zq'(z) \cdot \phi[q(z)] = h(z). \tag{19}$$

We have proved that we can use Lemma 1. By applying it, we have $p(z) \prec q(z)$, i.e.,
$$\frac{zF'(z)}{F(z)} \prec q(z) = \frac{1+z}{1-z}, \ z \in U. \tag{20}$$

Since $q(z) = \frac{1+z}{1-z}$ is a convex function and
$$\operatorname{Re} \frac{1+z}{1-z} > 0, \ z \in U,$$

the differential subordination in Equation (20) implies
$$\operatorname{Re} \frac{zF'(z)}{F(z)} > \operatorname{Re} q(z) > 0, \text{ hence } F \in S^*.$$

□

Example 1. Let $\lambda = \dfrac{1}{2}$, $n = 1$, $\beta = \dfrac{1}{2}$, $\gamma = 2$,

$$h(z) = \frac{1+z}{1-z} + \frac{4z}{(1-z)(7-3z)},\ f(z) = z + \frac{1}{4}z^2,\ z \in U,$$

$$S^1 f(z) = z f'(z) = z + \frac{1}{2}z^2,$$

$$L^1 f(z) = \int_0^z \frac{t + \frac{1}{4}t^2}{t}\,dt = z + \frac{1}{8}z^2,$$

$$I^1 f(z) = \frac{1}{2}S^1 f(z) + \frac{1}{2}L^1 f(z) = z + \frac{5}{16}z^2,$$

$$F(z) = \left(z^{\frac{1}{2}} + \frac{25}{112}z^{\frac{3}{2}}\right)^2 = z + \frac{25}{56}z^2 + \frac{625}{12544}z^3,$$

and

$$\frac{zF'(z)}{F(z)} = \frac{1 + \frac{25}{28}z + \frac{1875}{12544}z^2}{1 + \frac{25}{56}z + \frac{625}{12544}z^2},\ q(z) = \frac{1+z}{1-z}$$

$$\frac{1}{\beta} \cdot \frac{z(I^1 f(z))'}{I^1 f(z)} + \frac{\beta - 1}{\beta} = \frac{4(8 + 5z)}{16 + 5z} - 1.$$

From Theorem 1, we have:

If $f \in A$, and

$$\frac{4(8 + 5z)}{16 + 5z} - 1 \prec \frac{1+z}{1-z} + \frac{4z}{(1-z)(7-3z)},\ z \in U$$

then

$$p(z) = \frac{zF'(z)}{F(z)} = \frac{1 + \frac{25}{28}z + \frac{1875}{12544}z^2}{1 + \frac{25}{56}z + \frac{625}{12544}z^2} \prec \frac{1+z}{1-z},$$

meaning that $F(z) = z + \dfrac{25}{56}z^2 + \dfrac{625}{12544}z^3$ is a starlike function.

3. Conclusions

A new differential-integral operator is introduced proving that this operator is starlike. An example is given to show how the result can be applied in finding such operators. As it is the case for most operators, special classes of univalent functions could be introduced using it and this is subject to further studies. Another problem that can be studied is related to the parameters β and γ used in the definition of the operator. In this paper they are positive but the case of β and γ being complex numbers could be subject of further investigation. Starlikeness of certain order $0 \leq \alpha < 1$ can also be further studied both for the case of β and γ positive and for β and γ complex numbers.

Author Contributions: Conceptualization, M.A. and G.O.; methodology, M.A. and G.O.; software, M.A.; validation, M.A. and G.O.; formal analysis, M.A. and G.O.; investigation, M.A. and G.O.; resources, M.A. and G.O.; data curation, M.A. and G.O.; writing–original draft preparation, M.A. and G.O.; writing–review and editing, M.A.; visualization, M.A. and G.O.; supervision, G.O.; project administration, M.A. All authors have read and agreed to the published version of the manuscript.

Funding: This research received no external funding.

Conflicts of Interest: The authors declare no conflict of interest.

References

1. Alexander, J.W. Functions which map the interior of the unit circle upon simple regions. *Ann. Math.* **1915**, *17*, 12–22. [CrossRef]
2. Sălăgean, G.S. Subclass of univalent functions. In *Lectures Notes in Math. 1013*; Springer: Berlin, Germany, 1983; pp. 362–372.
3. Mocanu, P.T.; Miller, S.S. Differential Subordinations. In *Theory and Applications*; Marcel Dekker Inc.: New York, NY, USA; Basel, Switzerland, 2000.
4. Oros, G.I.; Alb-Lupaş, A. Sufficient conditions for univalence obtained by using Briot-Bouquet differential subordination. *Math. Stat.* **2020**, *8*, 126–136. [CrossRef]
5. Páll-Szabó, Á.O. On a class of univalent functions defined by Sălăgean integro-differential operator. *Miskolc Math. Notes* **2018**, *19*, 1095–1106. [CrossRef]
6. Oros, G.I.; Oros, G.; Diaconu, R. Differential subordination obtained with some new integral operator. *J. Comput. Appl.* **2015**, *19*, 904–910.
7. Bernardi, S.D. Convex and starlike univalent functions. *Trans. Am. Math. Soc.* **1969**, *135*, 429–446. [CrossRef]
8. Libera, R.J. Some classes of regular univalent functions. *Proc. Am. Math. Soc.* **1965**, *16*, 755–758. [CrossRef]
9. Mocanu, P.T. Starlikeness of certain integral operators. *Mathematica* **1994**, *36*, 179–184.

© 2020 by the authors. Licensee MDPI, Basel, Switzerland. This article is an open access article distributed under the terms and conditions of the Creative Commons Attribution (CC BY) license (http://creativecommons.org/licenses/by/4.0/).

Article

On The Third-Order Complex Differential Inequalities of ζ-Generalized-Hurwitz–Lerch Zeta Functions

Hiba Al-Janaby [1], Firas Ghanim [2] and Maslina Darus [3],*

[1] Department of Mathematics, College of Science, University of Baghdad, Baghdad 10071, Iraq; fawzihiba@yahoo.com
[2] Department of Mathematics, College of Science, University of Sharjah, Sharjah, UAE; fgahmed@sharjah.ac.ae
[3] Department of Mathematical Sciences, Faculty of Science and Technology, Universiti Kebangsaan Malaysia, Bangi 43600, Selangor, Malaysia
* Correspondence: maslina@ukm.edu.my

Received: 31 March 2020; Accepted: 2 May 2020; Published: 23 May 2020

Abstract: In the z- domain, differential subordination is a complex technique of geometric function theory based on the idea of differential inequality. It has formulas in terms of the first, second and third derivatives. In this study, we introduce some applications of the third-order differential subordination for a newly defined linear operator that includes ζ-Generalized-Hurwitz–Lerch Zeta functions (GHLZF). These outcomes are derived by investigating the appropriate classes of admissible functions.

Keywords: holomorphic function; univalent function; *p*-valent function; convolution product; ζ-Generalized Hurwitz–Lerch Zeta function; differential subordination; admissible functions

MSC: 30C45; 33C10; 30C80

---

1. Introduction and Terminology

Complex Function Theory (CFT) is a mathematical branch dating back to the 18th century. It investigates the functions of complex numbers. This branch has attracted the concern of several researchers. Among the remarkable names are Euler, Gauss, Riemann, Cauchy and others. It has numerous implementations in diverse fields of mathematics and science. These functions have many interesting properties that are not owned by real-valued functions. For instance, infinitely differentiable functions, holomorphic functions, every holomorphic function in the open unit disk can be represented as a Taylor series, conformal functions (that is, they preserve angles when $f'(z) \neq 0$), line integrals, and all types of handy formulas. The considerable area in CFT is the Geometric Function Theory (GFT). The study of GFT includes investigating the interaction between the analytical properties of the complex holomorphic function and the geometrical properties of the image domain. Riemann [1] in 1851 introduced the first major result in GFT named the Riemann Mapping Theorem. Later, in 1907, Koebe [2] was a prominent scientist who studied the univalent functions in the open unit disk. Thereafter, in 1912, Koebe [3] presented a modified version of the Riemann's mapping theorem by utilized univalent functions. The theory tends towards the principle of "univalent" and "holomorphic", Riemann's mapping theorem plays a significant role in the collection of both principles. This synthesis interprets the formula of a domain where the complex functions being defined, for details see [1,4].

On the other hand, differential inequality theory (inequalities including derivatives of functions) impacted the development of GFT due to it giving much information regarding the behavior of the holomorphic function. Further, there are many differential implications in which characterization of a holomorphic function is settled by a differential condition. For instance, the Noshiro–Warschawski theorem states that for a holomorphic function in the unit disk, $\Re(f'(z)) > 0$ implies that f is univalent function in the unit disk. Most of the known differential implications dealt with real-value inequalities that involved the absolute value, the imaginary part, or the real part of a complex function [4].

The principle of subordination is central to the theory of differential subordination of complex-valued function which is the generalizing the formula of differential inequality of real-valued function. Its origins back to Lindelöf [4] in 1909, though Littlewood ([5,6]) and Rogosinski ([7,8]) posed the term and examined the basic outcomes regarding subordination. This principle, as an advantageous tool, displays its importance to unify the presentation of several geometric classes in addition to achieving sandwich-type outcomes.

The methods of differential subordination are employed to study upper bounds for holomorphic functions in the unit disk. This technique inspired numerous researchers to work in GFT. The implementations and extensions of differential subordination theory have been developed in this theme and diverse other fields, such as functions of several complex variables, integral operator theory, meromorphic function theory, harmonic functions theory, differential equations and partial differential equations. Many papers handled the first-order and second-order differential subordination methods,but few articles handled the third-order differential subordination method. In 1935, Goluzin [9] studied the first significant outcome that includes the first-order differential subordination. Afterwards, Suffridge [10] in 1970 and Robinson [11] in 1947 discussed further successive investigations into first-order differential subordination. Later, in 1981, Miller and Mocanu [12] provided a systematic study of the theory of differential subordination. In 1985 [13] and 1987 [14], they evolved and studied several interesting outcomes on this theory. Next, numerous important studies were presented by Miller and Mocanu ([15–17]). In 1992, Ponnusamy and Juneja [18] considered the third-order inequalities and subordination. After that, in 2000, Miller and Mocano in their monograph [19] provided a marvelous and extensive discussion on this theory with numerous implementations.

In 2011, Antonino and Miller [20] investigated and extended the second-order differential subordination to the third-order case. Several authors provided fruitful implementations in the same direction of study. For their contributions, Tang et al. [21] considered some third-order differential subordination outcomes for meromorphically p-valent functions associated with the certain linear operator. At the same time, Tang and Deniz [22] studied a similar problem for holomorphic functions, involving the generalized Bessel functions. In 2015, Farzana et al. [23] introduced several third-order differential subordination outcomes for holomorphic functions associated with the fractional derivative operator. Related to this period, Tang et al. [24] used third-order differential subordination methods of holomorphic functions associated with generalized Bessel functions to yield sandwich-type outcomes containing this operator. In the same year, Ibrahim et al. [25] established some third-order differential subordination outcomes for holomorphic functions associated with a fractional integral operator (Carlson–Shaffer operator type). Subsequently, the problems of the third-order differential subordination were studied by El-Ashwah and Hassan [26], El-Ashwah and Hassan [27], Attiya et al. ([28,29]), Srivastava et al. [30] and Gochhayat and Prajapati [31]. Many of the studies have not yet been investigated utilizing third-order differential subordination technique. In this investigation, we impose a new generalized Noor-type linear integral operator $\mathcal{M}_p^\ell \vartheta(z)$ on the class $\mathcal{A}_p$ of p-valent functions by utilizing ξ-Generalized Hurwitz–Lerch Zeta functions (GHLZF). Some outcomes concerning an application of the third-order differential subordination for multivalent functions including operator $\mathcal{M}_p^\ell \vartheta(z)$ are studied.

Denote by $\mathbb{D} = \{z \in \mathbb{C} : |z| < 1\}$ the open unit disc in the complex plane $\mathbb{C}$, and $\mathcal{H}(\mathbb{D})$ the class of holomorphic functions in $\mathbb{D}$. For $\alpha \in \mathbb{C}$, $\jmath \in \mathbb{N} = \{1, 2, 3, ...\}$, let

$$\mathcal{H}[\alpha, \jmath] = \{\vartheta \in \mathcal{H}(\mathbb{D}) : \vartheta(z) = \alpha + \alpha_\jmath z^\jmath + \alpha_{\jmath+1} z^{\jmath+1} + ...\}, \tag{1}$$

and suppose that $\mathcal{H}_0 \equiv \mathcal{H}[0,1]$ and $\mathcal{H}_1 \equiv \mathcal{H}[1,1]$. Let $\mathcal{A}$ denote the class of all holomorphic functions ϑ in $\mathbb{D}$, normalized by the conditions $\vartheta(0) = \vartheta'(0) - 1 = 0$, and of the formula

$$\vartheta(z) = z + \sum_{j=2}^{\infty} \alpha_j z^j, \quad (z \in \mathbb{D}). \tag{2}$$

The subclass of $\mathcal{A}$ involving holomorphic univalent function is denoted by $\mathcal{S}$, [1]. In [4] the concept of subordination between holomorphic functions given as: for two functions ϑ_1 and ϑ_2, holomorphic in $\mathbb{D}$, the function ϑ_1 is said to be subordinate to ϑ_2, or ϑ_2 superordinate to ϑ_1 in $\mathbb{D}$, written $\vartheta_1 \prec \vartheta_2$, if there is a holomorphic function $\hbar$ in $\mathbb{D}$ with $\hbar(0) = 0$ and $|\hbar(z)| < 1$ for all $z \in \mathbb{D}$, such that $\vartheta_1(z) = \vartheta_2(\hbar(z))$. In particular, if the function ϑ_2 is univalent in $\mathbb{D}$, then the following characterization for subordination is gained as:

$$\vartheta_1 \prec \vartheta_2 \text{ if and only if } \vartheta_1(0) = \vartheta_2(0) \text{ and } \vartheta_1(\mathbb{D}) \subset \vartheta_2(\mathbb{D}).$$

The natural generalization of holomorphic univalent function is a p-valent (multivalent) function, that is, if for each w, the equation $\vartheta(z) = w$ has at most p roots in a domain $\mathfrak{D} \subset \mathbb{C}$, and if there is w_0 such that the equation $\vartheta(z) = w_0$ has exactly p roots in a Domain $\mathfrak{D}$. Let $\mathcal{A}_p$ ($p \in \mathbb{N} = \{1,2,3,...\}$) denote the class involves all p-valent functions in $\mathbb{D}$ of the form

$$\vartheta(z) = z^p + \sum_{j=p+1}^{\infty} \alpha_j z^j, \quad (z \in \mathbb{D}). \tag{3}$$

If ϑ is the p-valent function with $p = 1$, then ϑ is the holomorphic univalent function, [4].

As one of the most remarkable tools, namely Hadamard (convolution) product, utilizes to formulate assorted operators: differential, integral and convolution operators. The term "Hadamard product" is attributed to Hadamard in 1899 [1] and defined as: for two functions $\vartheta_\ell \in \mathcal{A}$ of the form $\vartheta_\ell(z) = z + \sum_{j=2}^{\infty} \alpha_{j,\ell} z^j$, $\ell = 1,2$, their convolution, $\vartheta_1 * \vartheta_2$, is given by

$$(\vartheta_1 * \vartheta_2)(z) = z + \sum_{j=2}^{\infty} \alpha_{j,1} \alpha_{j,2} z^j, \quad (z \in \mathbb{D}). \tag{4}$$

More generally, the convolution product of two functions $\vartheta_\ell \in \mathcal{A}_p$ of the formula $\vartheta_\ell(z) = z^p + \sum_{j=p+1}^{\infty} \alpha_{j,\ell} z^j$, $\ell = 1,2$, $p \in \mathbb{N}$, is the function $\vartheta_1 * \vartheta_2$ given by

$$(\vartheta_1 * \vartheta_2)(z) = z^p + \sum_{j=p+1}^{\infty} \alpha_{j,1} \alpha_{j,2} z^j, \quad (z \in \mathbb{D}). \tag{5}$$

In 1915, Alexander [32] was the first to introduce a linear integral operator which drafted in terms of the convolution, namely "Alexander operator" as follows: let $\vartheta \in \mathcal{A}$ and $I_A : \mathcal{A} \to \mathcal{A}$ be defined as

$$I_A \vartheta(z) = \int_0^z \frac{\vartheta(t)}{t} dt = -\log(1-z) * \vartheta(z)$$
$$= z + \sum_{j=2}^{\infty} \frac{\alpha_j}{j} z^j. \tag{6}$$

Later on, in 1965, Libera [33] given another linear integral operator so-called "Libera operator" $I_L : \mathcal{A} \to \mathcal{A}$ by the formula

$$I_L \vartheta(z) = \frac{2}{z} \int_0^z \vartheta(t)\, dt = \frac{2\,[z + \log(1-z)]}{z} * \vartheta(z)$$

$$= z + \sum_{j=2}^{\infty} \frac{2}{j+1}\, a_j\, z^j. \tag{7}$$

In 1969, Bernardi [34] imposed a more general linear integral operator $I_\varepsilon : \mathcal{A} \to \mathcal{A}$, for $\vartheta \in \mathcal{A}$ and $\varepsilon > -1$, as

$$I_\varepsilon \vartheta(z) = \frac{1+\varepsilon}{z^\varepsilon} \int_0^z \vartheta(t)\, t^\varepsilon\, dt = z + \sum_{j=2}^{\infty} \left(\frac{\varepsilon+1}{\varepsilon+j}\right) z^j * \vartheta(z)$$

$$= z + \sum_{j=2}^{\infty} \left(\frac{\varepsilon+1}{\varepsilon+j}\right) a_j\, z^j. \tag{8}$$

The operator I_ε is called the generalized Bernardi–Libera–Livingston integral operator. For $\varepsilon = 0$, the operator I_ε reduces to the Alexander operator I_A given by Equation (6) and for $\varepsilon = 1$, it reduces to the Libera operator I_L defined by Equation (7).

Utilizing the convolution technique, in 1975, Ruscheweyh [35] proposed a linear operator as: let $\vartheta \in \mathcal{A}$, $\wp > -1$ and $D^\wp : \mathcal{A} \to \mathcal{A}$ be defined by

$$D^\wp \vartheta(z) = \frac{z}{(1-z)^{\wp+1}} * \vartheta(z) = z + \sum_{j=2}^{\infty} \frac{\Gamma(\wp+j)}{\Gamma(j)\,\Gamma(\wp+1)}\, a_j\, z^j. \tag{9}$$

For $\wp = \varpi \in \mathbb{N}_0 = \mathbb{N} \cup \{0\}$, yields

$$D^\varpi \vartheta(z) = \frac{z(z^{\varpi-1}\vartheta(z))^\varpi}{\varpi!}. \tag{10}$$

Further, $D^0 \vartheta(z) = \vartheta(z)$ and $D'\vartheta(z) = z\vartheta'(z)$, $z \in \mathbb{D}$. The operator D^ϖ is called the Ruscheweyh derivative of ϖth order of ϑ.

Corresponding to the Ruscheweyh operator D^ϖ, $\varpi \in \mathbb{N}_0$ given by Equation (10), in 1999, Noor [36] considered the following linear operator: let $\vartheta \in \mathcal{A}$, $\varpi \in \mathbb{N}_0$ and $I_\varpi : \mathcal{A} \to \mathcal{A}$ be defined as

$$I_\varpi \vartheta(z) = \vartheta_\varpi^{(-1)}(z) * \vartheta(z) = \left[\frac{z}{(1-z)^{\varpi+1}}\right]^{-1} * \vartheta(z)$$

$$= z + \sum_{j=2}^{\infty} \frac{\Gamma(j+1)\,\Gamma(\varpi+1)}{\Gamma(\varpi+j)}\, a_j\, z^j, \tag{11}$$

such that

$$\vartheta_\varpi(z) * \vartheta_\varpi^{(-1)}(z) = \frac{z}{(1-z)^2}.$$

Evidently, $I_0 \vartheta(z) = z\vartheta'(z)$, $I_1 \vartheta(z) = \vartheta(z)$, $z \in \mathbb{D}$. This reverse relationship between the operators I_ϖ and D^ϖ gives a a cause for naming the Noor operator an integral operator. The operator I_ϖ is called as the Noor integral operator of ϖth order of ϑ.

Analogous to $D^{\wp}$, $\wp > -1$ written by Equation (9), in 2002, Choi, Saigo and Srivastava [37] defined the linear operator $I_{\wp,\mathfrak{F}} : \mathcal{A} \to \mathcal{A}$, for $\vartheta \in \mathcal{A}$, $\wp > -1$ and $\mathfrak{F} > 0$ by

$$I_{\wp,\mathfrak{F}}\vartheta(z) = \vartheta_{\wp}^{-1}(z) * \vartheta(z) = \left[\frac{z}{(1-z)^{\wp+1}}\right]^{-1} * \vartheta(z)$$

$$= z + \sum_{j=2}^{\infty} \frac{\Gamma(\mathfrak{F}+j-1)\,\Gamma(\wp+1)}{\Gamma(\mathfrak{F})\,\Gamma(\wp+j)}\, \alpha_j\, z^j, \quad (12)$$

such that

$$\vartheta_{\wp}(z) * \vartheta_{\wp}^{-1}(z) = \frac{z}{(1-z)^{\mathfrak{F}}}.$$

The operator $I_{\wp,\mathfrak{F}}$ is called the Choi–Saigo–Srivastava operator. For $\wp = \varpi$ and $\mathfrak{F} = 2$ reduces to the Noor integral operator I_{ϖ} of Equation (11).

In 2002, Liu and Noor [38] provided a linear operator as: for $\vartheta \in \mathcal{A}_p$, $\wp > -p$ and $I_{\wp+p} : \mathcal{A}_p \to \mathcal{A}_p$ defined by

$$I_{\wp+p}\vartheta(z) = \vartheta_{\wp+p}^{(-1)}(z) * \vartheta(z) = \left[\frac{z^p}{(1-z)^{\wp+p}}\right]^{-1} * \vartheta(z)$$

$$= z^p + \sum_{j=p+1}^{\infty} \frac{\Gamma(1+j)\,\Gamma(\wp+p)}{\Gamma(1+p)\,\Gamma(\wp+j)}\, \alpha_j\, z^j, \quad (13)$$

such that

$$\varphi_{\wp+p}(z) * \varphi_{\wp+p}^{(-1)}(z) = \frac{z^p}{(1-z)^{p+1}}.$$

Obviously, $I_{0+p}\vartheta(z) = z\vartheta'(z)/p$ and $I_{1+p}\vartheta(z) = \vartheta(z)$. The operator $I_{\wp+p}$ is an extended Noor integral operator I_{ϖ} of Equation (11). In addition, the operator $I_{\wp+p}$ is closely related to the Choi–Saigo–Srivastava operator $I_{\wp,\mathfrak{F}}$ of Equation (12).

The Theory of Hypergeometric Functions (HFT) has been incorporated in GFT. Employing hypergeometric functions in the proof of the famed problem "Bieberbach conjecture" by de Branges in 1984 [39] has given complex analysts a renewed attention to study the role of special functions. In this regard a lot of implementations and generalizations are found. The study of this theory gained an independent status. The Gauss Hypergeometric Function (GHF), denoted by $\mathcal{F}(\mu, \nu; \tau; \omega)$, was first introduced by Gauss in 1812 [39]. It is given as follows: for μ, ν and τ be complex numbers with τ other than $0, -1, -2, ...$, and

$$\mathcal{F}(\mu,\nu;\tau;z) = \sum_{j=0}^{\infty} \frac{(\mu)_j (\nu)_j}{(\tau)_j (1)_j} z^j = 1 + \frac{\mu\nu}{\tau} z + \frac{\mu(\mu+1)\nu(\nu+1)}{\tau(\tau+1)} \frac{z^2}{2!} + ... \quad (14)$$

where $(\varrho)_j$ is the Pochhammer symbol given by

$$(\varrho)_j := \frac{\Gamma(\varrho+j)}{\Gamma(\varrho)} = \begin{cases} 1, & (j=0), \\ \varrho(\varrho+1)(\varrho+2)...(\varrho+j-1), & (j \in \mathbb{N} = \{1,2,3,...\}). \end{cases}$$

Another important special function related to GHF is the incomplete beta function $\varphi_p(\mu, \tau; z)$ defined (for $\mu \in \mathbb{R}, \tau \in \mathbb{R}\setminus\mathbb{Z}_0^-, \mathbb{Z}_0^- = \{\ldots, -2, -1, 0\}$) by

$$\varphi_p(\mu, \tau; z) = z\mathcal{F}(\mu, 1; \tau; z) = \sum_{j=0}^{\infty} \frac{(\mu)_j}{(\tau)_j} z^{j+p}. \tag{15}$$

Other generalized Noor-type linear integral operators between classes of holomorphic functions associated with hypergeometric functions and its generalizations have been posed by authors. For instance, Al-Janaby et al. ([40,41]).

Recently, the theory of Hurwitz–Lerch Zeta functions has a fruitful role in the study operators. This theory is developed with numerous implementations and generalizations by various researchers. One may refer to Al-Janaby et al. [42,43], Ghanim [44], Ghanim and Darus [45], Ghanim and Al-Janaby [46], Răducanu and Srivastava [47], Srivastava and Attiya [48], Srivastava et al. [49,50], Xing and Jose [51], Choi and Srivastava [52], Milovanovic and Rassias [53] and Rassias and Yang [54–57].

In terms of the Hurwitz–Lerch Zeta function $\Phi(z, \gamma, \eta)$ defined by (see, for example, [58–60])

$$\Phi(z, \gamma, \eta) := \sum_{j=0}^{\infty} \frac{z^j}{(j+\eta)^\gamma} \tag{16}$$

$(\eta \in \mathbb{C}\setminus\mathbb{Z}_0^-, \gamma \in \mathbb{C}$ when $|z| < 1, 1 < \Re(\gamma)$ when $|z| = 1)$.

The following new family of the (GHLZF) was considered systematically by Srivastava [61]:

$$\Phi_{\mu_1,\cdots,\mu_r;\nu_1,\cdots,\nu_s}^{(\rho_1,\cdots,\rho_r,\sigma_1,\cdots,\sigma_s)}(z, \gamma, \eta; \zeta, \xi)$$

$$= \frac{1}{\xi \Gamma(\gamma)} \sum_{j=0}^{\infty} \frac{\prod_{i=1}^{r}(\mu_i)_{j\rho_i}}{(\eta+j)^\gamma \cdot \prod_{i=1}^{s}(\nu_i)_{j\sigma_i}} H_{0,2}^{2,0}\left[(\eta+j)\zeta^{\frac{1}{\xi}} \middle| \overline{(\gamma,1), \left(0, \frac{1}{\xi}\right)}\right] \frac{z^j}{j!} \tag{17}$$

$(0 < \min\{\Re(\eta), \Re(\gamma)\}, 0 < \Re(\zeta); 0 < \xi),$

where

$$\left(\mu_i \in \mathbb{C} \ (i=1,\cdots,p), \ \nu_i \in \mathbb{C}\setminus\mathbb{Z}_0^- \ (i=1,\cdots,s), \ 0 < \rho_i \ (i=1,\cdots,r),\right.$$

$$\left.0 < \sigma_i \ (i=1,\cdots,q), \text{ and } 0 \leqq 1 + \sum_{i=1}^{s}\sigma_i - \sum_{i=1}^{r}\rho_i\right)$$

and the equality in the convergence condition holds true for suitably bounded values of $|z|$ given by

$$|z| < \nabla := \left(\prod_{i=1}^{r} \rho_i^{-\rho_i}\right)\left(\prod_{i=1}^{s}\sigma_i^{\sigma_i}\right).$$

Definition 1. *The H-function involved in the right-hand side of Equation (17) is the well-known Fox's H-function ([62], Definition 1.1) (see also [30,63]) defined by*

$$H_{p,q}^{m,n}(z) = H_{p,q}^{m,n}\left[z \middle| \begin{array}{c}(a_1, A_1), \cdots, (a_p, A_p) \\ (b_1, B_1), \cdots, (b_q, B_q)\end{array}\right]$$

$$= \frac{1}{2\pi i}\int_{\mathcal{L}} \Xi(\gamma) z^{-\gamma} d\gamma \quad (z \in \mathbb{C}\setminus\{0\}, |\arg(z)| < \pi), \tag{18}$$

an empty product is interpreted as 1, m, n, p and q are integers such that

$$1 \leqq m \leqq q \quad \text{and} \quad 0 \leqq n \leqq p,$$

$$0 < A_i \quad (i = 1, \cdots, \mathfrak{p}) \quad \text{and} \quad 0 < B_i \quad (i = 1, \cdots, \mathfrak{q}),$$
$$a_i \in \mathbb{C} \quad (i = 1, \cdots, \mathfrak{p}) \quad \text{and} \quad b_i \in \mathbb{C} \quad (i = 1, \cdots, \mathfrak{q})$$

and $\mathcal{L}$ is a suitable Mellin–Barnes type contour separating the poles of the gamma functions

$$\{\Gamma(b_i + B_i\gamma)\}_{i=1}^m$$

from the poles of the gamma functions

$$\{\Gamma(1 - a_i + A_i\gamma)\}_{i=1}^n.$$

It is worthy of mention here that, by using the fact that ([61], p. 1496, Remark 7)

$$\lim_{\zeta \to 0} \left\{ H_{0,2}^{2,0} \left[(\eta + j)\zeta^{\frac{1}{\xi}} \middle| \overline{(\gamma, 1), \left(0, \frac{1}{\xi}\right)} \right] \right\} = \xi \, \Gamma(\gamma) \quad (0 < \xi),$$

Equation (17) reduces to the following form:

$$\Phi_{\mu_1, \cdots, \mu_r; \nu_1, \cdots, \nu_s}^{(\rho_1, \cdots, \rho_r, \sigma_1, \cdots, \sigma_s)}(z, \gamma, \eta; 0, \xi) := \Phi_{\mu_1, \cdots, \mu_r; \nu_1, \cdots, \nu_s}^{(\rho_1, \cdots, \rho_r, \sigma_1, \cdots, \sigma_s)}(z, \gamma, \eta)$$

$$= \sum_{j=0}^{\infty} \frac{\prod_{i=1}^{r}(\mu_i)_{j\rho_i}}{(\eta + j)^\gamma \cdot \prod_{i=1}^{s}(\nu_i)_{j\sigma_i}} \frac{z^j}{j!}. \tag{19}$$

Definition 2. *The function* $\Phi_{\mu_1, \cdots, \mu_r; \nu_1, \cdots, \nu_s}^{(\rho_1, \cdots, \rho_r, \sigma_1, \cdots, \sigma_s)}(z, \gamma, \eta)$ *involved in Equation* (19) *is the multiparameter extension and generalization of the Hurwitz–Lerch Zeta function* $\Phi(z, \gamma, \eta)$ *introduced by Srivastava et al.* ([64], *p. 503, Equation (6.2)) defined by*

$$\Phi_{\mu_1, \cdots, \mu_r; \nu_1, \cdots, \nu_s}^{(\rho_1, \cdots, \rho_r, \sigma_1, \cdots, \sigma_s)}(z, \gamma, \eta) = \sum_{j=0}^{\infty} \frac{\prod_{i=1}^{r}(\mu_i)_{j\rho_i}}{(\eta + j)^\gamma \cdot \prod_{i=1}^{s}(\nu_i)_{j\sigma_i}} \frac{z^j}{j!} \tag{20}$$

$$\left(r, s \subset \mathbb{N}_0; \, \mu_j \in \mathbb{C} \; (j-1, \cdots, r); \, \eta, \nu_j \in \mathbb{C} \setminus \mathbb{Z}_0^- \; (j = 1, \cdots, s); \right.$$
$$\rho_i, \sigma_i \in \mathbb{R}^+ \; (i = 1, \cdots, r; \, i = 1, \cdots, s);$$
$$-1 < \Delta \text{ when } \gamma, z \in \mathbb{C};$$
$$\Delta = -1 \text{ and } \gamma \in \mathbb{C} \text{ when } |z| < \nabla^*;$$
$$\left. \Delta = -1 \text{ and } \frac{1}{2} < \Re(\Xi) \text{ when } |z| = \nabla^* \right)$$

with

$$\nabla^* := \left(\prod_{i=1}^{r} \rho_i^{-\rho_i} \right) \cdot \left(\prod_{i=1}^{s} \sigma_i^{\sigma_i} \right),$$

$$\Delta := \sum_{i=1}^{s} \sigma_i - \sum_{i=1}^{r} \rho_i \quad \text{and} \quad \Xi := t + \sum_{i=1}^{s} \nu_i - \sum_{i=1}^{r} \mu_i + \frac{r-s}{2}.$$

In GFT, the third-order differential subordination methodology for holomorphic functions is indicated by Antonion and Miller [20], which is required in this investigation.

Definition 3 ([20], Definition 2, p. 441). *Let $\mathcal{J}$ denote the set of holomorphic functions ω that are univalent on the set $\overline{\mathbb{D}}\backslash\mathcal{G}(\omega)$, where*

$$\mathcal{G}(\omega) = \{\chi \in \partial\mathbb{D} : \lim_{z \to \chi} \omega(z) = \infty\},$$

is such that

$$\min |\omega'(\chi)| = \delta > 0$$

for $\chi \in \partial\mathbb{D}\backslash\mathcal{G}(\omega)$. Further, let $\mathcal{J}(\alpha) = \{\omega(z) \in \mathcal{J} : \omega(0) = \alpha\}$, $\mathcal{J}(0) = \mathcal{J}_0$ and $\mathcal{J}(1) = \mathcal{J}_1$.

Definition 4 ([20], Definition 1, p. 440). *Let $\Gamma : \mathbb{C}^4 \times \mathbb{D} \longrightarrow \mathbb{C}$ and the function $\pi(z)$ be univalent in $\mathbb{D}$. If the function $v(z)$ is holomorphic in $\mathbb{D}$ and satisfies the following third-order differential subordination:*

$$\Gamma(v(z), zv'(z), z^2v''(z), z^3v'''(z); z) \prec \pi(z), \tag{21}$$

then $v(z)$ is called a solution of the differential subordination. A univalent function $\omega(z)$ is called a dominant of the solutions of the differential subordination, or, more simply, a dominant if $v(z) \prec \omega(z)$ for all $v(z)$ achieving Equation (21). A dominant $\tilde{\omega}(z)$ that achieves $\tilde{\omega}(z) \prec \omega(z)$ for all dominants $\omega(z)$ of Equation (21) is said to be the best dominant.

The class of admissible functions related to differential subordination is presented next.

Definition 5 ([20], Definition 2, p. 449). *Let Λ be a set in $\mathbb{C}$, $\omega \in \mathcal{J}$ and $\jmath \in \mathbb{N}\backslash\{1\}$. The class of admissible functions denoted by $\Omega_\jmath[\Lambda, \omega]$ consists of those functions $\Gamma : \mathbb{C}^4 \times \mathbb{D} \longrightarrow \mathbb{C}$ that achieves the following admissibility condition:*

$$\Gamma(f, q, x, y; z) \notin \Lambda$$

whenever

$$f = \omega(\chi), \quad q = \kappa\zeta\omega'(\chi), \quad \Re\left(\frac{x}{q}+1\right) \geq \kappa\Re\left(\frac{\zeta\omega''(\chi)}{\omega'(\chi)}+1\right),$$

and

$$\Re\left(\frac{y}{q}\right) \geq \kappa^2\Re\left(\frac{\chi^2\omega'''(\chi)}{\omega'(\chi)}\right),$$

where $z \in \mathbb{D}$, $\chi \in \partial\mathbb{D}\backslash\mathcal{G}(\omega)$, and $\kappa \geq \jmath$.

The following theorem is a key outcome in third-order differential subordination.

Theorem 1 ([20], Definition 2, p. 449). *Let $v \in \mathcal{H}[\alpha, \jmath]$ with $\jmath \geq 2$, and let $\omega \in \mathcal{J}(\alpha)$ and achieve the following conditions:*

$$\Re\left(\frac{\chi\omega''(\chi)}{\omega'(\chi)}\right) \geq 0, \text{ and } \left|\frac{zv'(z)}{\omega'(\chi)}\right| \leq \kappa,$$

where $z \in \mathbb{D}$, $\chi \in \partial\mathbb{D}\setminus\mathcal{G}(\omega)$ and $\kappa \geq \jmath$. If Λ is a set in $\mathbb{C}$, $\Gamma \in \Omega_\jmath[\Lambda, \omega]$ and

$$\Gamma\left(v(z), zv'(z), z^2 v''(z), z^3 v'''(z); z\right) \in \Lambda,$$

then

$$v(z) \prec \omega(z) \quad (z \in \mathbb{D}).$$

2. Imposed Linear Integral Operator $\mathcal{M}_p^\ell \vartheta(z)$

This section considers a new generalized Noor-type linear integral operator $\mathcal{M}_p^\ell \vartheta(z)$ for p-valent functions associated with the GHLZF in $\mathbb{D}$ defined in Equation (17). Setting $\rho_1 = \cdots, \rho_r = \sigma_1 = \cdots = \sigma_s = 1$, and $\mu_i \in \mathbb{C} \setminus \mathbb{Z}_0^-$ $(i = 1, \cdots, r)$ as follows:

$$\Phi_{\mu_1,\cdots,\mu_r;\nu_1,\cdots,\nu_s}^{(1,\cdots,1,1,\cdots,1)}(z, \gamma, \eta; \zeta, \xi)$$

$$= \frac{1}{\xi \Gamma(\gamma)} \sum_{\jmath=0}^{\infty} \frac{\prod_{i=1}^{r}(\mu_i)_\jmath}{(\eta + \jmath)^\gamma \cdot \prod_{i=1}^{s}(\nu_i)_\jmath} H_{0,2}^{2,0}\left[(\eta + \jmath)\zeta^{\frac{1}{\xi}} \Big| \overline{(\gamma, 1), \left(0, \frac{1}{\xi}\right)}\right] \frac{z^\jmath}{\jmath!} \tag{22}$$

$$= \frac{1}{\xi \Gamma(\gamma)} \sum_{\jmath=p}^{\infty} \frac{\prod_{i=1}^{r}(\mu_i)_{\jmath-p}}{(\eta + (\jmath-p))^\gamma \cdot \prod_{i=1}^{s}(\nu_i)_{\jmath-p}} H_{0,2}^{2,0}\left[(\eta + (\jmath-p))\zeta^{\frac{1}{\xi}} \Big| \overline{(\gamma, 1), \left(0, \frac{1}{\xi}\right)}\right] \frac{z^{\jmath-p}}{(\jmath-p)!}.$$

Thus, from Equation (22), we derive a new function as:

$$\Psi_{\mu_1,\cdots,\mu_r}^{\nu_1,\cdots,\nu_s}(\gamma, \eta; \zeta, \xi, \wp) := \xi \Gamma(\gamma) \, Y \left[z^p \Phi_{\mu_1,\cdots,\mu_r;\nu_1,\cdots,\nu_s}^{(1,\cdots,1,1,\cdots,1)}(z, \gamma, \eta; \zeta, \xi) \right]$$

$$= z^p + \sum_{\jmath=p+1}^{\infty} \frac{Y \prod_{i=1}^{r}(\mu_i)_{\jmath-p}}{(\eta + (\jmath-p))^\gamma \cdot \prod_{i=1}^{s}(\nu_i)_{\jmath-p}} H_{0,2}^{2,0}\left[(\eta + (\jmath-p))\zeta^{\frac{1}{\xi}} \Big| \overline{(\gamma, 1), \left(0, \frac{1}{\xi}\right)}\right] \frac{z^\jmath}{(\jmath-p)!}, \tag{23}$$

where Y is defined as:

$$Y = \frac{\eta^\gamma}{H_{0,2}^{2,0}\left[\eta \zeta^{\frac{1}{\xi}} \Big| \overline{(\gamma, 1), \left(0, \frac{1}{\xi}\right)}\right]}. \tag{24}$$

By employing the principle of convolution product of ℓth order of GHLZF, we yield

$$\mathcal{N}_{p,\mu_1,\ldots,\mu_r}^{\ell,v_1,\ldots,v_s}(\gamma,\eta;\zeta,\xi,\wp) := \underbrace{\Psi_{\mu_1,\ldots,\mu_r}^{v_1,\ldots,v_s}(\gamma,\eta;\zeta,\xi,\wp) * \ldots * \Psi_{\mu_1,\ldots,\mu_r}^{v_1,\ldots,v_s}(\gamma,\eta;\zeta,\xi,\wp)}_{\ell-\text{times}}$$

$$= z^p + \sum_{j=p+1}^{\infty} \left[\frac{\prod_{i=1}^{r}(\mu_i)_{j-p} H_{0,2}^{2,0}\left[(\eta+(j-p))\zeta^{\frac{1}{\zeta}} \bigg| \overline{(\gamma,1),\left(0,\frac{1}{\zeta}\right)}\right]}{(\eta+(j-p))^{\gamma} \prod_{i=1}^{s}(v_i)_{j-p}(j-p)!} \right]^{\ell} z^j. \quad (25)$$

Next, we present a new function $\left(\mathcal{N}_{p,\mu_1,\ldots,\mu_r}^{\ell,v_1,\ldots,v_s}(\gamma,\eta;\zeta,\xi,\wp)\right)^{-1}$ given by

$$\left(\mathcal{N}_{p,\mu_1,\ldots,\mu_r}^{\ell,v_1,\ldots,v_s}(\gamma,\eta;\zeta,\xi,\wp)\right)^{-1}$$

$$= z^p + \sum_{j=p+1}^{\infty} \left[\frac{(\eta+(j-p))^{\gamma} \prod_{i=1}^{s}(v_i)_{j-p}(j-p)!}{\prod_{i=1}^{r}(\mu_i)_{j-p} H_{0,2}^{2,0}\left[(\eta+(j-p))\zeta^{\frac{1}{\zeta}} \bigg| \overline{(\gamma,1),\left(0,\frac{1}{\zeta}\right)}\right]} \right]^{\ell} \left[\frac{(\wp+p)_{j-p}}{(j-p)!} \right] z^j, \quad (26)$$

such that,

$$\left(\mathcal{N}_{p,\mu_1,\ldots,\mu_r}^{\ell,v_1,\ldots,v_s}(\gamma,\eta;\zeta,\xi,\wp)\right) * \left(\mathcal{N}_{p,\mu_1,\ldots,\mu_r}^{\ell,v_1,\ldots,v_s}(\gamma,\eta;\zeta,\xi,\wp)\right)^{-1} = \frac{z^p}{(1-z)^{\wp+p}}$$

$$= z^p + \sum_{j=p+1}^{\infty} \frac{(\wp+p)_{j-p}}{(j-p)!} z^j, \quad (\wp > -1).$$

Therefore, from Equation (26), we consider the following linear operator: $\mathcal{M}_{p,\mu_1,\ldots,\mu_r}^{\ell,v_1,\ldots,v_s}(\gamma,\eta;\zeta,\xi,\wp)$: $\mathcal{A}_p \longrightarrow \mathcal{A}_p$, which is defined by

$$\mathcal{M}_{p,\mu_1,\ldots,\mu_r}^{\ell,v_1,\ldots,v_s}(\gamma,\eta;\zeta,\xi,\wp)\vartheta(z) = \left(\mathcal{N}_{p,\mu_1,\ldots,\mu_r}^{\ell,v_1,\ldots,v_s}(\gamma,\eta;\zeta,\xi,\wp)\right)^{-1} * \vartheta(z)$$

$$= z^p + \sum_{j=p+1}^{\infty} \left[\frac{(\eta+(j-p))^{\gamma} \prod_{i=1}^{s}(v_i)_{j-p}(j-p)!}{\prod_{i=1}^{r}(\mu_i)_{j-p} H_{0,2}^{2,0}\left[(\eta+(j-p))\zeta^{\frac{1}{\zeta}} \bigg| \overline{(\gamma,1),\left(0,\frac{1}{\zeta}\right)}\right]} \right]^{\ell} \left[\frac{(\wp+p)_{j-p}}{(j-p)!} \right] a_j z^j. \quad (27)$$

Remark 1. *For suitably specializing the parameters of ℓ, p, ζ, ξ, γ, η, s, r, v_i and μ_i, the operator $\mathcal{M}_p^{\ell}\vartheta(z)$ defined in Equation (27) can be reduced to various operators previously mentioned. Thus, we have the following special cases:*

1. *For $\ell = p = \gamma = \eta = 1$, $\zeta = 0$, $s = 1$, $v_1 = 1$, $r = 2$, $\mu_1 = 2$ and $\mu_2 = 1$ in Equation (27), we yield the Ruscheweyh operator given in Equation (9).*
2. *For $\ell = p = \gamma = \eta = 1$, $\zeta = 0$, $s = 2$, $v_1 = v_2 = 1$, $r = 3$, $\mu_1 = 2$ and $\mu_2 = \mu_3 = \wp + 1$, the operator Equation (27) reduce to the Noor operator defined by Equation (11).*
3. *For $\ell = \gamma = 1$, $\zeta = 0$, $s = 2$, $v_1 = \eta$, $v_2 = 1$, $r = 3$, $\mu_1 = \eta + 1$ $\mu_2 = \mu_2 = \wp + p$, the operator Equation (27), we have the extended Noor operator given by Equation (13).*
4. *For $\ell = 1$, $\zeta = 0$, $s = 2$, $\gamma = 1$, $v_1 = \eta$, $v_2 = \tau$, $r = 3$, $\mu_1 = \eta + 1$, $\mu_2 = 1$, and $\mu_3 = \mu$, the operator Equation (27) provides the Noor-type integral operator defined by [65].*

5. For $\ell = p = \gamma = \eta = 1$, $\zeta = 0$, $s = 2$, $\nu_1 = 1$, $\nu_2 = \tau$, $r = 3$, $\mu_1 = 2$, $\mu_2 = \mu$, and $\mu_3 = \nu$, the operator Equation (27) provides the Noor integral operator given in [66].
6. For $\ell = p = \gamma = \eta = 1$, $\zeta = 0$, $s = 1$, $\nu_1 = \tau$, $r = 3$, $\mu_1 = 2$, $\mu_2 = \mu$, and $\mu_3 = \nu$, the operator Equation (27) reduce to the generalized Noor-type linear integral operator defined in [67].
7. For $\ell = p = \gamma = \eta = 1$, $\zeta = 0$ $s = 2$, $\nu_1 = \nu_2 = 1$, $r = 3$, $\mu_1 = \mu_2 = 2$ and $\mu_3 = \wp + 1$, the operator Equation (27) reduce to Alexander operator given in Equation (6).
8. For $\ell = p = \gamma = \eta = 1$, $\zeta = 0$ $s = 1$, $\nu_1 = 1$, $r = 2$, $\mu_1 = 2$ and $\mu_2 = \wp + 1$, the operator Equation (27) is reduced to $\vartheta(z)$ given by Equation (2).

For convenience, Equation (27) is written as

$$\mathcal{M}_p^\ell \vartheta(z) \equiv \mathcal{M}_{p,\mu_1,\ldots,\mu_r}^{\ell,\nu_1,\ldots,\nu_r}(\gamma, \eta; \zeta, \xi, \wp) \vartheta(z). \tag{28}$$

This operator achieves the differential recurrence relation

$$\frac{Az}{p}\left[\mathcal{M}_p^\ell \vartheta(z)\right]' = \mathcal{M}_p^{\ell+1} \vartheta(z) - (1 - A)\mathcal{M}_p^\ell \vartheta(z), \tag{29}$$

where $A = \dfrac{p\left[(\eta+(\jmath-p))^\gamma \prod_{i=1}^s (\nu_i)_{\jmath-p} (\jmath-p)! - Y \prod_{i=1}^r (\mu_i)_{\jmath-p} H_{0,2}^{2,0}\left[(\eta+(\jmath-p))\zeta^{\frac{1}{\xi}} \middle| \overline{(\gamma,1), \left(0, \frac{1}{\xi}\right)}\right]\right]}{(\jmath-p) Y \prod_{i=1}^r (\mu_i)_{\jmath-p} H_{0,2}^{2,0}\left[(\eta+(\jmath-p))\zeta^{\frac{1}{\xi}} \middle| \overline{(\gamma,1), \left(0, \frac{1}{\xi}\right)}\right]}$. Throughout this paper, the generalized Noor-type linear integral operator will be denoted by $\mathcal{M}_p^\ell \vartheta(z)$.

3. Differential Subordination with $\mathcal{M}_p^\ell \vartheta(z)$

This section introduces certain appropriate class of admissible functions and studies some third-order differential subordination outcomes for the operator $\mathcal{M}_p^\ell \vartheta(z)$ defined by Equation (27).

Definition 6. *Let $\mathfrak{A}$ be a set in $\mathbb{C}$, $\omega \in \mathcal{J}_0$ and $\jmath \in \mathbb{N}\backslash\{1\}$. The class of admissible functions $\Sigma_\mathcal{M}[\mathfrak{A}, \omega]$ consists of those functions $\maltese : \mathbb{C}^4 \times \mathbb{D} \longrightarrow \mathbb{C}$ that satisfy the following admissibility condition:*

$$\maltese(u_1, u_2, u_3, u_4; z) \notin \mathfrak{A}$$

whenever

$$u_1 = \omega(\chi), \quad u_2 = \frac{\kappa\chi\omega'(\chi) + \frac{p(1-A)}{A}\omega(\chi)}{\frac{p}{A}},$$

$$\Re\left(\frac{p\left[u_3 - 2(1-A)u_2 + (1-A)^2 u_1\right]}{A[u_2 - (1-A)u_1]}\right) \geq \kappa\Re\left(\frac{\zeta\omega''(\chi)}{\omega'(\chi)} + 1\right),$$

and

$$\Re\left(\frac{p^2[u_4 - (1-A)^3 u_1] - p[3A + 3p(1-A)][u_3 - 2(1-A)u_2 + (1-A)^2 u_1]}{A^2[u_2 - (1-A)u_1]} - \frac{3p^2(1-A)^2}{A^2} + 2\right)$$
$$\geq \kappa^2 \Re\left(\frac{\chi^2 \omega'''(\chi)}{\omega'(\chi)}\right),$$

where $z \in \mathbb{D}$, $\chi \in \partial\mathbb{D}\backslash\mathcal{G}(\omega)$, and $\kappa \geq \jmath$.

Theorem 2. Let $\Psi \in \Sigma_\mathcal{M}[\mathfrak{A}, \omega]$. If $\vartheta \in \mathcal{A}_p$ and $\omega \in \mathcal{J}_0$ achieve the following conditions:

$$\Re\left(\frac{\chi\omega''(\chi)}{\omega'(\chi)}\right) \geq 0, \quad \left|\frac{\mathcal{M}_p^{\ell+1}\vartheta(z) - (1-A)\mathcal{M}_p^\ell \vartheta(z)}{\omega'(\chi)}\right| \leq \frac{|A|\kappa}{p}, \tag{30}$$

and

$$\{\Psi\left(\mathcal{M}_p^\ell \vartheta(z), \mathcal{M}_p^{\ell+1}\vartheta(z), \mathcal{M}_p^{\ell+2}\vartheta(z), \mathcal{M}_p^{\ell+3}\vartheta(z); z\right) : z \in \mathbb{D}\} \subset \mathfrak{A}, \tag{31}$$

then

$$\mathcal{M}_p^\ell \vartheta(z) \prec \omega(z), \quad (z \in \mathbb{D}). \tag{32}$$

Proof. Define the following holomorphic function $v(z)$ in $\mathbb{D}$ by

$$v(z) = \mathcal{M}_p^\ell \vartheta(z). \tag{33}$$

From Equations (29) and (33), we have

$$\mathcal{M}_p^{\ell+1}\vartheta(z) = \frac{zv'(z) + \frac{p(1-A)}{A}v(z)}{\frac{p}{A}}. \tag{34}$$

Further computations show that

$$\mathcal{M}_p^{\ell+2}\vartheta(z) = \frac{z^2 v''(z) + \left[1 + \frac{2p(1-A)}{A}\right]zv'(z) + \frac{p^2(1-A)^2}{A^2}v(z)}{\frac{p^2}{A^2}}, \tag{35}$$

and

$$\mathcal{M}_p^{\ell+3}\vartheta(z) = \frac{z^3 v'''(z) + \left[3 + \frac{3p(1-A)}{A}\right]z^2 v''(z) + \left[1 + \frac{3p(1-A)}{A} + \frac{3p^2(1-A)^2}{A^2}\right]zv'(z) + \frac{p^3(1-A)^3}{A^3}v(z)}{\frac{p^3}{A^3}}. \tag{36}$$

Define the parameters u_1, u_2, u_3 and u_4 as:

$$u_1 = f, \quad u_2 = \frac{g + \frac{p(1-A)}{A}f}{\frac{p}{A}}, \tag{37}$$

$$u_3 = \frac{h + \left[1 + \frac{2p(1-A)}{A}\right]g + \frac{p^2(1-A)^2}{A^2}f}{\frac{p^2}{A^2}}, \tag{38}$$

and

$$u_4 = \frac{t + \left[3 + \frac{3p(1-A)}{A}\right]h + \left[1 + \frac{3p(1-A)}{A} + \frac{3p^2(1-A)^2}{A^2}\right]g + \frac{p^3(1-A)^3}{A^3}f}{\frac{p^3}{A^3}}. \tag{39}$$

Now, we define the transformation $\Gamma : \mathbb{C}^4 \times \mathbb{D} \longrightarrow \mathbb{C}$ as follows:

$$\Gamma(f,g,h,t;z) = ¥(u_1, u_2, u_3, u_4; z)$$

$$= ¥\left(f, \frac{g + \frac{p(1-A)}{A}f}{\frac{p}{A}}, \frac{h + \left[1 + \frac{2p(1-A)}{A}\right]g + \frac{p^2(1-A)^2}{A^2}f}{\frac{p^2}{A^2}}, \frac{t + 3\left[1 + \frac{p(1-A)}{A}\right]h + \left[1 + \frac{3p(1-A)}{A} + \frac{3p^2(1-A)^2}{A^2}\right]g + \frac{p^3(1-A)^3}{A^3}f}{\frac{p^3}{A^3}}; z\right). \quad (40)$$

By utilizing Theorem 1 and Equations (33) to (36), and from Equation (40), we yield

$$\Gamma\left(v(z), zv'(z), z^2v''(z), z^3v'''(z); z\right) = ¥\left(\mathcal{M}_p^\ell\vartheta(z), \mathcal{M}_p^{\ell+1}\vartheta(z), \mathcal{M}_p^{\ell+2}\vartheta(z), \mathcal{M}_p^{\ell+3}\vartheta(z); z\right). \quad (41)$$

Therefore, Equation (31) becomes

$$\Gamma\left(v(z), zv'(z), z^2v''(z), z^3v'''(z); z\right) \in \mathfrak{A}.$$

A computation utilizing Equations (37), (38) and (39) acquire

$$\frac{h}{q} + 1 = \frac{p\left[u_3 - 2(1-A)u_2 + (1-A)^2 u_1\right]}{A[u_2 - (1-A)u_1]},$$

and

$$\frac{t}{q} = \frac{p^2[u_4 - (1-A)^3 u_1] - p[3A + 3p(1-A)][u_3 - 2(1-A)u_2 + (1-A)^2 u_1]}{A^2[u_2 - (1-A)u_1]} - \frac{3p^2(1-A)^2}{A^2} + 2.$$

We also note that

$$\left|\frac{zv'(z)}{\omega'(\chi)}\right| = \left|\frac{\frac{p}{A}\left[\mathcal{M}^{\ell+1}(z) - (1-A)\mathcal{M}^\ell(z)\right]}{\omega'(\chi)}\right| \leq \kappa.$$

Hence, the admissibility condition for $¥ \in \Sigma_\mathcal{M}[\mathfrak{A}, \omega]$ in Definition 8 is equivalent to the admissibility condition of $\Gamma \in \Omega_2[\mathfrak{A}, \omega]$ as given in Definition 5 and by Theorem 1, we obtain

$$\mathcal{M}_p^\ell\vartheta(z) \prec \omega(z).$$

The proof of Theorem 2 is complete. □

If $\mathfrak{A} \neq \mathbb{C}$ is a simply connected domain, then $\mathfrak{A} = \hbar(\mathbb{D})$ for some conformal mapping $\hbar(z)$ of $\mathbb{D}$ onto $\mathfrak{A}$. In this case the class $\Sigma_\mathcal{M}[\hbar(\mathbb{D}), \omega]$ is written as $\Sigma'_\mathcal{M}[\hbar, \omega]$. The following outcome is a directly consequence of Theorem 2.

Theorem 3. *Let $¥ \in \Sigma_\mathcal{M}[\hbar, \omega]$. If $\vartheta \in \mathcal{A}_p$ and $\omega \in \mathcal{J}_0$ achieve the following condition (28) given as follows:*

$$\Re\left(\frac{\chi\omega''(\chi)}{\omega'(\chi)}\right) \geq 0, \quad \left|\frac{\mathcal{M}_p^{\ell+1}\vartheta(z) - (1-A)\mathcal{M}_p^\ell\vartheta(z)}{\omega'(\chi)}\right| \leq \frac{|A|\kappa}{p},$$

and
$$¥\left(\mathcal{M}_p^\ell \vartheta(z), \mathcal{M}_p^{\ell+1}\vartheta(z), \mathcal{M}_p^{\ell+2}\vartheta(z), \mathcal{M}_p^{\ell+3}\vartheta(z); z\right) \prec h(z), \tag{42}$$

then
$$\mathcal{M}_p^\ell \vartheta(z) \prec \omega(z), \quad (z \in \mathbb{D}). \tag{43}$$

The next outcome is an extension of Theorem 3 to the case where the behavior of $\omega(z)$ on $\partial \mathbb{D}$ is not known.

Corollary 1. *Let $\mathfrak{A} \subset \mathbb{C}$ and let $\omega(z)$ be univalent in $\mathbb{D}$, $\omega(0) = 0$. Let $¥ \in \Sigma_\mathcal{M}[\mathfrak{A}, \omega_\varepsilon]$ for some $\varepsilon \in (0,1)$ where $\omega_\varepsilon(z) = \omega(\varepsilon z)$. If $\vartheta \in \mathcal{A}_p$ achieves*

$$\Re\left(\frac{\chi \omega_\varepsilon''(\chi)}{\omega_\varepsilon'(\chi)}\right) \geq 0, \quad \left|\frac{\mathcal{M}_p^{\ell+1}\vartheta(z) - (1-A)\mathcal{M}_p^\ell \vartheta(z)}{\omega_\varepsilon'(\chi)}\right| \leq \frac{|A|\kappa}{p},$$

and

$$¥\left(\mathcal{M}_p^\ell \vartheta(z), \mathcal{M}_p^{\ell+1}\vartheta(z), \mathcal{M}_p^{\ell+2}\vartheta(z), \mathcal{M}_p^{\ell+3}\vartheta(z); z\right) \in \mathfrak{A},$$

then

$$\mathcal{M}_p^\ell \vartheta(z) \prec \omega(z),$$

where $z \in \mathbb{D}$ and $\chi \in \partial \mathbb{D} \setminus \mathcal{G}(\omega_\varepsilon)$.

Proof. By utilizing Theorem 3, we have $\mathcal{M}_p^\ell \vartheta(z) \prec \omega_\varepsilon(z)$. Then we get the outcome from $\omega_\varepsilon(z) \prec \omega(z)$. □

The next outcome is an immediate consequence of Corollary 1.

Corollary 2. *Let $\mathfrak{A} \subset \mathbb{C}$ and let $\omega(z)$ be univalent in $\mathbb{D}$, $\omega(0) = 0$. Let $¥ \in \Sigma_\mathcal{M}[\hbar, \omega_\varepsilon]$ for some $\varepsilon \in (0,1)$ where $\omega_\varepsilon(z) = \omega(\varepsilon z)$. If $\vartheta \in \mathcal{A}_p$ achieves*

$$\Re\left(\frac{\chi \omega_\varepsilon''(\chi)}{\omega_\varepsilon'(\chi)}\right) \geq 0, \quad \left|\frac{\mathcal{M}_p^{\ell+1}\vartheta(z) - (1-A)\mathcal{M}_p^\ell \vartheta(z)}{\omega_\varepsilon'(\chi)}\right| \leq \frac{|A|\kappa}{p},$$

and

$$¥\left(\mathcal{M}_p^\ell \vartheta(z), \mathcal{M}_p^{\ell+1}\vartheta(z), \mathcal{M}_p^{\ell+2}\vartheta(z), \mathcal{M}_p^{\ell+3}\vartheta(z); z\right) \prec \hbar(z), \tag{44}$$

then

$$\mathcal{M}^\ell(z) \prec \omega(z),$$

where $z \in \mathbb{D}$ and $\chi \in \partial \mathbb{D} \setminus \mathcal{G}(\omega_\varepsilon)$.

The following outcome gives the best dominant of the differential subordination of Equation (40).

Theorem 4. Let $\hbar(z)$ be univalent in $\mathbb{D}$. Let $\yen : \mathbb{C}^4 \times \mathbb{D} \longrightarrow \mathbb{C}$. Suppose that the differential equation:

$$\yen\left(\omega(z), \frac{z\omega'(z) + \frac{p(1-A)}{A}\omega(z)}{\frac{p}{A}}, \frac{z^2\omega''(z) + \left[1 + \frac{2p(1-A)}{A}\right]z\omega'(z) + \frac{p^2(1-A)^2}{A^2}\omega(z)}{\frac{p^2}{A^2}},\right.$$

$$\left.\frac{z^3\omega'''(z) + 3\left[1 + \frac{p(1-A)}{A}\right]z\omega''(z) + \left[1 + \frac{3p(1-A)}{A} + \frac{3p^2(1-A)^2}{A^2}\right]z\omega'(z) + \frac{p^3(1-A)^3}{A^3}\omega(z)}{\frac{p^3}{A^3}}; z\right) = \hbar(z),$$

(45)

has a solution $\omega(z)$ with $\omega(0) = 0$ which achieves Equation (30). If $\vartheta \in A_p$ achieves Equation (44) and

$$\yen\left(\mathcal{M}_p^\ell\vartheta(z), \mathcal{M}_p^{\ell+1}\vartheta(z), \mathcal{M}_p^{\ell+2}\vartheta(z), \mathcal{M}_p^{\ell+3}\vartheta(z); z\right),\qquad(46)$$

is holomorhic in $\mathbb{D}$, then

$$\mathcal{M}_p^\ell\vartheta(z) \prec \omega(z),\qquad(47)$$

and $\omega(z)$ is the best dominant.

Proof. By utilizing Theorem 3 that $\omega(z)$ is a dominant of Equation (44). Since $\omega(z)$ achieves Equation (45), it is also a solution of Equation (44) and therefore $\omega(z)$ will be dominated by all dominants. Thus $\omega(z)$ is the best dominant. □

In the case $\omega(z) = \mathcal{Q}z$ ($\mathcal{Q} > 0$) and in view of Definition 8, the class of admissible functions $\Sigma_\mathcal{M}[\mathfrak{A}, \omega]$ denoted by $\Sigma_\mathcal{M}[\mathfrak{A}, \mathcal{Q}]$ is defined below:

Definition 7. Let $\mathfrak{A}$ be a set in $\mathbb{C}$ and $\mathcal{Q} > 0$. The class of admissible functions $\Sigma_\mathcal{M}[\mathfrak{A}, \mathcal{Q}]$ consists of those functions $\yen : \mathbb{C}^4 \times \mathbb{D} \longrightarrow \mathbb{C}$ that achieve the admissibility condition

$$\yen\left(\mathcal{Q}e^{i\theta}, \left[\frac{\kappa A}{p} + (1-A)\right]\mathcal{Q}e^{i\theta}, \frac{\mathcal{L} + \left[\left[1 + \frac{2p(1-A)}{A}\right]\kappa + \frac{p^2(1-A)^2}{A^2}\right]\mathcal{Q}e^{i\theta}}{\frac{p^2}{A^2}},\right.$$

$$\left.\frac{\mathcal{V} + \left[1 + \frac{p(1-A)}{A}\right]3\mathcal{L} + \left[\left[1 + \frac{3p(1-A)}{A} + \frac{3p^2(1-A)^2}{A^2}\right]\kappa + \frac{p^3(1-A)^3}{A^3}\right]\mathcal{Q}e^{i\theta}}{\frac{p^3}{A^3}}\right) \notin \mathfrak{A},$$

(48)

where $z \in \mathbb{D}$, $\Re(\mathcal{L}e^{-i\theta}) \geq (\kappa - 1)\kappa\mathcal{Q}$ and $\Re(\mathcal{V}e^{-i\theta}) \geq 0$ for all real θ and $\kappa \in \mathbb{N}\setminus\{1\}$.

Corollary 3. Let $\yen \in \Sigma_\mathcal{M}[\mathfrak{A}, \mathcal{Q}]$. If $\vartheta \in A_p$ achieves the following conditions:

$$\left|\mathcal{M}_p^{\ell+1}\vartheta(z) - (1-A)\mathcal{M}_p^\ell\vartheta(z)\right| \leq \frac{|A|\kappa\mathcal{Q}}{p},\qquad(49)$$

and

$$\yen\left(\mathcal{M}_p^\ell\vartheta(z), \mathcal{M}_p^{\ell+1}\vartheta(z), \mathcal{M}_p^{\ell+2}\vartheta(z), \mathcal{M}_p^{\ell+3}\vartheta(z); z\right) \in \mathfrak{A},\qquad(50)$$

then

$$\left|\mathcal{M}^\ell(z)\right| < \mathcal{Q}(z).$$

In the case $\mathfrak{A} = \omega(\mathbb{D}) = \{z_1 : |z_1| < \mathcal{Q}\}$, $(\mathcal{Q} > 0)$, for simplification we denote by $\Sigma_\mathcal{M}[\mathcal{Q}]$ to the class $\Sigma_\mathcal{M}[\mathfrak{A}, \mathcal{Q}]$.

Corollary 4. *Let* $\yen \in \Sigma_\mathcal{M}[\mathcal{Q}]$. *If* $\vartheta \in \mathcal{A}_p$ *achieves the following conditions*

$$\left| \mathcal{M}_p^{\ell+1} \vartheta(z) - (1-A) \mathcal{M}_p^\ell \vartheta(z) \right| \leq \frac{|A|\kappa \mathcal{Q}}{p},$$

and

$$\left| \yen \left(\mathcal{M}_p^\ell \vartheta(z), \mathcal{M}^{\ell+1}(z), \mathcal{M}_p^{\ell+2} \vartheta(z), \mathcal{M}_p^{\ell+3} \vartheta(z); z \right) \right| < \mathcal{Q}, \tag{51}$$

then

$$\left| \mathcal{M}_p^\ell \vartheta(z) \right| < \mathcal{Q}.$$

Corollary 5. *If* $\kappa \geq 2$, $\mathcal{Q} > 0$. *If* $\vartheta \in \mathcal{A}_p$ *achieves*

$$\left| \mathcal{M}_p^{\ell+1} \vartheta(z) - \mathcal{M}_p^\ell \vartheta(z) \right| < \frac{(\kappa - p)|A| \mathcal{Q}}{p},$$

then

$$\left| \mathcal{M}_p^\ell \vartheta(z) \right| < \mathcal{Q}.$$

Proof. Let $\yen(u_1, u_2, u_3, u_4; z) = u_2 - u_1$. Utilizing Corollary 3 with $\mathfrak{A} = \hbar(\mathbb{D})$ and

$$\hbar(z) = \frac{(\kappa - p)|A| \mathcal{Q}}{p} z, \quad (\mathcal{Q} > 0, z \in \mathbb{D}).$$

We have to find the condition so that $\yen \in \Sigma_\mathcal{M}[\mathfrak{A}, \mathcal{Q}]$, that is, the admissibility condition of Equation (48) is achieved. This follows since

$$\left| \yen \left(\mathcal{Q} e^{i\theta}, \left[\frac{\kappa A}{p} + (1-A) \right] \mathcal{Q} e^{i\theta}, \frac{\mathcal{L} + \left[1 + \frac{2p(1-A)}{A} \right] \kappa + \frac{p^2(1-A)^2}{A^2} \right] \mathcal{Q} e^{i\theta}}{\frac{p^2}{A^2}}, \right. \right.$$

$$\left. \frac{\mathcal{V} + \left[1 + \frac{p(1-A)}{A} \right] 3\mathcal{L} + \left[\left[1 + \frac{3p(1-A)}{A} + \frac{3p^2(1-A)^2}{A^2} \right] \kappa + \frac{p^3(1-A)^3}{A^3} \right] \mathcal{Q} e^{i\theta}}{\frac{p^3}{A^3}} \right) \right|$$

$$= \left| \left[\frac{\kappa}{p} - 1 \right] A \mathcal{Q} e^{i\theta} \right| = \frac{(\kappa - p)|A| \mathcal{Q}}{p}.$$

The required outcome is obtained. □

Corollary 6. *If* $\kappa \geq 2$, $\mathcal{Q} > 0$. *If* $\vartheta \in \mathcal{A}_p$ *achieves*

$$\left| \mathcal{M}_p^{\ell+2} \vartheta(z) - \mathcal{M}_p^{\ell+1} \vartheta(z) \right| < \frac{\left[2\left(1 + \left| 1 + \frac{p(1-2A)}{A} \right| \right) + \left| \frac{p^2(A-1)}{A} \right| \right] \mathcal{Q}}{\frac{p^2}{|A|^2}},$$

then
$$\left|\mathcal{M}_p^\ell \vartheta(z)\right| < \mathcal{Q}.$$

Proof. Let $¥(u_1, u_2, u_3, u_4; z) = u_3 - u_2$. Utilizing Corollary 3 with $\mathfrak{A} = \hbar(\mathbb{D})$ and

$$\hbar(z) = \frac{\left[2\left(1 + \left|1 + \frac{p(1-2A)}{A}\right|\right) + \left|\frac{p^2(A-1)}{A}\right|\right]\mathcal{Q}}{\frac{p^2}{|A|^2}} z, \quad (\mathcal{Q} > 0, z \in \mathbb{D}).$$

It is enough to show that $¥ \in \Sigma_\mathcal{M}[\mathfrak{A}, \mathcal{Q}]$, that is, the admissibility condition of Equation (48) is achieved. This follows since

$$\left|¥\left(\mathcal{Q}e^{i\theta}, \left[\frac{\kappa A}{p} + (1-A)\right]\mathcal{Q}e^{i\theta}, \frac{\mathcal{L} + \left[\left[1 + \frac{2p(1-A)}{A}\right]\kappa + \frac{p^2(1-A)^2}{A^2}\right]\mathcal{Q}e^{i\theta}}{\frac{p^2}{A^2}},\right.\right.$$

$$\left.\left.\frac{\mathcal{V} + \left[1 + \frac{p(1-A)}{A}\right]3\mathcal{L} + \left[\left[1 + \frac{3p(1-A)}{A} + \frac{3p^2(1-A)^2}{A^2}\right]\kappa + \frac{p^3(1-A)^3}{A^3}\right]\mathcal{Q}e^{i\theta}}{\frac{p^3}{A^3}}\right)\right|$$

$$= \left|\frac{\mathcal{L} + \left[1 + \frac{p(1-2A)}{A}\right]\kappa\mathcal{Q}e^{i\theta} + \left[\frac{p^2(A-1)}{A}\right]\mathcal{Q}e^{i\theta}}{\frac{p^2}{A^2}}\right| = \left|\frac{\mathcal{L}e^{-i\theta} + \left[1 + \frac{p(1-2A)}{A}\right]\kappa\mathcal{Q} + \left[\frac{p^2(A-1)}{A}\right]\mathcal{Q}}{\frac{p^2}{A^2}e^{-i\theta}}\right|$$

$$\geq \frac{\Re(\mathcal{L}e^{-i\theta}) + \left|1 + \frac{p(1-2A)}{A}\right|\kappa\mathcal{Q} + \left|\frac{p^2(A-1)}{A}\right|\mathcal{Q}}{\frac{p^2}{|A|^2}} \geq \frac{\left[2\left(1 + \left|1 + \frac{p(1-2A)}{A}\right|\right) + \left|\frac{p^2(A-1)}{A}\right|\right]\mathcal{Q}}{\frac{p^2}{|A|^2}}.$$

This completes the proof. □

Corollary 7. *If $\kappa \geq 2$, $\mathcal{Q} > 0$. If $\vartheta \in \mathcal{A}_p$ achieves*

$$\left|\mathcal{M}_p^{\ell+2}\vartheta(z) - \mathcal{M}_p^{\ell+1}\vartheta(z)\right| < \frac{\left[2\left(\left|2 + \frac{3p(1-A)}{A}\right| + \left|\frac{p(1-A)}{A}\right|\left(1 + \frac{3p(1-A)}{A}\right)\right|\right) + \left|\frac{p^2(1-A)^2}{A^2}\left(\frac{p(1-A)}{A} - 1\right)\right|\right]\mathcal{Q}}{\frac{p^3}{|A|^3}},$$

then
$$\left|\mathcal{M}_p^\ell \vartheta(z)\right| < \mathcal{Q}.$$

Proof. Let $¥(u_1, u_2, u_3, u_4; z) = u_4 - \frac{A}{p}u_3$. Using Corollary 3 with $\mathfrak{A} = \hbar(\mathbb{D})$ and

$$\hbar(z) = \frac{\left[2\left(\left|2 + \frac{3p(1-A)}{A}\right| + \left|\frac{p(1-A)}{A}\right|\left(1 + \frac{3p(1-A)}{A}\right)\right|\right) + \left|\frac{p^2(1-A)^2}{A^2}\left(\frac{p(1-A)}{A} - 1\right)\right|\right]\mathcal{Q}}{\frac{p^3}{|A|^3}} z, \quad (\mathcal{Q} > 0, z \in \mathbb{D}).$$

It is adequate to show that $\mathscr{Y} \in \Sigma_{\mathcal{M}}[\mathfrak{A}, \mathcal{Q}]$, that is, the admissibility condition of Equation (48) is achieved. This follows since

$$\left| \mathscr{Y}\left(\mathcal{Q}e^{i\theta}, \left[\frac{\kappa A}{p}+(1-A)\right]\mathcal{Q}e^{i\theta}, \frac{\mathcal{L}+\left[\left[1+\frac{2p(1-A)}{A}\right]\kappa+\frac{p^2(1-A)^2}{A^2}\right]\mathcal{Q}e^{i\theta}}{\frac{p^2}{A^2}}, \right.\right.$$

$$\left.\left. \frac{\mathcal{V}+\left[1+\frac{p(1-A)}{A}\right]3\mathcal{L}+\left[\left[1+\frac{3p(1-A)}{A}+\frac{3p^2(1-A)^2}{A^2}\right]\kappa+\frac{p^3(1-A)^3}{A^3}\right]\mathcal{Q}e^{i\theta}}{\frac{p^3}{A^3}} \right)\right|$$

$$= \left| \frac{\mathcal{V}+\left[2+\frac{3p(1-A)}{A}\right]\mathcal{L}+\left[\frac{p(1-A)}{A}\left(1+\frac{3p(1-A)}{A}\right)\right]\kappa\mathcal{Q}e^{i\theta}+\left[\frac{p^2(1-A)^2}{A^2}\left(\frac{p(1-A)}{A}-1\right)\right]\mathcal{Q}e^{i\theta}}{\frac{p^3}{A^3}} \right|$$

$$= \left| \frac{\mathcal{V}e^{-i\theta}+\left[2+\frac{3p(1-A)}{A}\right]\mathcal{L}e^{-i\theta}+\left[\frac{p(1-A)}{A}\left(1+\frac{3p(1-A)}{A}\right)\right]\kappa\mathcal{Q}+\left[\frac{p^2(1-A)^2}{A^2}\left(\frac{p(1-A)}{A}-1\right)\right]\mathcal{Q}}{\frac{p^3}{A^3}e^{-i\theta}} \right|$$

$$\geq \frac{\Re(\mathcal{V}e^{-i\theta})+\left|2+\frac{3p(1-A)}{A}\right|\Re(\mathcal{L}e^{-i\theta})+\left|\frac{p(1-A)}{A}\left(1+\frac{3p(1-A)}{A}\right)\right|\kappa\mathcal{Q}+\left|\frac{p^2(1-A)^2}{A^2}\left(\frac{p(1-A)}{A}-1\right)\right|\mathcal{Q}}{\frac{p^3}{|A|^3}}$$

$$\geq \frac{(\kappa-1)\kappa\mathcal{Q}\left|2+\frac{3p(1-A)}{A}\right|+\left|\frac{p(1-A)}{A}\left(1+\frac{3p(1-A)}{A}\right)\right|\kappa\mathcal{Q}+\left|\frac{p^2(1-A)^2}{A^2}\left(\frac{p(1-A)}{A}-1\right)\right|\mathcal{Q}}{\frac{p^3}{|A|^3}}$$

$$\geq \frac{\left[2\left(\left|2+\frac{3p(1-A)}{A}\right|+\left|\frac{p(1-A)}{A}\left(1+\frac{3p(1-A)}{A}\right)\right|\right)+\left|\frac{p^2(1-A)^2}{A^2}\left(\frac{p(1-A)}{A}-1\right)\right|\right]\mathcal{Q}}{\frac{p^3}{|A|^3}}.$$

The required outcome is derived. □

4. Conclusions and Future Directions

In the terms of the $\check{\varsigma}$-Generalized Hurwitz–Lerch Zeta functions (GHLZF) in the z- domain, a new generalized Noor-type linear integral operator is introduced. This operator was utilized to study new classes of holomorphic functions in $\mathbb{D}$. In addition, new applications of the third-order differential subordination outcome that involves this new operator were investigated. The third-order differential inequalities were imposed in this work to show the uppercase of this new generalized Noor-type linear integral operator in $\mathbb{D}$.

Author Contributions: H.A.-J. writing the original draft; F.G. and M.D. writing review and editing. All authors have read and agreed to the published version of the manuscript.

Funding: This research was funded by Universiti Kebangsaan Malaysia, grant number GUP-2019-032.

Conflicts of Interest: The authors declare no conflict of interest.

References

1. Duren, P.L. *Univalent Functions*; Springer: New York, NY, USA, 1983.
2. Koebe, P. Über die Uniformisierung beliebiger analytischer kurven. *Nachrichten Ges. Wiss. Göttingen. Math. Phys. Kl.* **1907**, *1907*, 191–210.
3. Koebe, P. Über eine neue Methode der konformen Abbildung und Uniformisierung. *Nachrichten von der Gesellschaft der Wissenschaften zu Göttingen Mathematisch-Physikalische Klasse* **1912**, *1912*, 844–848.
4. Goodman, A.W. *Univalent Functions, I*; Mariner: Tampa, FL, USA, 1983.
5. Littlewood, J.E. On equalities in the theory of functions. *Proc. Lond. Math. Soc.* **1925**, *23*, 481–519. [CrossRef]
6. Littlewood, J.E. *Lectures on the Theory of Functions*; Oxford University Press: Oxford, UK; London, UK, 1944.
7. Rogosinski, W. On subordinate functions. *Proc. Camb. Philos. Soc.* **1939**, *35*, 1–26. [CrossRef]
8. Rogosinski, W. On the coefficients of subordinate functions. *Proc. London Math. Soc.* **1945**, *48*, 48–82. [CrossRef]
9. Goluzin, G.M. On the majorization principle in function theory (Russian). *Dokl. Akad. Nauk. SSSR* **1953**, *42*, 647–650.
10. Suffridge, T.J. Some remarks on convex maps of the unit disk. *Duke Math. J.* **1970**, *37*, 775–777. [CrossRef]
11. Robinson, R.M. Univalent majorants. *Trans. Am. Math. Soc.* **1947**, *61*, 1–35. [CrossRef]
12. Miller, S.S.; Mocanu, P.T. Differential subordinations and univalent functions. *Michigan Math. J.* **1981**, *28*, 157–171. [CrossRef]
13. Miller, S.S.; Mocanu, P.T. On some classes of first order differential subordinations. *Michigan Math. J.* **1985**, *32*, 185–195. [CrossRef]
14. Miller, S.S.; Mocanu, P.T. Differential subordinations and inequalities in the complex plane. *J. Differ. Eqns.* **1987**, *67*, 199–211. [CrossRef]
15. Miller, S.S.; Mocanu, P.T. The theory and applications of second-order differential subordinations. *Studia Univ. Babeş-Bolyai Math.* **1989**, *34*, 3–33.
16. Miller, S.S.; Mocanu, P.T. A Special Differential Subordination and Its Application to Univalency Conditions. In *Current Topics in Analytic Function Theory*; World Scientific: Singapore; London, UK, 1992; pp. 171–185.
17. Miller, S.S.; Mocanu, P.T. Briot-Bouquet differential equations and differential subordinations. *Complex Var.* **1997**, *33*, 217–237. [CrossRef]
18. Ponnusamy, S.; Juneja, O.P. Third-Order Differential Inequalities in the Complex Plane. In *Current Topics in Analytic Function Theory*; World Scientific: Singapore; London, UK, 1992.
19. Miller, S.S.; Mocanu, P.T. *Differential Subordinations: Theory and Applications*; Marcel Dekker: New York, NY, USA, 2000.
20. Antonino, J.A.; Miller, S.S. Third-order differential inequalities and subordinations in the complex plane. *Complex Var. Appl.* **2011**, *56*, 439–454. [CrossRef]
21. Tang, H.; Srivastiva, H.M.; Li, S.; Ma, L. Third-order differential subordinations and superordination results for meromorphically multivalent functions associated with the Liu-Srivastava Operator. *Abstr. Appl. Anal.* **2014**, *2014*, 1–11. [CrossRef]
22. Tang, H.; Deniz, E. Third-order differential subordinations results for analytic functions involving the generalized Bessel functions. *Acta Math. Sci.* **2014**, *6*, 1707–1719. [CrossRef]
23. Farzana, H.A.; Stephen, B.A.; Jeyaraman, M.P. Certain third-order differential subordination and superordination results of meromorphic multivalent functions. *Asia Pacific J. Math.* **2015**, *2*, 76–87.
24. Tang, H.; Srivastiva, H.M.; Deniz, E.; Li, S.-H. Third-order differential superordination involving the generalized Bessel functions. *Bull. Malay. Math. Soc.* **2015**, *38*, 1669–1688. [CrossRef]
25. Ibrahim, R.W.; Ahmad, M.Z.; Al-Janaby, H.F. The Third-Order Differential Subordination and Superordination involving a fractional operator. *Open Math.* **2015**, *13*, 706–728. [CrossRef]
26. El-Ashwah, R.M.; Hassan, A.H. Some third-order differential subordination and superordination results of some meromorphic functions using a Hurwitz-Lerech Zeta type operator. *Ilirias J. Math.* **2015**, *4*, 1–15.
27. El-Ashwah, R.M.; Hassan, A.A. Third-order differential subordination and superordination results by using Fox-Wright generalized hypergeometric function. *Func. Anal. TMA* **2016**, *2*, 34–51.
28. Attiya, A.A.; Kwon, O.S.; Hyang, P.J.; Cho, N.E. An integrodifferential operator for meromorphic functions associated with the Hurwitz–Lerch Zeta function. *Filomat* **2016**, *30*, 2045–2057. [CrossRef]

29. Răducanu, D. Third-order differential subordinations for analytic functions associated with generalized Mittag-Leffler functions. *Mediterr. J. Math.* **2017**, *14*, 1–18. [CrossRef]
30. Srivastava, H.M.; Prajapati, A.; Gochhayat, P. Third-order differential subordination and differential superordination results for analytic functions involving the Srivastava-Attiya operator. *Appl. Math. Inf. Sci.* **2018**, *12*, 469–481. [CrossRef]
31. Gochhayat, P.; Prajapati, A. Applications of third order differential subordination and superordination involving generalized Struve function. *Filomat* **2019**, *33*, 3047–3059. [CrossRef]
32. Alexander, J.W. Functions which map the interior of the unit circle upon simple regions. *Ann. Math.* **1915**, *17*, 12–22. [CrossRef]
33. Libera, R.J. Some classes of regular univalent functions. *Proc. Am. Math. Soc.* **1965**, *16*, 755–758. [CrossRef]
34. Bernardi, S.D. Convex and starlike univalent functions. *Trans. Am. Math. Soc.* **1969**, *135*, 429–446. [CrossRef]
35. Ruscheweyh, S. New criteria for univalent functions. *Proc. Am. Math. Soc.* **1975**, *49*, 109–115. [CrossRef]
36. Noor, K.I. On new classes of integral operators. *J. Nat. Geom.* **1999**, *16*, 71–80.
37. Cho, N.E.; Saigo, M.; Srivastava, H.M. Some inclusion properties of a certain family of integral operators. *J. Math. Anal. Appl.* **2002**, *276*, 432–445. [CrossRef]
38. Liu, J.-L.; Noor, K.I. Some properties of noor integral operator. *J. Nat. Geom.* **2002**, *21*, 81–90.
39. Branges, L.D. A proof of the Bieberbach conjecture. *Acta Math.* **1984**, *154*, 137–152. [CrossRef]
40. Al-Janay, H.F.; Ghanim, F. On Subclass Noor-Type Harmonic Multivalent Functions Based on Hypergeometric Functions. *Kragujev. J. Math.* **2020**, *45*, 499–519.
41. Al-Janaby, H.F.; Ghanim, F.; Ahmad, M.Z. Harmonic Multivalent Functions Associated with an Extended Generalized Linear Operator of Noor-type. *J. Nonlinear Funct. Anal. Appl.* **2019**, *24*, 269–292.
42. Al-Janaby, H.F.; Ghanim, F.; Agarwal, P. Geometric Studies on Inequalities of Harmonic Functions in a Complex Field Based on ξ-Generalized Hurwitz–Lerch Zeta Function *Iran. J. Math. Sci. Inform.* **2020**, accepted.
43. Al-Janay, H.F.; Ghanim, F.; Darus, M. Some Geometric Properties of Integral Operators Proposed by Hurwitz–Lerch Zeta Function. *IOP Conf. Ser. J. Phys. Conf. Ser.* **2019**, *1212*, 1–6.
44. Ghanim, F. Study of a certain subclass of Hurwitz–Lerch zeta function related to a linear operator. *Abstr. Appl. Anal.* **2013**, *2013*, 1–7. [CrossRef]
45. Ghanim, F.; Darus, M. New result of analytic functions related to Hurwitz-zeta function. *Sci. World J.* **2013**, *2013*, 1–5. [CrossRef]
46. Ghanim, F.; Al-Janaby, H.F. A Certain Subclass of Univalent Meromorphic Functions Defined by a Linear Operator Associated with the Hurwitz–Lerch Zeta Function. *Rad Hrvat. Akad. Znan. Umjet. Mat. Znan.* **2019**, *23*, 71–83. [CrossRef]
47. Răducanu, D.; Srivastava, H.M. A new class of analytic functions defined by means of a convolution operator involving the Hurwitz–Lerch zeta function. *Integr. Trans. Spec. Funct.* **2007**, *18*, 933–943. [CrossRef]
48. Srivastava, H.M.; Attiya, A.A. An integral operator associated with the Hurwitz–Lerch zeta function and differential subordination. *Integr. Trans. Spec. Funct.* **2007**, *18*, 207–216. [CrossRef]
49. Srivastava, H.M.; Gaboury, S.A.; Ghanim, F. Certain subclasses of meromorphically univalent functions defined by a linear operator associated with the λ−generalized Hurwitz–Lerch zeta function. *Integr. Transf. Spec. Funct.* **2015**, *26*, 258–272. [CrossRef]
50. Srivastava, H.M.; Gaboury, S.A.; Ghanim, F. Some further properties of a linear operator associated with the λ−generalized Hurwitz–Lerch zeta function related to the class of meromorphically univalent functions. *Appl. Math. Comput.* **2015**, *259*, 1019–1029.
51. Xing, S.C.; Jose, L.L. A note on the asymptotic expansion of the Lerch's transcendent. *Integr. Trans. Spec. Funct.* **2019**, *30*, 844–855.
52. Choi, J.; Srivastava, H.M. The Multiple Hurwitz Zeta Function and the Multiple Hurwitz-Euler Eta Function. *Taiwan J. Math.* **2011**, *15*, 501–522. [CrossRef]
53. Milovanovic, G.V.; Rassias, M.T. (Eds.) *Analytic Number Theory, Approximation Theory and Special Functions—In Honor of Hari M. Srivastava*; Springer: Basel, Switzerland, 2014.
54. Rassias, M.T.; Yang, B. On an Equivalent Property of a Reverse Hilbert-Type Integral Inequality Related to the Extended Hurwitz-Zeta Function. *J. Math. Inequal.* **2019**, *13*, 315–334. [CrossRef]
55. Rassias, M.T.; Yang, B. On a Hilbert-type integral inequality related to the extended Hurwitz zeta function in the whole plane. *Acta Appl. Math.* **2019**, *160*, 67–80. [CrossRef]

56. Rassias, M.T.; Yang, B. Equivalent properties of a Hilbert-type integral inequality with the best constant factor related to the Hurwitz zeta function. *Ann. Funct. Anal.* **2018**, *9*, 282–295. [CrossRef]
57. Rassias, M.T.; Yang, B.; Raigorodskii, A. Two Kinds of the Reverse Hardy-Type Integral Inequalities with the Equivalent Forms Related to the Extended Riemann Zeta Function. *Appl. Anal. Discret. Math.* **2018**, *12*, 273–296. [CrossRef]
58. Srivastava, H.M.; Choi, J. *Series Associated with Zeta and Related Functions*; Kluwer Academic Publishers: Dordrecht, Germany, 2001.
59. Srivastava, H.M. Some formulas for the Bernoulli and Euler polynomials at rational arguments. *Math. Proc. Camb. Philos. Soc.* **2000**, *129*, 77–84. [CrossRef]
60. Srivastava, H.M.; Choi, J. *Zeta and q−zeta Functions and Associated Series and Integrals*; Elsevier Science Publishers: Amsterdam, The Netherland, 2012.
61. Srivastava, H.M. A new family of the λ−generalized Hurwitz–Lerch zeta functions with applications. *Appl. Math. Inf. Sci.* **2014**, *8*, 1485–1500. [CrossRef]
62. Mathai, A.M.; Saxena, R.K.; Haubold, H.J. *The H-function: Theory and applications*; Springer: New York, NY, USA, 2010.
63. Srivastava, H.M.; Gupta, K.C.; Goyal, S.P. *The H−functions of One and Two Variables with Applications*; South Asian Publishers: New Delhi, India, 1982.
64. Srivastava, H.M.; Gaboury, S.; Tremblay, R. New relations involving an extended multiparameter Hurwitz–Lerch zeta function with applications. *Int. J. Anal.* **2014**, *2014*, 1–14. [CrossRef]
65. Cho, N.E.; Kwon, O.S.; Srivastava, H.M. Inclusion relationships and argument properties for certain subclasses of multivalent functions associated with a family of linear operators. *J. Math. Anal. Appl.* **2004**, *292*, 470–483. [CrossRef]
66. Noor, K.L. Integral operators defined by convolution with hypergeometric functions. *Appl. Math. Comput.* **2006**, *182*, 1872–1881. [CrossRef]
67. Darus, M.; Ibrahim, R.W. Integral operator defined by convolution product of hypergeometric functions. *Int. J. Nonlinear Sci.* **2012**, *13*, 153–157.

© 2020 by the authors. Licensee MDPI, Basel, Switzerland. This article is an open access article distributed under the terms and conditions of the Creative Commons Attribution (CC BY) license (http://creativecommons.org/licenses/by/4.0/).

Article

Subclasses of Bi-Univalent Functions Defined by Frasin Differential Operator

Ibtisam Aldawish [1], Tariq Al-Hawary [2] and B. A. Frasin [3,*]

[1] Department of Mathematics and Statistics, College of Science, IMSIU (Imam Mohammed Ibn Saud Islamic University), P.O. Box 90950, Riyadh 11623, Saudi Arabia; imaldawish@imamu.edu.sa
[2] Department of Applied Science, Ajloun College, Al-Balqa Applied University, Ajloun 26816, Jordan; tariq_amh@bau.edu.jo
[3] Faculty of Science, Department of Mathematics, Al al-Bayt University, Mafraq 25113, Jordan
* Correspondence: bafrasin@yahoo.com

Received: 17 April 2020; Accepted: 9 May 2020; Published: 13 May 2020

Abstract: Let Ω denote the class of functions $f(z) = z + a_2 z^2 + a_3 z^3 + \cdots$ belonging to the normalized analytic function class $\mathcal{A}$ in the open unit disk $\mathbb{U} = \{z : |z| < 1\}$, which are bi-univalent in $\mathbb{U}$, that is, both the function f and its inverse f^{-1} are univalent in $\mathbb{U}$. In this paper, we introduce and investigate two new subclasses of the function class Ω of bi-univalent functions defined in the open unit disc $\mathbb{U}$, which are associated with a new differential operator of analytic functions involving binomial series. Furthermore, we find estimates on the Taylor–Maclaurin coefficients $|a_2|$ and $|a_3|$ for functions in these new subclasses. Several (known or new) consequences of the results are also pointed out.

Keywords: analytic functions; univalent functions; bi-univalent functions; Taylor–Maclaurin series

MSC: 30C45

1. Introduction and Definitions

Let $\mathcal{A}$ be the class of all analytic functions f in the open unit disk $\mathbb{U} = \{z : |z| < 1\}$, normalized by the conditions $f(0) = 0$ and $f'(0) = 1$ of the form

$$f(z) = z + \sum_{n=2}^{\infty} a_n z^n. \tag{1}$$

Further, by $\mathcal{S}$ we shall denote the class of all functions in $\mathcal{A}$ which are univalent in $\mathbb{U}$.

A function $f \in \mathcal{A}$ is said to be starlike if $f(\mathbb{U})$ is a starlike domain with respect to the origin; i.e., the line segment joining any point of $f(\mathbb{U})$ to the origin lies entirely in $f(\mathbb{U})$ and a function $f \in \mathcal{A}$ is said to be convex if $f(\mathbb{U})$ is a convex domain; i.e., the line segment joining any two points in $f(\mathbb{U})$ lies entirely in $f(\mathbb{U})$. Analytically, $f \in \mathcal{A}$ is starlike, denoted by $\mathcal{S}^*$, if and only if $\operatorname{Re}(zf'(z)/f(z)) > 0$, whereas $f \in \mathcal{A}$ is convex, denoted by $\mathcal{K}$, if and only if $\operatorname{Re}(1 + zf''(z)/f'(z)) > 0$. The classes $\mathcal{S}^*(\alpha)$ and $\mathcal{K}(\alpha)$ of starlike and convex functions of order $\alpha (0 \leq \alpha < 1)$, are respectively characterized by

$$\operatorname{Re}\left(\frac{zf'(z)}{f(z)}\right) > \alpha \qquad (z \in \mathbb{U}), \tag{2}$$

and

$$\operatorname{Re}\left(1 + \frac{zf''(z)}{f'(z)}\right) > \alpha \qquad (z \in \mathbb{U}). \tag{3}$$

For a function f in $\mathcal{A}$, and making use of the binomial series

$$(1-\lambda)^m = \sum_{j=0}^{m} \binom{m}{j}(-1)^j \lambda^j \quad (m \in \mathbb{N} = \{1,2,\cdots\}, j \in \mathbb{N}_0 = \mathbb{N} \cup \{0\}),$$

Frasin [1] (see also [2–4]) introduced the differential operator $D_{m,\lambda}^{\zeta} f(z)$ defined as follows:

$$D^0 f(z) = f(z), \quad (4)$$
$$D_{m,\lambda}^1 f(z) = (1-\lambda)^m f(z) + (1-(1-\lambda)^m) z f'(z) = D_{m,\lambda} f(z), \; \lambda > 0; m \in \mathbb{N}, \quad (5)$$
$$D_{m,\lambda}^{\zeta} f(z) = D_{m,\lambda}(D_{m,\lambda}^{\zeta-1} f(z)) \quad (\zeta \in \mathbb{N}). \quad (6)$$

If f is given by Equation (1), then from Equations (5) and (6) we see that

$$D_{m,\lambda}^{\zeta} f(z) = z + \sum_{n=2}^{\infty} \left(1 + (n-1) \sum_{j=1}^{m} \binom{m}{j}(-1)^{j+1} \lambda^j \right)^{\zeta} a_n z^n, \; \zeta \in \mathbb{N}_0. \quad (7)$$

Using the relation in Equation (7), it is easily verified that

$$C_j^m(\lambda) z (D_{m,\lambda}^{\zeta} f(z))' = D_{m,\lambda}^{\zeta+1} f(z) - (1 - C_j^m(\lambda)) D_{m,\lambda}^{\zeta} f(z) \quad (8)$$

where $C_j^m(\lambda) := \sum_{j=1}^{m} \binom{m}{j}(-1)^{j+1} \lambda^j$.

We observe that for $m = 1$, we obtain the differential operator $D_{1,\lambda}^{\zeta}$ defined by Al-Oboudi [5] and for $m = \lambda = 1$, we get Sălăgean differential operator D^{ζ} [6].

In [7], Frasin defined the subclass $S(\alpha, s, t)$ of analytic functions f satisfying the following condition

$$\operatorname{Re}\left\{\frac{(s-t)zf'(sz)}{f(sz) - f(tz)}\right\} > \alpha, \quad (9)$$

for some $0 \leq \alpha < 1$ $s, t \in \mathbb{C}$ with $|s| \leq 1$; $|t| \leq 1$; $s \neq t$ and for all $z \in \mathbb{U}$. We also denote by $\mathcal{T}(\alpha, s, t)$ the subclass of $\mathcal{A}$ consisting of all functions $f(z)$ such that $zf'(z) \in S(\alpha, s, t)$. The class $S(\alpha, 1, t)$ was introduced and studied by Owa et al. [8]. When $t = -1$, the class $S(\alpha, 1, -1) \equiv S_s(\alpha)$ was introduced by Sakaguchi [9] and is called Sakaguchi function of order α (see [10,11]), where as $S_s(0) = S_s$ is the class of starlike functions with respect to symmetrical points in $\mathbb{U}$. In addition, we note that $S(\alpha, 1, 0) \equiv S^*(\alpha)$ and $\mathcal{T}(\alpha, 1, 0) = \mathcal{K}(\alpha)$.

Determination of the bounds for the coefficients a_n is an important problem in geometric function theory as they give information about the geometric properties of these functions. For example, the bound for the second coefficient a_2 of functions in S gives the growth and distortion bounds as well as covering theorems. It is well known that the n-th coefficient a_n is bounded by n for each $f \in S$.

In this paper, we estimate the initial coefficients $|a_2|$ and $|a_3|$ coefficient problem for certain subclasses of bi-univalent functions.

The Koebe one-quarter theorem [12] proves that the image of $\mathbb{U}$ under every univalent function $f \in S$ contains the disk of radius $\frac{1}{4}$. Therefore, every function $f \in S$ has an inverse f^{-1}, defined by

$$f^{-1}(f(z)) = z \quad (z \in \mathbb{U})$$

and

$$f(f^{-1}(w)) = w, \quad \left(|w| < r_0(f), \; r_0(f) \geq \frac{1}{4}\right),$$

where
$$f^{-1}(w) = h(w) = w + \sum_{n=2}^{\infty} A_n w^n. \tag{10}$$

A simple computation shows that
$$w = f(h(w)) = w + (A_2 + a_2)w^2 + (A_3 - 2a_2^2 + a_3)w^3 + (A_4 + 5a_2^3 - 5a_2 a_3 + a_4)w^4 + \cdots. \tag{11}$$

Comparing the initial coefficients in Equation (11), we find that $A_2 = -a_2$, $A_3 = 2a_2^2 - a_3$ and $A_4 = 5a_2^3 + 5a_2 a_3 - a_4$.

By putting these values in the Equation (10), we get
$$f^{-1}(w) = w - a_2 w^2 + (2a_2^2 - a_3)w^3 - (5a_2^3 - 5a_2 a_3 + a_4)w^4 + \cdots.$$

A function $f \in \mathcal{A}$ is said to be bi-univalent in the open unit disk $\mathbb{U}$ if both the function f and its inverse f^{-1} are univalent there. Let Ω denote the class of bi-univalent functions defined in the univalent unit disk $\mathbb{U}$. Examples of functions in the class Ω are
$$\frac{z}{1-z}, \quad \log \frac{1}{1-z}, \quad \log \sqrt{\frac{1+z}{1-z}}.$$

However, the familiar Koebe function is not a member of Ω. Other common examples of functions in $\mathbb{U}$ such as
$$\frac{2z - z^2}{2} \quad \text{and} \quad \frac{z}{1 - z^2}$$
are not members of Ω either.

Finding bounds for the coefficients of classes of bi-univalent functions dates back to 1967 (see Lewin [13]). Brannan and Taha [14] (see also [15]) introduced certain subclasses of the bi-univalent function class Ω similar to the familiar subclasses $\mathcal{S}^*(\alpha)$ and $\mathcal{K}(\alpha)$ (see [16]). Thus, following Brannan and Taha [14] (see also [15]), a function $f \in \mathcal{A}$ is in the class $\mathcal{S}^*_\Omega[\alpha]$ of strongly bi-starlike functions of order α ($0 < \alpha \le 1$) if each of the following conditions are satisfied:
$$f \in \Omega \text{ and } \left| \arg \left(\frac{z f'(z)}{f(z)} \right) \right| < \frac{\alpha \pi}{2} \quad (0 < \alpha \le 1,\ z \in \mathbb{U})$$
and
$$\left| \arg \left(\frac{z g'(w)}{g(w)} \right) \right| < \frac{\alpha \pi}{2} \quad (0 < \alpha \le 1,\ w \in \mathbb{U}),$$
where g is the extension of f^{-1} to $\mathbb{U}$. The classes $\mathcal{S}^*_\Omega(\alpha)$ and $\mathcal{K}_\Omega(\alpha)$ of bi-starlike functions of order α and bi-convex functions of order α, corresponding (respectively) to the function classes defined by Equations (2) and (3), were also introduced analogously. For each of the function classes $\mathcal{S}^*_\Omega(\alpha)$ and $\mathcal{K}_\Omega(\alpha)$, they found non-sharp estimates on the first two Taylor–Maclaurin coefficients $|a_2|$ and $|a_3|$ (for details, see [14,15]).

Motivated by the earlier works of Srivastava et al. [17] and Frasin and Aouf [18] (see also [10,12,13,19–33]) in the present paper we introduce two new subclasses $\mathcal{B}^\zeta_\Omega(\lambda, \alpha, s, t)$ and $\mathcal{B}^\zeta_\Omega(\lambda, \beta, s, t)$ of the function class Ω, that generalize the previous defined classes. This subclass is defined with the aid of the new differential operator $D^\zeta_{m,\lambda}$ of analytic functions involving binomial series in the open unit disk $\mathbb{U}$. In addition, upper bounds for the second and third coefficients for functions in this new subclass are derived.

In order to derive our main results, we have to recall the following lemma [34].

Lemma 1. *If* $\mathbb{P} \in \mathcal{P}$ *then*
$$|c_k| \le 2 \quad (k \in \mathbb{N}),$$

where $\mathcal{P}$ is the family of all functions $\mathbb{P}$, analytic in $\mathbb{U}$, for which

$$\text{Re}(\mathbb{P}(z)) > 0 \quad (z \in \mathbb{U}),$$

where $\mathbb{P}(z) = 1 + c_1 z + c_2 z^2 + c_3 z^3 + \cdots \ (z \in \mathbb{U})$.

Unless otherwise mentioned, we presume throughout this paper that

$$\lambda > 0; m \in \mathbb{N}, s, t \in \mathbb{C} \text{ with } |s| \leq 1; |t| \leq 1; s \neq t; \zeta \in \mathbb{N}_0.$$

2. Coefficient Bounds for the Function Class $\mathcal{B}_\Omega^\zeta(\lambda, \alpha, s, t)$

Definition 1. *A function $f(z)$ given by Equation (1) is said to be in the class $\mathcal{B}_\Omega^\zeta(\lambda, \alpha, s, t)$ if the following conditions are satisfied:*

$$f \in \Omega \text{ and } \left| \arg \left(\frac{(s-t)z(D_{m,\lambda}^\zeta f(z))'}{D_{m,\lambda}^\zeta f(sz) - D_{m,\lambda}^\zeta f(tz)} \right) \right| < \frac{\alpha \pi}{2} \quad (0 < \alpha \leq 1, z \in \mathbb{U}) \tag{12}$$

and

$$\left| \arg \left(\frac{(s-t)w(D_{m,\lambda}^\zeta g(w))'}{D_{m,\lambda}^\zeta g(sw) - D_{m,\lambda}^\zeta g(tw)} \right) \right| < \frac{\alpha \pi}{2} \quad (0 < \alpha \leq 1, w \in \mathbb{U}) \tag{13}$$

where the function g is given by

$$g(w) = w - a_2 w^2 + (2a_2^2 - a_3)w^3 - (5a_2^3 - 5a_2 a_3 + a_4)w^4 + \cdots. \tag{14}$$

We begin by finding the estimates on the coefficients $|a_2|$ and $|a_3|$ for functions in the class $\mathcal{B}_\Omega^\zeta(\lambda, \alpha, s, t)$.

Theorem 1. *Let $f(z)$ given by (1) be in the class $\mathcal{B}_\Omega^\zeta(\lambda, \alpha, s, t)$. Then*

$$|a_2| \leq \frac{2\alpha}{\sqrt{\left| \alpha(6 - 2s^2 - 2t^2 - 2ts)\left(1 + 2C_j^m(\lambda)\right)^\zeta - \left(1 + C_j^m(\lambda)\right)^{2\zeta} [2\alpha(2s + 2t - t^2 - s^2 - 2ts) + (\alpha - 1)(2 - s - t)^2] \right|}} \tag{15}$$

and

$$|a_3| \leq \frac{4\alpha^2}{|(2-s-t)^2| \left(1 + C_j^m(\lambda)\right)^{2\zeta}} + \frac{2\alpha}{|(3 - s^2 - t^2 - ts)| \left(1 + 2C_j^m(\lambda)\right)^\zeta}. \tag{16}$$

Proof. From Equations (12) and (13), we have

$$\frac{(s-t)z \left(D_{m,\lambda}^\zeta f(z)\right)'}{D_{m,\lambda}^\zeta f(sz) - D_{m,\lambda}^\zeta f(tz)} = [p(z)]^\alpha \tag{17}$$

and

$$\frac{(s-t)w \left(D_{m,\lambda}^\zeta g(w)\right)'}{D_{m,\lambda}^\zeta g(sw) - D_{m,\lambda}^\zeta g(tw)} = [q(w)]^\alpha, \tag{18}$$

where $p(z)$ and $q(w)$ in $\mathcal{P}$ and have the forms

$$p(z) = 1 + p_1 z + p_2 z^2 + p_3 z^3 + \cdots \tag{19}$$

and

$$q(w) = 1 + q_1 w + q_2 w^2 + q_3 w^3 + \cdots . \tag{20}$$

This yields the following relations:

$$(2 - s - t)\left(1 + C_j^m(\lambda)\right)^\zeta a_2 = \alpha p_1, \tag{21}$$

$$(3 - s^2 - t^2 - ts)\left(1 + 2C_j^m(\lambda)\right)^\zeta a_3 - (2s + 2t - s^2 - 2ts - t^2)\left(1 + C_j^m(\lambda)\right)^{2\zeta} a_2^2 \tag{22}$$
$$= \alpha p_2 + \frac{\alpha(\alpha - 1)}{2} p_1^2,$$

$$-(2 - s - t)\left(1 + C_j^m(\lambda)\right)^\zeta a_2 = \alpha q_1 \tag{23}$$

and

$$\left[(6 - 2s^2 - 2t^2 - 2ts)\left(1 + 2C_j^m(\lambda)\right)^\zeta - (2s + 2t - s^2 - t^2 - 2ts)\left(1 + C_j^m(\lambda)\right)^{2\zeta}\right] a_2^2 \tag{24}$$
$$-(3 - s^2 - t^2 - ts)\left(1 + 2C_j^m(\lambda)\right)^\zeta a_3 = \alpha q_2 + \frac{\alpha(\alpha - 1)}{2} q_1^2.$$

From Equations (21) and (23), we obtain

$$p_1 = -q_1 \tag{25}$$

and

$$2(2 - s - t)^2 \left(1 + C_j^m(\lambda)\right)^{2\zeta} a_2^2 = \alpha^2 (p_1^2 + q_1^2). \tag{26}$$

Now by adding Equation (22) and Equation (24), we deduce that

$$\left[(6 - 2s^2 - 2t^2 - 2ts)\left(1 + 2C_j^m(\lambda)\right)^\zeta - 2(2s + 2t - t^2 - s^2 - 2ts)\left(1 + C_j^m(\lambda)\right)^{2\zeta}\right] a_2^2 \tag{27}$$
$$= \alpha(p_2 + q_2) + \frac{\alpha(\alpha - 1)}{2}(p_1^2 + q_1^2).$$

From Equations (27) and (26), we have

$$\alpha \left[(6 - 2s^2 - 2t^2 - 2ts)\left(1 + 2C_j^m(\lambda)\right)^\zeta - 2(2s + 2t - t^2 - s^2 - 2ts)\left(1 + C_j^m(\lambda)\right)^{2\zeta}\right] a_2^2 \tag{28}$$
$$= \alpha^2(p_2 + q_2) + (\alpha - 1)(2 - s - t)^2 \left(1 + C_j^m(\lambda)\right)^{2\zeta} a_2^2.$$

Therefore, we have

$$a_2^2 = \frac{\alpha^2(p_2 + q_2)}{\left|\alpha(6 - 2s^2 - 2t^2 - 2ts)\left(1 + 2C_j^m(\lambda)\right)^\zeta - 2\alpha(2s + 2t - t^2 - s^2 - 2ts)\left(1 + C_j^m(\lambda)\right)^{2\zeta} - (\alpha - 1)(2 - s - t)^2 \left(1 + C_j^m(\lambda)\right)^{2\zeta}\right|}.$$

Applying Lemma 1 for the coefficients p_2 and q_2, we immediately have

$$|a_2| \leq \frac{2\alpha}{\sqrt{\left|\alpha(6 - 2s^2 - 2t^2 - 2ts)\left(1 + 2C_j^m(\lambda)\right)^\zeta - \left(1 + C_j^m(\lambda)\right)^{2\zeta}\left[2\alpha(2s + 2t - t^2 - s^2 - 2ts) + (\alpha - 1)(2 - s - t)^2\right]\right|}}$$

which gives us the desired estimate on $|a_2|$ as asserted in Equation (15).

Next in order to find the bound on $|a_3|$, by subtracting Equation (24) from Equation (22), we get

$$2(3 - s^2 - t^2 - ts)\left(1 + 2C_j^m(\lambda)\right)^\zeta a_3 - (6 - 2s^2 - 2t^2 - 2ts)\left(1 + 2C_j^m(\lambda)\right)^\zeta a_2^2$$
$$= \alpha(p_2 - q_2) + \frac{\alpha(\alpha - 1)}{2}(p_1^2 - q_1^2). \tag{29}$$

From Equations (25), (26) and (29), we obtain

$$2(3 - s^2 - t^2 - ts)\left(1 + 2C_j^m(\lambda)\right)^\zeta a_3$$
$$= (6 - 2s^2 - 2t^2 - 2ts)\left(1 + 2C_j^m(\lambda)\right)^\zeta \frac{\alpha^2(p_1^2 + q_1^2)}{2(2 - s - t)^2\left(1 + C_j^m(\lambda)\right)^{2\zeta}} + \alpha(p_2 - q_2)$$

or, equivalently,

$$a_3 = \frac{\alpha^2(p_1^2 + q_1^2)}{2(2 - s - t)^2\left(1 + C_j^m(\lambda)\right)^{2\zeta}} + \frac{\alpha(p_2 - q_2)}{2(3 - s^2 - t^2 - ts)\left(1 + 2C_j^m(\lambda)\right)^\zeta}.$$

Applying Lemma 1 for the coefficients p_1, p_2, q_1 and q_2, we have

$$|a_3| \leq \frac{4\alpha^2}{|(2 - s - t)^2|\left(1 + C_j^m(\lambda)\right)^{2\zeta}} + \frac{2\alpha}{|(3 - s^2 - t^2 - ts)|\left(1 + 2C_j^m(\lambda)\right)^\zeta}.$$

We get desired estimate on $|a_3|$ as asserted in Equation (16). □

Putting $\zeta = 0$ in Theorem 1, we get the following consequence.

Corollary 1. *Let $f(z)$ given by Equation (1) be in the class $B_\Omega^0(\alpha, s, t)$, $0 < \alpha \leq 1$. Then*

$$|a_2| \leq \frac{2\alpha}{\sqrt{|\alpha(6 - 2s^2 - 2t^2 - 2ts) - [2\alpha(2s + 2t - t^2 - s^2 - 2ts) + (\alpha - 1)(2 - s - t)^2]|}}$$

and

$$|a_3| \leq \frac{4\alpha^2}{|(2 - s - t)^2|} + \frac{2\alpha}{|(3 - s^2 - t^2 - ts)|}.$$

Putting $s = 1$ and $t = -1$ in Corollary 1, we immediately have the following result.

Corollary 2. *Let $f(z)$ given by Equation (1) be in the class $B_\Omega^0(\alpha, 1, -1)$, $0 < \alpha \leq 1$. Then*

$$|a_2| \leq \alpha$$

and
$$|a_3| \leq \alpha(\alpha+1).$$

If we put $s = 1$ and $t = 0$ in Corollary 1, we obtain well-known the class $\mathcal{S}_\Omega^*[\alpha]$ of strongly bi-starlike functions of order α and get the following corollary.

Corollary 3. *Let $f(z)$ given by Equation (1) be in the class $\mathcal{S}_\Omega^*[\alpha]$, $0 < \alpha \leq 1$. Then*
$$|a_2| \leq \frac{2\alpha}{\sqrt{\alpha+1}}$$

and
$$|a_3| \leq \alpha(4\alpha+1).$$

3. Coefficient Bounds for the Function Class $\mathcal{B}_\Omega^\zeta(\lambda, \beta, s, t)$

Definition 2. *A function $f(z)$ given by Equation (1) is said to be in the class $\mathcal{B}_\Omega^\zeta(\lambda, \beta, s, t)$ if the following conditions are satisfied:*

$$f \in \Omega \text{ and } \mathrm{Re}\left(\frac{(s-t)z(D_{m,\lambda}^\zeta f(z))'}{D_{m,\lambda}^\zeta f(sz) - D_{m,\lambda}^\zeta f(tz)}\right) > \beta \qquad (0 \leq \beta < 1,\ z \in \mathbb{U}) \tag{30}$$

and

$$\mathrm{Re}\left(\frac{(s-t)w(D_{m,\lambda}^\zeta g(w))'}{D_{m,\lambda}^\zeta g(sw) - D_{m,\lambda}^\zeta g(tw)}\right) > \beta \qquad (0 \leq \beta < 1,\ w \in \mathbb{U}) \tag{31}$$

where the function g is given by Equation (14).

Theorem 2. *Let $f(z)$ given by Equation (1) be in the class $\mathcal{B}_\Omega^\zeta(\lambda, \beta, s, t)$. Then*

$$|a_2| \leq \sqrt{\frac{2(1-\beta)}{|(3-s^2-t^2-ts)\left(1+2C_j^m(\lambda)\right)^\zeta - (2s+2t-t^2-s^2-2ts)\left(1+C_j^m(\lambda)\right)^{2\zeta}|}} \tag{32}$$

and

$$|a_3| \leq \frac{4(1-\beta)^2}{|(2-s-t)^2|\left(1+C_j^m(\lambda)\right)^{2\zeta}} + \frac{2(1-\beta)}{|(3-s^2-t^2-ts)|\left(1+2C_j^m(\lambda)\right)^\zeta}. \tag{33}$$

Proof. It follows from Equations (30) and (31) that there exist p and $q \in \mathcal{P}$ such that

$$\frac{(s-t)z\left(D_{m,\lambda}^\zeta f(z)\right)'}{D_{m,\lambda}^\zeta f(sz) - D_{m,\lambda}^\zeta f(tz)} = \beta + (1-\beta)p(z) \tag{34}$$

and

$$\frac{(s-t)w\left(D_{m,\lambda}^\zeta g(w)\right)'}{D_{m,\lambda}^\zeta g(sw) - D_{m,\lambda}^\zeta g(tw)} = \beta + (1-\beta)q(w) \tag{35}$$

where $p(z)$ and $q(w)$ in $\mathcal{P}$ given by Equations (19) and (20).
This yields the following relations:

$$(2-s-t)\left(1+C_j^m(\lambda)\right)^\zeta a_2 = (1-\beta)p_1, \tag{36}$$

39

$$(3-s^2-t^2-ts)\left(1+C_j^m(\lambda)\right)^\zeta a_3 - (2s+2t-s^2-2ts-t^2)\left(1+C_j^m(\lambda)\right)^{2\zeta} a_2^2$$
$$= (1-\beta)p_2, \tag{37}$$

$$-(2-s-t)\left(1+C_j^m(\lambda)\right)^\zeta a_2 = (1-\beta)q_1 \tag{38}$$

and

$$\left[(6-2s^2-2t^2-2ts)\left(1+C_j^m(\lambda)\right)^\zeta - (2s+2t-s^2-t^2-2ts)\left(1+C_j^m(\lambda)\right)^{2\zeta}\right]a_2^2$$
$$-(3-s^2-t^2-ts)\left(1+C_j^m(\lambda)\right)^\zeta a_3 = (1-\beta)q_2. \tag{39}$$

From Equations (36) and (38), we obtain

$$p_1 = -q_1 \tag{40}$$

and

$$2(2-s-t)^2\left(1+C_j^m(\lambda)\right)^{2\zeta} a_2^2 = (1-\beta)^2(p_1^2+q_1^2). \tag{41}$$

Now by adding Equation (37) and Equation (39), we deduce that

$$\left[(6-2s^2-2t^2-2ts)\left(1+C_j^m(\lambda)\right)^\zeta - 2(2s+2t-s^2-t^2-2ts)\left(1+C_j^m(\lambda)\right)^{2\zeta}\right]a_2^2$$
$$= (1-\beta)(p_2+q_2). \tag{42}$$

Thus, we have

$$|a_2^2| \leq \frac{(1-\beta)(|p_2|+|q_2|)}{|(6-2s^2-2t^2-2ts)\left(1+C_j^m(\lambda)\right)^\zeta - 2(2s+2t-t^2-s^2-2ts)\left(1+C_j^m(\lambda)\right)^{2\zeta}|}$$
$$= \frac{2(1-\beta)}{|(3-s^2-t^2-ts)\left(1+C_j^m(\lambda)\right)^\zeta - (2s+2t-t^2-s^2-2ts)\left(1+C_j^m(\lambda)\right)^{2\zeta}|}$$

which gives us the desired estimate on $|a_2|$ as asserted in Equation (32). Next in order to find the bound on $|a_3|$, by subtracting Equation (39) from Equation (37), we get

$$2(3-s^2-t^2-ts)\left(1+C_j^m(\lambda)\right)^\zeta a_3 - (6-2s^2-2t^2-2ts)\left(1+2C_j^m(\lambda)\right)^\zeta a_2^2$$
$$= (1-\beta)(p_2-q_2). \tag{43}$$

From Equations (40), (41) and (43), we obtain

$$2(3-s^2-t^2-ts)\left(1+2C_j^m(\lambda)\right)^\zeta a_3$$
$$= (1-\beta)(p_2-q_2) + (6-2s^2-2t^2-2ts)\left(1+2C_j^m(\lambda)\right)^\zeta \frac{(1-\beta)^2(p_1^2+q_1^2)}{2(2-s-t)^2\left(1+C_j^m(\lambda)\right)^{2\zeta}}$$

or, equivalently,

$$a_3 = \frac{(1-\beta)^2(p_1^2+q_1^2)}{2(2-s-t)^2\left(1+C_j^m(\lambda)\right)^{2\zeta}} + \frac{(1-\beta)(p_2-q_2)}{2(3-s^2-t^2-ts)\left(1+2C_j^m(\lambda)\right)^\zeta}.$$

Applying Lemma 1 for the coefficients p_1, p_2, q_1 and q_2, we have

$$|a_3| \leq \frac{4(1-\beta)^2}{|(2-s-t)^2|\left(1+C_j^m(\lambda)\right)^{2\zeta}} + \frac{2(1-\beta)}{|(3-s^2-t^2-ts)|\left(1+2C_j^m(\lambda)\right)^{\zeta}}.$$

We get desired estimate on $|a_3|$ as asserted in Equation (33). □

It is worth to mention that a similar technique in the real space has been used in the study of random environments, see [35].

Putting $\zeta = 0$ in Theorem 2, we have the following corollary.

Corollary 4. *Let $f(z)$ given by Equation (1) be in the class $\mathcal{B}_{\Omega}^0(\beta, s, t)$. Then*

$$|a_2| \leq \sqrt{\frac{2(1-\beta)}{|3+st-2(s+t)|}}$$

and

$$|a_3| \leq \frac{4(1-\beta)^2}{|(2-s-t)^2|} + \frac{2(1-\beta)}{|(3-s^2-t^2-ts)|}.$$

Putting $s = 1$ and $t = -1$ in Corollary 4, we immediately have the following result.

Corollary 5. *Let $f(z)$ given by Equation (1) be in the class $\mathcal{B}_{\Omega}^0(\beta, 1, -1), 0 \leq \beta < 1$. Then*

$$|a_2| \leq \sqrt{1-\beta}$$

and

$$|a_3| \leq (1-\beta)(2-\beta).$$

If we take $s = 1$ and $t = 0$ in Corollary 4, we obtain well-known the class $\mathcal{S}_{\Omega}^*(\beta)$ of strongly bi-starlike functions of order β and get the following corollary.

Corollary 6. *Let $f(z)$ given by Equation (1) be in the class $\mathcal{S}_{\Omega}^*(\beta), 0 \leq \beta < 1$. Then*

$$|a_2| \leq \sqrt{2(1-\beta)}$$

and

$$|a_3| \leq (1-\beta)(5-4\beta).$$

4. Conclusions

In this paper, two new subclasses of bi-univalent functions related to a new differential operator $D_{m,\lambda}^{\zeta}$ of analytic functions involving binomial series in the open unit disk $\mathbb{U}$ were introduced and investigated. Furthermore, we obtained the second and third Taylor–Maclaurin coefficients of functions in these classes. The novelty of our paper consists of the fact that the operator used by defining the new subclasses of Ω is a very general operator that generalizes two important differential operators, Sălăgean differential operator D^{ζ} and Al-Oboudi differential operator $D_{1,\lambda}^{\zeta}$. These operators are playing an important role in geometric function theory to define new generalized subclasses of analytic univalent functions and then study their properties. The special cases taken from the main results confirm the validity of these results. We mentioned that all the above estimates for the coefficients $|a_2|$ and $|a_3|$ for the function classes $\mathcal{B}_{\Omega}^{\zeta}(\lambda, \alpha, s, t)$ and $\mathcal{B}_{\Omega}^{\zeta}(\lambda, \beta, s, t)$ are not sharp. To find the sharp upper bounds for the above estimations, it is still an interesting open problem, as well as for $|a_n|$, $n \geq 4$.

Author Contributions: Conceptualization, I.A. and B.A.F.; methodology, B.A.F.; validation, I.A., T.A.-H. and B.A.F.; formal analysis, T.A.-H.; writing—review and editing, T.A.-H. and B.A.F.; project administration, B.A.F. All authors have read and agreed to the published version of the manuscript.

Funding: This research received no external funding.

Acknowledgments: The authors would like to thank the referees for their helpful comments and suggestions.

Conflicts of Interest: The authors declare no conflict of interest.

References

1. Frasin, B.A. A new differential operator of analytic functions involving binomial series. *Bol. Soc. Paran. Mat.* **2020**, *38*, 205–213. [CrossRef]
2. Al-Hawary, T.; Frasin, B.A.; Yousef, F. Coefficient estimates for certain classes of analytic functions of complex order. *Afr. Mat.* **2018**, *29*, 1265–1271. [CrossRef]
3. Wanas, A.K.; Frasin, B.A. Strong differential sandwich results for Frasin operator. *Earthline J. Math. Sci.* **2020**, *3*, 95–104. [CrossRef]
4. Yousef, F.; Al-Hawary, T.; Murugusundaramoorthy, G. Fekete-Szegö functional problems for some subclasses of bi-univalent functions defined by Frasin differential operator. *Afr. Mat.* **2019**, *30*, 495–503. [CrossRef]
5. Al-Oboudi, F. M. On univalent functions defined by a generalized Sălăgean operator. *Int. J. Math. Math. Sci.* **2004**, *2004*, 1429–1436. [CrossRef]
6. Sălăgean, G. Subclasses of univalent functions. In *Complex Analysis—Fifth Romanian-Finnish Seminar*; Springer: Berlin/Heidelberg, Germany, 1983; pp. 362–372.
7. Frasin, B.A. Coefficient inequalities for certain classes of Sakaguchi type functions. *Int. J. Nonlinear Sci.* **2010**, *10*, 206–211.
8. Owa, S.; Sekine, T.; Yamakawa, R. On Sakaguchi type functions. *App. Math. Comp* **2007**, 356–361. [CrossRef]
9. Sakaguchi, K. On a certain univalent mapping. *J. Math. Soc. Jpn.* **1959**, *11*, 72–75. [CrossRef]
10. Cho, N.E.; Kwon, O.S.; Owa, S. Certain subclasses of Sakaguchi functions. *SEA Bull. Math.* **1993**, *17*, 121–126.
11. Owa, S.; Sekine, T.; Yamakawa, R. Notes on Sakaguchi functions. *RIMS Kokyuroku* **2005**, *1414*, 76–82.
12. Duren, P.L. *Univalent Functions, Grundlehren der Mathematischen Wissenschaften, Band 259*; Springer: New York, NY, USA; Berlin/Heidelberg Germany; Tokyo, Japan, 1983.
13. Lewin, M. On a coeffcient problem for bi-univalent functions. *Proc. Am. Math. Soc.* **1967**, *18*, 63–68. [CrossRef]
14. Brannan, D.A.; Taha, T.S. On some classes of bi-univalent functions. In *Mathematical Analysis and Its Applications*; Pergamon Press: Pergamon, Turkey, 1988; pp. 53–60.
15. Taha, T.S. Topics in Univalent Function Theory. Ph.D. Thesis, University of London, London, UK, 1981.
16. Brannan, D.A.; Clunie, J.; Kirwan, W.E. Coefficient estimates for a class of starlike functions. *Canad. J. Math.* **1970**, *22*, 476–485. [CrossRef]
17. Srivastava, H.M.; Mishra, A.K.; Gochhayat, P. Certain subclasses of analytic and biunivalent functions. *Appl. Math. Lett.* **2010**, *23*, 1188–1192. [CrossRef]
18. Frasin, B.A.; Aouf, M.K. New subclasses of bi-univalent functions. *Appl. Math. Lett.* **2011**, *24*, 1569–1573. [CrossRef]
19. Bulut, S. Coefficient estimates for a class of analytic and biunivalent functions. *Novi Sad J. Math.* **2013**, *43*, 59–65.
20. Cağlar, M.; Orhan, H.; Yağmur, N. Coefficient bounds for new subclasses of bi-univalent functions. *Filomat* **2013**, *27*, 1165–1171. [CrossRef]
21. Crisan, O. Coefficient estimates for certain subclasses of bi-univalent functions. *Gen. Math. Notes* **2013**, *16*, 93–102.
22. Deniz, E. Certain subclasses of bi-univalent functions satisfying subordinate conditions. *J. Class. Anal.* **2013**, *2*, 49–60. [CrossRef]
23. Frasin, B.A. Coefficient bounds for certain classes of bi-univalent functions. *Hacettepe J. Math. Stat.* **2014**, *43*, 383–389. [CrossRef]
24. Frasin, B.A.; Al-Hawary, T. Initial Maclaurin Coefficients Bounds for New Subclasses of Bi-univalent Functions. *Theory App. Math. Comp. Sci.* **2015**, *5*, 186–193.

25. Frasin, B.A.; Al-Hawary, T.; Yousef, F. Necessary and sufficient conditions for hypergeometric functions to be in a subclass of analytic functions. *Afr. Mat.* **2019**, *30*, 223–230. [CrossRef]
26. Goyal, S.P.; Goswami, P. Estimate for initial Maclaurin coefficients of bi-univalent functions for a class defined by fractional derivatives. *J. Egyp. Math. Soc.* **2012**, *20*, 179–182. [CrossRef]
27. Hayami, T.; Owa, S. Coefficient bounds for bi-univalent functions. *Pan Am. Math. J.* **2012**, *22*, 15–26.
28. Li, X-F; Wang, A-P. Two new subclasses of bi-univalent functions. *Int. Math. Forum.* **2012**, *7*, 1495–1504.
29. Magesh, N.; Yamini, J. Coefficient bounds for certain subclasses of bi-univalent functions. *Int. Math. Forum* **2013**, *8*, 1337–1344. [CrossRef]
30. Yousef, F.; Al-Hawary, T.; Frasin, B.A. Fekete-Szegö inequality for analytic and bi-univalent functions subordinate to Chebyshev polynomials. *Filomat* **2018**, *32*, 3229–3236. [CrossRef]
31. Yousef, F.; Alroud, S.; Illafe, M. A comprehensive subclass of bi-univalent functions associated with Chebyshev polynomials of the second kind. *Bol. Soc. Mat. Mex.* **2019**. [CrossRef]
32. Porwal, S.; Darus, M. On a new subclass of bi-univalent functions. *J. Egyp. Math. Soc.* **2013**, *21*, 190–193. [CrossRef]
33. Murugusundaramoorthy, G.; Magesh, N.; Prameela, V. Coefficient bounds for certain subclasses of bi-univalent functions. *Abs. App. Ana.* **2013**, *2013*, 573017. [CrossRef]
34. Pommerenke, C. *Univalent Functions*; Vandenhoeck and Rupercht: Gottingen, Germany, 1975.
35. Shang, Y. Emergence in random noisy environments. *Int. J. Math. Ana.* **2010**, *4*, 1205–1215.

© 2020 by the authors. Licensee MDPI, Basel, Switzerland. This article is an open access article distributed under the terms and conditions of the Creative Commons Attribution (CC BY) license (http://creativecommons.org/licenses/by/4.0/).

Article

Starlikness Associated with Cosine Hyperbolic Function

Abdullah Alotaibi [1], Muhammad Arif [2,*], Mohammed A. Alghamdi [1] and Shehzad Hussain [2]

[1] Operator Theory and Applications Research Group, Department of Mathematics, Faculty of Science, King Abdulaziz University, Jeddah 21589, Saudi Arabia; mathker11@hotmail.com (A.A.); Prof-malghamdi@hotmail.com (M.A.A.)
[2] Department of Mathematics, Abdul Wali Khan University Mardan, Mardan 23200, Pakistan; shehzad873822@gmail.com
* Correspondence: marifmaths@awkum.edu.pk

Received: 3 June 2020; Accepted: 6 July 2020; Published: 8 July 2020

Abstract: The main contribution of this article is to define a family of starlike functions associated with a cosine hyperbolic function. We investigate convolution conditions, integral preserving properties, and coefficient sufficiency criteria for this family. We also study the differential subordinations problems which relate the Janowski and cosine hyperbolic functions. Furthermore, we use these results to obtain sufficient conditions for starlike functions connected with cosine hyperbolic function.

Keywords: Janowski functions; subordination; cosine hyperbolic function

1. Introduction and Definitions

The aims of this particular section is to include some basic notions about the Geometric Function Theory that will help to understand our key findings in a clear way. In this regards, first we start to define the most basic family $\mathcal{A}$ which consists of holomorphic (or analytic) functions in $\mathcal{D} = \{z \in \mathbb{C} : |z| < 1\}$ by:

$$\mathcal{A} = \left\{ q : q \text{ is holomorphic in } \mathcal{D} \text{ with } q(z) = z + \sum_{k=2}^{\infty} a_k z^k \right\}.$$

Also the set $\mathcal{S} \subset \mathcal{A}$ describes the family of all univalent functions which is define here by the following set builder form:

$$\mathcal{S} = \{q \in \mathcal{A} : q \text{ is univalent in } \mathcal{D}\}.$$

Next we consider defining the idea of subordinations between holomorphic functions q_1 and q_2, indicated by $q_1 \prec q_2$, as; the functions $q_1, q_2 \in \mathcal{A}$ are connected by the relation of subordination, if there exists a holomorphic function v with the restrictions $v(0) = 0$ and $|v(z)| < |z|$ such that $q_1(z) = q_2(v(z))$. Moreover, if the function $q_2 \in \mathcal{S}$ in $\mathcal{D}$, then we obtain:

$$q_1 \prec q_2 \Leftrightarrow q_1(0) = q_2(0) \ \& \ q_1(\mathcal{D}) \subset q_2(\mathcal{D}). \tag{1}$$

Image domains are of primary significance in the analysis of analytical functions. Analytic functions are classified into various families based on geometry of image domains. In 1992, Ma and Minda [1] considered a holomorphic function Δ normalized by the conditions $\Delta(0) = 1$ and $\Delta'(0) > 0$ with $\text{Re}\Delta > 0$ in $\mathcal{D}$. The function Δ transforms the $\mathcal{D}$ disc into a region that is star-shaped about 1 and is symmetric on the real axis. In particular, if we take $\Delta(z) = \frac{1+Lz}{1+Mz}$ with $-1 \leq M < L \leq 1$, then it maps $\mathcal{D}$ to a disc which lies in the right-half plan with center on the real axis while $\frac{1-L}{1-M}$ and $\frac{1+L}{1+M}$ are its different end points of the diameter. This familiar function is recognized as Janowski function [2]. Some interesting problems such as convolution properties, coefficients inequalities, sufficient conditions, subordinates results, and integral preserving were discussed recently in [3–7] for some of the generalized families associated with circular domain. The image of the function $\Delta(z) = \sqrt{1+z}$ shows that the image domain is bounded by right-half plan of the Bernoullis lemniscate given by $|v^2 - 1| < 1$, see [8]. The function $\Delta(z) = 1 + \frac{4}{3}z + \frac{2}{3}z^2$ maps $\mathcal{D}$ into the image set bounded by the cardioid

$$(y^2 + x^2 - 2x + \frac{5}{9})^2 - 16(y^2 + x^2 - \frac{2}{3}x + \frac{1}{9}) = 0,$$

which was examined in [9] and further studied in [10]. The function $\Delta(z) = 1 + \sin z$ was established by Cho and his coauthors in [11] while $\Delta(z) = e^z$ is recently studied in [12,13]. Furthermore, many subfamilies of starlike functions have also been introduced recently in [14–18] by choosing some particular functions such as functions associated with Bell numbers, functions related with shell-like curve connected with Fibonacci numbers, functions connected with conic domains and rational functions instead of the function Δ.

Differential subordinations are natural generalizations in complex plane of differential inequalities on real line. Information obtained from derivative plays important role in studying properties of real valued functions. In complex plane, there are various differential implications, in which a function is characterized by using differential conditions. Noshiro-Warschawski theorem is an example of such differential implication which gives the univalency criterion for analytic functions. In numerous cases, properties of function are determined from the range of the combination of the derivatives of the function. For more details about differential subordinations, see [19].

Let h be a holomorphic function defined on $\mathcal{D}$ with $h(0) = 1$. Recently, Ali et al. have obtained sufficient conditions on λ such that

$$1 + \lambda z h'(z)/h^n(z) \prec \sqrt{1+z} \implies h(z) \prec \sqrt{1+z}, \quad \text{for } n = 0, 1, 2.$$

Similar type implications have been investigated in some of the recent papers by different researchers, for example see the articles contributed by Haq et al. [20], Kumar et al. [21,22], Paprocki and Sokół [23], Raza et al. [24], Sharma et al. [25] and Tuneski [26].

Now we establish the family $\mathcal{S}^*_{\cosh}$ of starlike functions connected with cosine hyperbolic function that are defined by:

$$\mathcal{S}^*_{\cosh} = \left\{ q \in \mathcal{A} : \frac{zq'(z)}{q(z)} \prec \cosh(z), \ (z \in \mathcal{D}) \right\}. \tag{2}$$

Geometrically, the function $\frac{zq'(z)}{q(z)}$ maps $\mathcal{D}$ onto an open disk symmetric with respect to the real axis with centre $\frac{\cosh(1)+\cos(1)}{2}$ and radius $\frac{\cosh(1)-\cos(1)}{2}$. It is interesting to see that the cosine and cosine hyperbolic functions have the same image domain in $\mathcal{D}$. For detail see [14].

Also, since $\cosh(z)$ maps the region $\mathcal{D}$ onto the image which is bounded by

$$\left| \ln\left(v + \sqrt{v^2 - 1}\right) \right| < 1.$$

Thus, the class $\mathcal{S}^*_{\cosh}$ can also be defined in a different way as; a function $q \in \mathcal{A}$ belongs to the class $\mathcal{S}^*_{\cosh}$ if and only if the following inequality will be true

$$\left| \ln\left(\frac{zq'(z)}{q(z)} + \sqrt{\left(\frac{zq'(z)}{q(z)}\right)^2 - 1} \right) \right| < 1.$$

We need to get the foregoing Lemma to establish our principal results.

Lemma 1. *[27] Let v be a holomorphic function in $\mathcal{D}$ with $v(0) = 0$. If*

$$|v(z_0)| = \max\{|v(z)| \text{ for } |z| \leq |z_0|\},$$

then a number l ($l \geq 1$) occurs in such a way that $z_0 v'(z_0) = l v(z_0)$.

To avoid repetitions, we assume the following restrictions

$$1 \leq M < L \leq 1,\ j \in \mathbb{N} = \{1,2,\ldots\},\ k \in \mathbb{N}_0 = \mathbb{N} \cup \{0\},$$

otherwise we will state it where different.

2. Sufficient Conditions Associated with Cosh

Theorem 1. *Let an analytic function h (with $h(0) = 1$) satisfying the relation of subordination*

$$1 + \lambda \left(zh'(z)\right)^j \prec \frac{1 + Lz}{1 + Mz}, \tag{3}$$

with the following limitation

$$|\lambda| \geq \frac{(L-M)}{\sin^j(1) - |M|\sinh^j(1)},\quad (\text{for } j \in \mathbb{N}). \tag{4}$$

Then

$$h(z) \prec \cosh(z). \tag{5}$$

Proof. Let us assume that

$$p(z) = 1 + \lambda \left(zh'(z)\right)^j. \tag{6}$$

Then the function p is holomorphic in $\mathcal{D}$ with $p(0) = 1$. Also consider

$$v(z) = \cosh^{-1}(h(z)), \tag{7}$$

where we selected the principle branches of the functions that are logarithmic and square root. Then v is clearly a holomorphic function in $\mathcal{D}$ with $v(0) = 0$. Also since

$$\cosh^{-1}(z) = \ln\left[z + \sqrt{z^2 - 1}\right].$$

To complete the proof of this result, we just need to prove $|v(z)| < 1$ in $\mathcal{D}$. By virtue of (7), we have

$$p(z) = 1 + \lambda \{zv'(z)\sinh(v(z))\}^j.$$

Therefore

$$\left|\frac{p(z) - 1}{L - Mp(z)}\right| = \left|\frac{\lambda \{zv'(z)\sinh(v(z))\}^j}{(L-M) - \lambda M \{zv'(z)\sinh(v(z))\}^j}\right|.$$

Now, we suppose that a point $z_0 \in \mathcal{D}$ occurs such that

$$\max_{|z| \leq |z_0|} |v(z)| = |v(z_0)| = 1.$$

Also, by Lemma 1, a number $l \geq 1$ exists with $z_0 v'(z_0) = lv(z_0)$. In addition, we also suppose that $v(z_0) = e^{i\theta}$ for $\theta \in [-\pi, \pi]$. Then we have

$$\left| \frac{p(z_0) - 1}{L - Mp(z_0)} \right| = \left| \frac{\lambda \{lv(z_0) \sinh(e^{i\theta})\}^j}{(L - M) - \lambda M \{lv(z_0) \sinh(e^{i\theta})\}^j} \right|$$

$$\geq \left| \frac{|\lambda| \{l |\sinh(e^{i\theta})|\}^j}{(L - M) + |\lambda| |M| \{l |\sinh(e^{i\theta})|\}^j} \right|. \tag{8}$$

If $|z| = r, -\pi \leq \theta \leq \pi$, then simple calculation illustrates that

$$\left|\cosh\left(e^{i\theta}\right)\right|^2 = \cosh^2(\cos\theta)\cos^2(\sin\theta) + \sinh^2(\cos\theta)\sin^2(\sin\theta) = \phi(\theta),$$

$$\left|\sinh\left(e^{i\theta}\right)\right|^2 = \sinh^2(\cos\theta)\cos^2(\sin\theta) + \cosh^2(\cos\theta)\sin^2(\sin\theta) = \mu(\theta).$$

A routine simplification ensures that $0, \pm\pi, \pm\frac{\pi}{2}$ are the roots of $\phi'(\theta) = 0$ and $\mu'(\theta) = 0$ in $[-\pi, \pi]$. Also, since

$$\phi(\theta) = \phi(-\theta),$$
$$\mu(\theta) = \mu(-\theta),$$

it is enough to conclude that $\theta \in [0, \pi]$ and thus we achieve

$$\max\{\phi(\theta)\} = \phi(0) = \phi(\pi) = \cosh^2(1),$$
$$\min\{\phi(\theta)\} = \phi\left(\frac{\pi}{2}\right) = \cos^2(1),$$
$$\max\{\mu(\theta)\} = \mu(0) = \mu(\pi) = \sinh^2(1),$$
$$\min\{\mu(\pi)\} = \mu\left(\frac{\pi}{2}\right) = \sin^2(1).$$

Thus, we have

$$\cos(1) \leq \left|\cosh\left(e^{i\theta}\right)\right| \leq \cosh(1), \tag{9}$$

$$\sin(1) \leq \left|\sinh\left(e^{i\theta}\right)\right| \leq \sinh(1). \tag{10}$$

Therefore, using (8)–(10), we attain

$$\left| \frac{p(z_0) - 1}{L - Mp(z_0)} \right| \geq \frac{|\lambda| l^j \sin^j(1)}{(L - M) + |\lambda| |M| l^j \sinh^j(1)}.$$

Now let

$$\varsigma(l) = \frac{|\lambda| l^j \sin^j(1)}{(L - M) + |\lambda| |M| l^j \sinh^j(1)}.$$

Then

$$\varsigma'(l) = \frac{|\lambda| (L - M) jl^{j-1} \sin^j(1)}{\left[(L - M) + |\lambda| |M| l^j \sinh^j(1)\right]^2} > 0.$$

This confirms that the function ς is increasing and therefore $\varsigma(l) \geq \varsigma(1)$, so

$$\left| \frac{p(z_0) - 1}{L - Mp(z_0)} \right| \geq \frac{|\lambda| \sin^j(1)}{(L - M) + |\lambda| |M| \sinh^j(1)}.$$

Now using (4), we achieve
$$\left|\frac{p(z_0)-1}{L-Mp(z_0)}\right| \geq 1,$$

and this contradicts the hypothesis
$$1+\lambda\left(zh'(z)\right)^j \prec \frac{1+Lz}{1+Mz}.$$

Hence the proof is completed. □

If we put $h(z) = \frac{zq'(z)}{q(z)}$ in (3), we achieve the below Corollary.

Corollary 1. *Let $q \in \mathcal{A}$ and justifying*
$$1+\lambda\left\{z\left(\frac{zq'(z)}{q(z)}\right)'\right\}^j \prec \frac{1+Lz}{1+Mz}. \quad (11)$$

with
$$|\lambda| \geq \frac{(L-M)}{\sin^j(1) - |M|\sinh^j(1)}.$$

Then $q \in \mathcal{S}^*_{\cosh}$.

If we choose $L = 1$, $M = 0$ in (11), we get the following result.

Corollary 2. *If $q \in \mathcal{A}$ and obeying the subordination*
$$1+\lambda\left\{z\left(\frac{zq'(z)}{q(z)}\right)'\right\}^j \prec 1+z.$$

with
$$|\lambda| \geq \frac{1}{\sin^j(1)}.$$

Then $q \in \mathcal{S}^*_{\cosh}$.

Theorem 2. *Let an analytic function h ($h(0) = 1$) satisfying the relation of subordination*
$$1+\lambda z\frac{h'(z)}{h^k(z)} \prec \frac{1+Lz}{1+Mz} \quad (\forall k \in \mathbb{N}_0), \quad (12)$$

with the following restriction
$$|\lambda| \geq \frac{(L-M)\cosh^k(1)}{\sin(1) - |M|\sinh(1)}. \quad (13)$$

Then
$$h(z) \prec \cosh(z).$$

Proof. Let us suppose
$$p(z) = 1+\lambda\frac{zh'(z)}{h^k(z)}.$$

Then the function p is holomorphic in $\mathcal{D}$ with $p(0) = 1$. Inserting (7), we have
$$p(z) = 1+\lambda\frac{zv'(z)\sinh(v(z))}{(\cosh v(z))^k}.$$

and so
$$\left|\frac{p(z)-1}{L-Mp(z)}\right| = \left|\frac{\lambda z v'(z)\sinh(v(z))}{(L-M)\cosh^k v(z) - \lambda M\{zv'(z)\sinh(v(z))\}}\right|.$$

By virtue of Lemma 1 along with (9) and (10), we have

$$\left|\frac{p(z_0)-1}{L-Mp(z_0)}\right| = \left|\frac{\lambda l v(z_0)\sinh(v(z_0))}{(L-M)\cosh^k(v(z_0)) - \lambda M\{lv(z_0)\sinh(v(z_0))\}}\right|,$$

$$\geq \frac{l|\lambda||\sinh(e^{i\theta})|}{(L-M)\left|\cosh^k(e^{i\theta})\right| + l|\lambda||M||\sinh(e^{i\theta})|},$$

$$\geq \frac{l|\lambda|\sin(1)}{(L-M)\cosh^k(1) + l|\lambda||M|\sinh(1)}.$$

Now let
$$\varsigma_1(l) = \frac{l|\lambda|\sin(1)}{(L-M)\cosh^k(1) + l|\lambda||M|\sinh(1)}.$$

Then
$$\varsigma_1'(l) = \frac{|\lambda|(L-M)\cosh^k(1)\sinh(1)}{\left\{(L-M)\cosh^k(1) + l|\lambda||M|\sinh(1)\right\}^2} > 0.$$

Applying (13), we have
$$\left|\frac{p(z_0)-1}{L-Mp(z_0)}\right| \geq 1.$$

A contradiction to the hypothesis occurs and hence the proof is completed. □

If we take $h(z) = \frac{zq'(z)}{q(z)}$ in (12), we obtain the below result.

Corollary 3. *If $q \in \mathcal{A}$ and obeying the subordination*

$$1 + \lambda z \left(\frac{q(z)}{zq'(z)}\right)^k \left(\frac{zq'(z)}{q(z)}\right)' \prec \frac{1+Lz}{1+Mz} \quad (14)$$

with
$$|\lambda| \geq \frac{(L-M)\cosh^k(1)}{\sin(1) - |M|\sinh(1)},$$

then the function $q \in \mathcal{S}^*_{\cosh}$.

If we choose $L = 1$, $M = 0$ in (14), we get the following result.

Corollary 4. *If $q \in \mathcal{A}$ and obeying the subordination*

$$1 + \lambda z \left(\frac{q(z)}{zq'(z)}\right)^k \left(\frac{zq'(z)}{q(z)}\right)' \prec 1 + z.$$

with
$$|\lambda| \geq \frac{\cosh^k(1)}{\sin(1)},$$

then $q \in \mathcal{S}^*_{\cosh}$.

Theorem 3. *Assume that*

$$|\lambda| \geq \frac{(L-M)\cosh^k(1)}{\sin^j(1) - |M|\sinh^j(1)}. \quad (15)$$

If h is a holomorphc function defined on $\mathcal{D}$ with $h(0) = 1$ and satisfying

$$1 + \lambda \frac{(zh'(z))^j}{h^k(z)} \prec \frac{1 + Lz}{1 + Mz}, \tag{16}$$

then

$$h(z) \prec \cosh(z).$$

Proof. Let us choose a function

$$p(z) = 1 + \lambda \frac{(zh'(z))^j}{h^k(z)}.$$

Then the function p is holomorphic in $\mathcal{D}$ with $p(0) = 1$. Applying some simple computation, we get

$$p(z) = 1 + \lambda \frac{\{zv'(z) \sinh(v(z))\}^j}{(\cosh v(z))^k}.$$

and so

$$\left| \frac{p(z) - 1}{L - Mp(z)} \right| = \left| \frac{\lambda \frac{\{zv'(z) \sinh(v(z))\}^j}{(\cosh(v(z)))^k}}{L - M \left\{ 1 + \lambda \frac{\{zv'(z) \sinh(v(z))\}^j}{(\cosh(v(z)))^k} \right\}} \right|,$$

$$= \left| \frac{\lambda \{zv'(z) \sinh(v(z))\}^j}{(L - M)(\cosh v(z))^k - \lambda M \{zv'(z) \sinh(v(z))\}^j} \right|.$$

By using Lemma 1, we have

$$\left| \frac{p(z_0) - 1}{L - Mp(z_0)} \right| = \left| \frac{\lambda \{lv(z_0) \sinh(v(z_0))\}^j}{(L - M) \cosh^k(v(z_0)) - \lambda M \{lv(z_0) \sinh(v(z_0))\}^j} \right|$$

$$\geq \frac{l^j |\lambda| \left|\sinh^j(e^{i\theta})\right|^j}{(L - M) \left|\cosh^k(e^{i\theta})\right| + l^j |\lambda| |M| \left|\sinh^j(e^{i\theta})\right|},$$

$$\geq \frac{l^j |\lambda| \sin^j(1)}{(L - M) \cosh^k(1) + l^j |\lambda| |M| \sinh^j(1)}.$$

Now, let

$$\varsigma_2(l) = \frac{l^j |\lambda| \sin^j(1)}{(L - M) \cosh^k(1) + l^j |\lambda| |M| \sinh^j(1)}.$$

Then,

$$\varsigma_2'(l) = \frac{jl^{j-1} |\lambda| (L - M) \sin^j(1) \cosh^k(1)}{\left\{ (L - M) \cosh^k(1) + l^j |\lambda| |M| \sinh^j(1) \right\}^2} > 0,$$

which shows that ς is an increasing function and it has its minimum value at $l = 1$, so

$$\left| \frac{p(z_0) - 1}{L - Mp(z_0)} \right| \geq \frac{|\lambda| \sin^j(1)}{(L - M) \cosh^k(1) + |\lambda| |M| \sinh^j(1)}.$$

Now by using (15), we have

$$\left| \frac{p(z_0) - 1}{L - Mp(z_0)} \right| \geq 1,$$

which yields a contradiction to our assumption. This completes the proof. □

If we put $h(z) = \frac{zq'(z)}{q(z)}$ in (16), we obtain the following result.

Corollary 5. *If $q \in \mathcal{A}$ and obeying the subordination*

$$1 + \lambda \left(\frac{q(z)}{zq'(z)}\right)^k \left\{z\left(\frac{zq'(z)}{q(z)}\right)'\right\}^j \prec \frac{1+Lz}{1+Mz}. \quad (17)$$

with

$$|\lambda| \geq \frac{(L-M)\cosh^k(1)}{\sin^j(1) - |M|\sinh^j(1)},$$

then $q \in \mathcal{S}^*_{\cosh}$.

If we choose $L = 1$, $M = 0$ in (17), we get the following result.

Corollary 6. *If $q \in \mathcal{A}$ and obeying the subordination*

$$1 + \lambda \left(\frac{q(z)}{zq'(z)}\right)^k \left\{z\left(\frac{zq'(z)}{q(z)}\right)'\right\}^j \prec 1 + z.$$

with

$$|\lambda| \geq \frac{\cosh^k(1)}{\sin^j(1)}.$$

Then $q \in \mathcal{S}^*_{\cosh}$.

3. Bernardi Integral Operator and Its Relationships

The role of operators in the field of functions theory is very crucial in exploring the nature of the geometry of analytic functions. Several differential and integral operators were introduced by using convolution of certain analytic functions. It is found that this formalism gives ease in more mathematical study and also allows explaining the geometrical properties of analytical and univalent functions. Alexander was the first, who started studying the operator back in 1916. Later Libera [28] and Bernardi [29] added several integral operators to study the classes of starlike, convex, and close-to-convex functions. Also, the mapping properties of these operators was discussed in [30].

The Bernardi [29] integral operator is defined by;

$$\mathcal{J}(z) = \frac{\xi+1}{z^\xi} \int_0^z t^{\xi-1} q(t)\, dt, \text{ for } \xi \geq 0. \quad (18)$$

In this part of the article, we analyze the mapping properties of functions belonging to the class $\mathcal{S}^*_{\cosh}$ under the integral operator described in (18) above. Some similar findings of this type are also discussed here.

Theorem 4. *Assume that*

$$|\lambda| \geq \frac{(L-M)(\cosh(1)+\xi)}{\sin(1) - |N|\sinh(1) - (1+|M|)\cosh(1)(\cosh(1)+\xi)}. \quad (19)$$

If

$$1 + \lambda z \left(\frac{zq'(z)}{q(z)}\right) \prec \frac{1+Lz}{1+Mz}, \quad (20)$$

then
$$\frac{z\mathcal{J}'(z)}{\mathcal{J}(z)} \prec \cosh(z),$$
where the operator $\mathcal{J}$ is given by (18).

Proof. Let a function v be defined by
$$v(z) = \cosh^{-1}\left(\frac{z\mathcal{J}'(z)}{\mathcal{J}(z)}\right). \tag{21}$$

where we have chosed the principle branches of the square root and logarithmic functions. Since $\cosh^{-1} z$ function is defined by
$$\cosh^{-1} z = \ln\left[z + (z^2 - 1)^{1/2}\right],$$

therefore v is an analytic function in $\mathcal{D}$ with $v(0) = 0$. To prove our result, we need only to show that $|v(z)| < 1$ in $\mathcal{D}$. From (21), we have
$$\frac{z\mathcal{J}'(z)}{\mathcal{J}(z)} = \cosh(v(z)).$$

Logarithmic differentiation of above relation yields
$$1 + \frac{z\mathcal{J}''(z)}{\mathcal{J}'(z)} - \frac{z\mathcal{J}'(z)}{\mathcal{J}(z)} = \frac{zv'(z)\sinh(v(z))}{\cosh(v(z))}.$$

Using (18), we have
$$(\xi + 1)q(z) = z\mathcal{J}'(z) + \xi\mathcal{J}(z).$$

Differentiating logarithmically, we have
$$\frac{zq'(z)}{q(z)} = \frac{z\mathcal{J}'(z)}{\mathcal{J}(z)}\left\{\frac{1 + \frac{z\mathcal{J}''(z)}{\mathcal{J}(z)} - \frac{z\mathcal{J}'(z)}{\mathcal{J}(z)}}{\frac{z\mathcal{J}'(z)}{\mathcal{J}(z)} + \xi} + 1\right\}$$
$$= \frac{zv'(z)\sinh(v(z)) + (\cosh(v(z)) + \xi)\cosh(v(z))}{\cosh(v(z)) + \xi}.$$

Now, we define a function
$$p(z) = 1 + \lambda z\left(\frac{zq'(z)}{q(z)}\right)$$
$$= 1 + \lambda z\left\{\frac{zv'(z)\sinh(v(z)) + \cosh(v(z))(\cosh(v(z)) + \xi)}{\cosh(v(z)) + \xi}\right\},$$

where p is analytic in $\mathcal{D}$ with $p(0) = 1$. Also
$$\left|\frac{p(z) - 1}{L - Mp(z)}\right| = \left|\frac{\lambda zv'(z)\sinh(v(z)) + (\cosh(v(z)) + \xi)\cosh(v(z))}{(L-M)(\cosh(v(z)) + \xi) - \lambda M z\{zv'(z)\sinh(v(z)) + (\cosh(v(z)) + \xi)\cosh(v(z))\}}\right|.$$

Suppose that there exists a point $z_0 \in \mathcal{D}$ such that
$$\max_{|z| \leq |z_0|} |v(z)| = |v(z_0)| = 1.$$

By using Lemma 5, there exists a number $l \geq 1$ such that $z_0 v'(z_0) = l v(z_0)$. We also suppose that $v(z_0) = e^{i\theta}$. Then we have

$$\left|\frac{p(z_0)-1}{L-Mp(z_0)}\right| = \left|\frac{\lambda z_0 \{z_0 v'(z_0) \sinh(v(z_0)) + \cosh(v(z_0))(\cosh(v(z_0)) + \bar{\zeta})\}}{(L-M)(\cosh(v(z_0)) + \bar{\zeta}) - \lambda M z_0 \{z_0 v'(z_0) \sinh(v(z_0)) + \cosh(v(z_0))(\cosh(v(z_0)) + \bar{\zeta})\}}\right|$$

$$= \left|\frac{\lambda z_0 \{l v(z_0) \sinh(e^{i\theta}) + \cosh(e^{i\theta})(\cosh(e^{i\theta}) + \bar{\zeta})\}}{(L-M)(\cosh(e^{i\theta}) + \bar{\zeta}) - \lambda M z_0 \{l v(z_0) \sinh(e^{i\theta}) + \cosh(e^{i\theta})(\cosh(e^{i\theta}) + \bar{\zeta})\}}\right|.$$

Let $|z| = r$, $-\pi \leq \theta \leq \pi$. Then a simple computation shows that

$$\left|\cosh\left(e^{i\theta}\right)\right| = |\cosh(\cos\theta)\cos(\sin\theta) + i\sinh(\cos\theta)\sin(\sin\theta)|$$

$$\left|\cosh\left(e^{i\theta}\right)\right|^2 = \cosh^2(\cos\theta)\cos^2(\sin\theta) + \sinh^2(\cos\theta)\sin^2(\sin\theta) = \phi(\theta).$$

A simple computation shows that the equation $\phi'(\theta) = 0$ has five roots in $[-\pi, \pi]$ namely $0, \pm\pi, \pm\frac{\pi}{2}$. Since $\phi(\theta) = \phi(-\theta)$, it is sufficient to consider $\theta \in [0, \pi]$ and this implies that

$$\max\{\phi(\theta)\} = \phi(0) = \phi(\pi) = \cosh^2(1),$$
$$\min\{\phi(\theta)\} = \phi\left(\frac{\pi}{2}\right) = \cosh^2(1).$$

Also, consider

$$\left|\sinh\left(e^{i\theta}\right)\right| = \sinh(\cos\theta)\cos(\sin\theta) + i\cosh(\cos\theta)\sin(\sin\theta),$$

$$\left|\sinh\left(e^{i\theta}\right)\right|^2 = \sinh^2(\cos\theta)\cos^2(\sin\theta) + \cosh^2(\cos\theta)\sin^2(\sin\theta) = \mu(\theta)$$

Similarly, after simple calculations the equation $\mu'(\theta) = 0$ has five roots in $[-\pi, \pi]$ namely $0, \pm\pi, \pm\frac{\pi}{2}$. Since $\mu(\theta) = \mu(-\theta)$, it is sufficient to consider those roots which lies in $[0, \pi]$ and we see that

$$\max\{\mu(\theta)\} = \mu(0) = \mu(\pi) = \cosh^2(1) - 1,$$
$$\min\{\mu(\pi)\} = \mu\left(\frac{\pi}{2}\right) = 1 - \cos^2(1).$$

Thus, we conclude that

$$\cos(1) \leq \left|\cosh\left(e^{i\theta}\right)\right| \leq \cosh(1),$$
$$\sin(1) \leq \left|\sinh\left(e^{i\theta}\right)\right| \leq \sinh(1).$$

Now

$$\left|\frac{p(z_0)-1}{L-Mp(z_0)}\right| \geq \frac{|\lambda|\{|e^{i\theta}|l|e^{i\theta}||\sinh(e^{i\theta})| - |\cosh(e^{i\theta})|(|\cosh(e^{i\theta})| + \bar{\zeta})\}}{(L-M)(|\cosh(e^{i\theta})| + \bar{\zeta}) + |\lambda||M||e^{i\theta}|\{l|e^{i\theta}||\sinh(e^{i\theta})| + |\cosh(e^{i\theta})|(|\cosh(e^{i\theta})| + \bar{\zeta})\}}$$

$$\geq \frac{|\lambda|\{l\sin(1) - \cosh(1)(\cosh(1) + \bar{\zeta})\}}{(L-M)(\cosh(1) + \bar{\zeta}) + |\lambda||M|\{l\sinh(1) + \cosh(1)(\cosh(1) + \bar{\zeta})\}}.$$

Now let

$$\Phi(l) = \frac{|\lambda|\{l\sin(1) - \cosh(1)(\cosh(1) + \bar{\zeta})\}}{(L-M)(\cosh(1) + \bar{\zeta}) + |\lambda||M|\{l\sinh(1) + \cosh(1)(\cosh(1) + \bar{\zeta})\}}.$$

Then

$$\Phi'(l) = \frac{|\lambda|[(L-M)(\sin(1)\cosh(1)+\tilde{\zeta}\sin(1))+|\lambda||M|\cosh(1)\{\sin(1)(1+\tilde{\zeta}+\cosh(1))\}]}{[(L-M)(\cosh(1)+\tilde{\zeta})+|\lambda||M|\{l\sinh(1)+(\cosh(1)+\tilde{\zeta})\cosh(1)\}]^2} > 0.$$

This shows that Φ is an increasing function and has its minimum value at $l = 1$, so

$$\left|\frac{p(z_0)-1}{L-Mp(z_0)}\right| \geq \frac{|\lambda|\{\sin(1)-\cosh(1)(\cosh(1)+\tilde{\zeta})\}}{(L-M)(\tilde{\zeta}+\cosh(1))+|\lambda||M|\{\sinh(1)+\cosh(1)(\cosh(1)+\tilde{\zeta})\}}.$$

Now by (19), we have

$$\left|\frac{p(z_0)-1}{L-Mp(z_0)}\right| \geq 1.$$

A contradiction to the hypothesis

$$1 + \lambda z \left(\frac{zq'(z)}{q(z)}\right) \prec \frac{1+Lz}{1+Mz}.$$

Hence we have the required result. □

Theorem 5. *Assume that*

$$|\lambda| \geq \frac{(L-M)(\tilde{\zeta}+1)}{\sin(1)-|M|\sinh(1)-(1+|M|)(1+\tilde{\zeta})\cosh(1)}. \tag{22}$$

If

$$1 + \lambda q(z) \prec \frac{1+Lz}{1+Mz},$$

then

$$\frac{J(z)}{z} \prec \cosh(z).$$

where J is the Bernardi integral operator defined in (18).

Proof. Let a function v be defined by

$$v(z) = \cosh^{-1}\left(\frac{J(z)}{z}\right), \tag{23}$$

where we have chosed the principle branches of the square root and logarithmic functions. Then v is analytic in $\mathcal{D}$ with $v(0) = 0$. We need only to show that $|v(z)| < 1$ in $\mathcal{D}$. From (23), we have

$$\frac{J(z)}{z} = \cosh(v(z)). \tag{24}$$

Also we define a function

$$p(z) = 1 + \lambda q(z), \tag{25}$$

where p is analytic in $\mathcal{D}$ with $p(0) = 1$. Now by using (18), (24) and (25), we have

$$\left|\frac{p(z)-1}{L-Mp(z)}\right| = \left|\frac{\lambda z\{zv'(z)\sinh(v(z))+(1+\tilde{\zeta})\cosh(v(z))\}}{(L-M)(1+\tilde{\zeta})-\lambda zM\{zv'(z)\sinh(v(z))+(1+\tilde{\zeta})\cosh(v(z))\}}\right|.$$

Suppose that there exists a point $z_0 \in \mathcal{D}$ such that

$$\max_{|z|\leq|z_0|}|v(z)| = |v(z_0)| = 1.$$

By using Lemma 5, there exists a number $l \geq 1$ such that $z_0 v'(z_0) = lv(z_0)$. We also suppose that $v(z_0) = e^{i\theta}$. Then we have

$$\left|\frac{p(z_0)-1}{L-Mp(z_0)}\right| = \left|\frac{\lambda z_0 \{z_0 v'(z_0) \sinh(v(z_0)) + (1+\xi)\cosh(v(z_0))\}}{(L-M)(1+\xi) - \lambda z_0 M \{z_0 v'(z_0) \sinh(v(z_0)) + (1+\xi)\cosh(v(z_0))\}}\right|$$

$$\geq \frac{|\lambda|\{l|\sinh(e^{i\theta})| - (1+\xi)|\cosh(e^{i\theta})|\}}{(L-M)(1+\xi) + |\lambda||M|\{l|\sinh(e^{i\theta})| - (1+\xi)|\cosh(e^{i\theta})|\}}$$

$$\geq \frac{|\lambda|\{l\sin(1) - (1+\xi)\cosh(1)\}}{(L-M)(1+\xi) + |\lambda||M|\{l\sinh(1) + (1+\xi)\cosh(1)\}}.$$

Now let

$$\Theta(l) = \frac{|\lambda|\{l\sin(1) - (1+\xi)\cosh(1)\}}{(L-M)(1+\xi) + |\lambda||M|\{l\sinh(1) + (1+\xi)\cosh(1)\}}.$$

Then

$$\Theta'(l) = \frac{|\lambda|\{((L-M)(1+\xi)\sin(1) + |\lambda||M|(\sin(1)(1+\xi)) + \sinh(1))\}}{(L-M)(1+\xi) + |\lambda||M|\{l\sinh(1) + (1+\xi)\cosh(1)\}^2} > 0,$$

which shows that Θ is an increasing function and it has its minimum value at $l = 1$, so

$$\left|\frac{p(z_0)-1}{L-Mp(z_0)}\right| \geq \frac{|\lambda|\{\sin(1) - (1+\xi)\cosh(1)\}}{(L-M)(1+\xi) + |\lambda||M|\{\sinh(1) + (1+\xi)\cosh(1)\}}.$$

Now by (22), we have

$$\left|\frac{p(z_0)-1}{L-Mp(z_0)}\right| \geq 1.$$

A contradiction to the hypothesis

$$1 + \lambda q(z) \prec \frac{1+Lz}{1+Mz}.$$

Hence we have the required result. □

Theorem 6. *Assume that*

$$|\lambda| \geq \frac{(L-M)(1+\xi)}{\sin(1) - |M|\sinh(1) - (1+|M|)(1+\xi)\cosh(1)}. \tag{26}$$

If

$$1 + \lambda z q'(z) \prec \frac{1+Lz}{1+Mz}, \tag{27}$$

then

$$\mathcal{J}'(z) \prec \cosh(z),$$

where $\mathcal{J}$ is the Bernardi integral operator defined in (18).

Proof. Using the same steps as used in the last result, one can easily complete this proof. □

4. Convolution Conditions and Its Consequences

The technique of convolution (or Hadamard product) is extremely important in the solution of various function theory problems and due to this facts this concept becomes the major part of this field. The main goal of this portion is to analyze the properties of convolution and its implications for the family $\mathcal{S}^*_{\cosh}$ of starlike functions subordinated with cosine hyperbolic function. For $q_1, q_2 \in \mathcal{A}$, the convolution, denoted by $(q_1 * q_2)(z)$, is defined by

$$(q_1 * q_2)(z) = z + \sum_{k=2}^{\infty} a_k b_k z^k, \quad (z \in \mathcal{D}).$$

Also, the following facts will be true only if $q \in \mathcal{A}$;

$$q(z) * \frac{z}{1-z} = q(z) \text{ and } q(z) * \frac{z}{(1-z)^2} = zq'(z). \tag{28}$$

Now using these concepts we now start to state and prove our first result.

Theorem 7. Let $q \in \mathcal{A}$. Then $q \in \mathcal{S}^*_{\cosh}$ if and only if

$$\frac{1}{z}\left[q(z) * \frac{z - \delta z^2}{(1-z)^2}\right] \neq 0, \quad (z \in \mathcal{D}), \tag{29}$$

for all $\delta = \delta_\theta = \frac{\cosh e^{i\theta}}{\cosh e^{i\theta} - 1}$ and also for $\delta = 1$.

Proof. Since given that $q \in \mathcal{S}^*_{\cosh}$ is holomorphic in $\mathcal{D}$, it follows that $q(z) \neq 0$ for all $z \in \mathcal{D}^* = \mathcal{D}\setminus\{0\}$. That is $\frac{q(z)}{z} \neq 0$ for $z \in \mathcal{D}$ which is equivalent to (29) for $\delta = 1$. Thus, the proof is completed for $\delta = 1$. Now from (2), a holomorphic function v occurs with the property that $v(0) = 0$ and $|v(z)| < 1$ so that

$$\frac{zq'(z)}{q(z)} = \cosh(v(z)),$$

and it is equivalent to

$$\frac{zq'(z)}{q(z)} \neq \cosh\left(e^{i\theta}\right), \text{ for } \theta \in [0, 2\pi]. \tag{30}$$

Using (28), we can easily obtain

$$q(z) * \frac{1}{(1-z)^2} - \cosh\left(e^{i\theta}\right)\left(\frac{q(z)}{z} * \frac{1}{(1-z)}\right) \neq 0,$$

and then by simple computation, we have

$$\frac{1}{z}\left[q(z) * \frac{z - \delta z^2}{(1-z)^2}\right] \neq 0, \quad (z \in \mathcal{D}),$$

which is the needed relationship.

For the converse part let assume that (29) hold for $\delta = 1$, it implies that $\frac{q(z)}{z} \neq 0$ for all $z \in \mathcal{D}$. Thus, the function $\mathfrak{h}(z) = \frac{zq'(z)}{q(z)}$ is holomorphic in $\mathcal{D}$ with $\mathfrak{h}(0) = 1$. Also, let us take $\mathcal{H}(z) = \cosh\left(e^{i\theta}\right)$ for $z \in \mathcal{D}$ and since we have proven that (29) and (30) are identical, thus forming the relationship (30), it is evident that $\mathcal{H}(\partial\mathcal{D}) \cap \mathfrak{h}(\mathcal{D}) = \phi$. Hence, a connected part of $\mathbb{C}\setminus\mathcal{H}(\partial\mathcal{D})$ contains the simply connected domain $\mathfrak{h}(\mathcal{D})$. The univalence of the function $\mathfrak{h}$, together with the fact $\mathcal{H}(0) = \mathfrak{h}(0) = 1$, illustrates that $\mathfrak{h} \prec \mathcal{H}$ and it implies that $q \in \mathcal{S}^*_{\cosh}$. □

Theorem 8. Let $q \in \mathcal{A}$. Then a neccesary and sufficient condition $q \in \mathcal{S}^*_{\cosh}$ is that

$$1 - \sum_{n=2}^{\infty} \frac{n - \cosh\left(e^{i\theta}\right)}{\cosh\left(e^{i\theta}\right) - 1} a_n z^{n-1} \neq 0, \quad (z \in \mathcal{D}). \tag{31}$$

Proof. In the last theorem, we have proved that $q \in S^*_{\cosh}$ if and only if the relation (29) held. We can rewrite (29) as

$$0 \neq \frac{1}{z}\left[q(z) * \frac{z - \delta z^2}{(1-z)^2}\right]$$

$$= \frac{1}{z}\left[q(z) * \left(\frac{z}{(1-z)^2} - \delta \frac{z^2}{(1-z)^2}\right)\right]$$

$$= \frac{q(z)}{z} * \left(1 + \sum_{n=2}^{\infty} nz^{n-1} - L\sum_{n=2}^{\infty}(n-1)z^{n-1}\right)$$

$$= 1 - \sum_{n=2}^{\infty}((\delta - 1)n - \delta) a_n z^{n-1}$$

$$= 1 - \sum_{n=2}^{\infty} \frac{n - \cosh(e^{i\theta})}{\cosh(e^{i\theta}) - 1} a_n z^{n-1},$$

and this completes the proof. □

Theorem 9. *If the function $q \in A$ satisfies the following inequality*

$$\sum_{n=2}^{\infty} \left|\frac{n - \cosh(e^{i\theta})}{\cosh(e^{i\theta}) - 1}\right| |a_n| < 1, \tag{32}$$

then $q \in S^*_{\cosh}$.

Proof. To establish this result, we need to prove the relationship (31). For this consider

$$\left|1 - \sum_{n=2}^{\infty}((\delta-1)n - \delta) a_n z^{n-1}\right| > 1 - \sum_{n=2}^{\infty}\left|((\delta-1)n - \delta) a_n z^{n-1}\right|$$

$$= 1 - \sum_{n=2}^{\infty}|((\delta-1)n - \delta)| |a_n| |z|^{n-1}$$

$$> 1 - \sum_{n=2}^{\infty}|((\delta-1)n - \delta)| |a_n|$$

$$= 1 - \sum_{n=2}^{\infty}\left|\frac{n - \cosh(e^{i\theta})}{\cosh(e^{i\theta}) - 1}\right| |a_n| > 0,$$

where we have used inequality (32). Thus, by virtue of Theorem 8, the proof is completed. □

5. Conclusions

In the present research article, we examined some interesting properties of starlike functions associated with the cosine hyperbolic function which is symmetric about the real axis. These results included convolutions properties, Bernardi integral preserving problems and coefficient sufficiency criteria. In addition to that we also calculated some conditions on λ so that; if for each $j \in \mathbb{N}$, $k \in \mathbb{N}_0 = \mathbb{N} \cup \{0\}$

$$1 + \lambda \frac{(zh'(z))^j}{h^k(z)} \prec \frac{1 + Lz}{1 + Mz} \Rightarrow h(z) \prec \cosh(z), \ (z \in \mathcal{D}).$$

Furthermore, these results are used to find sufficiency criterion for the function belongs to the newly defined family $S^*_{\cosh}$. Moreover, some other problems like coefficient bounds, Hankel determinant, partial sum inequalities, and many more can be discussed for this class as a future work.

Author Contributions: Conceptualization, A.A. and M.A.; Formal analysis, M.A. and S.H.; Funding acquisition, M.A.A.; Investigation, M.A.; Methodology, S.H. and A.A.; Software, M.A.A.; Supervision, M.A., A.A. and S.H.; Visualization, M.A.A.; Writing—original draft, M.A. and S.H.; Writing—review & editing, M.A., A.A. and S.H. All authors have read and agreed to the published version of the manuscript.

Funding: This project was funded by the Deanship of Scientific Research (DSR) at King Abdulaziz University, Jeddah, under grant no. (RG-84-130-38).

Acknowledgments: This project was funded by the Deanship of Scientific Research (DSR) at King Abdulaziz University, Jeddah, under grant no. (RG-84-130-38). The authors, therefore, acknowledge with thanks DSR for technical and financial support.

Conflicts of Interest: The authors declare no conflict of interest.

References

1. Ma, W.; Minda, D. A unified treatment of some special classes of univalent functions. In *Proceeding of the Conference on Complex Analysis*; Li, Z., Ren, F., Yang, L., Zhang, S., Eds.; International Press: Vienna, Austria, 1994; pp. 157–169.
2. Janowski, W. Extremal problems for a family of functions with positive real part and for some related families. *Ann. Pol. Math.* **1970**, *23*, 159–177. [CrossRef]
3. Ahmad, K.; Arif, M.; Liu, J.-L. Convolution properties for a family of analytic functions involving q-analogue of Ruscheweyh differential operator. *Turk. J. Math.* **2019**, *43*, 1712–1720. [CrossRef]
4. Arif, M.; Ahmad, K.; Liu, J. L.; Sokół, J. A new class of analytic functions associated with Sălăgean operator. *J. Funct. Spaces* **2019**, *2019*, 5157394. [CrossRef]
5. Shi, L.; Khan, Q.; Srivastava, G.; Liu, J.-L.; Arif, M. A study of multivalent q-starlike functions connected with circular domain. *Mathematics* **2019**, *7*, 670. [CrossRef]
6. Srivastava, H.M.; Khan, B.; Khan, N.; Ahmad, Q.Z. Coeffcient inequalities for q-starlike functions associated with the Janowski functions. *Hokkaido Math. J.* **2019**, *48*, 407–425. [CrossRef]
7. Srivastava, H.M.; Tahir, M.; Khan, B.; Ahmad, Q.Z.; Khan, N. Some general classes of q-starlike functions associated with the Janowski functions. *Symmetry* **2019**, *11*, 292. [CrossRef]
8. Sokół, J.; Stankiewicz, J. Radius of convexity of some subclasses of strongly starlike functions. *Zeszyty Nauk. Politech. Rzeszowskiej Mat* **1996**, *19*, 101–105.
9. Sharma, K.; Jain, N.K.; Ravichandran, V. Starlike functions associated with a cardioid. *Afrika Matematika* **2016**, *27*, 923–939. [CrossRef]
10. Shi, L.; Ali, I.; Arif, M.; Cho, N.E.; Hussain, S.; Khan, H. A study of third Hankel determinant problem for certain subfamilies of analytic functions involving cardioid domain. *Mathematics* **2019**, *7*, 418. [CrossRef]
11. Cho, N.E.; Kumar, V.; Kumar, S.S.; Ravichandran, V. Radius problems for starlike functions associated with the sine function. *Bull. Iran. Math. Soc.* **2019**, *45*, 213–232. [CrossRef]
12. Mendiratta, R.; Nagpal, S.; Ravichandran, V. On a subclass of strongly starlike functions associated with exponential function. *Bull. Malays. Math. Sci. Soc.* **2015**, *38*, 365–386. [CrossRef]
13. Shi, L.; Srivastava, H.M.; Arif, M.; Hussain, S.; Khan H. An investigation of the third Hankel determinant problem for certain subfamilies of univalent functions involving the exponential function. *Symmetry* **2019**, *11*, 598. [CrossRef]
14. Bano, K.; Raza, M. Starlike functions associated with cosine functions. *Bull. Iran. Math. Soc.* **2020**, revised.
15. Cho, N.E.; Kumar, S.; Kumar, V.; Ravichandran, V.; Srivastava, H.M. Starlike functions related to the Bell numbers. *Symmetry* **2019**, *11*, 219. [CrossRef]
16. Dziok, J.; Raina, R.K.; Sokół, J. On a class of starlike functions related to a shell-like curve connected with Fibonacci numbers. *Math. Comput. Model.* **2013**, *57*, 1203–1211. [CrossRef]
17. Kanas, S.; Răducanu, D. Some classes of analytic functions related to conic domains. *Math. Slovaca* **2014**, *64*, 1183–1196. [CrossRef]
18. Kumar, S.; Ravichandran, V. A subclass of starlike functions associated with a rational function. *Southeast Asian Bull. Math.* **2016**, *40*, 199–212.
19. Miller, S.S.; Mocanu, P.T. *Differential Subordinations Theory and Its Applications*; Marcel Dekker Inc.: New York, NY, USA; Basel, Switzerland, 2000.
20. Haq, M.; Raza, M.; Arif, M.; Khan, Q.; Tang, H. Q-analogue of differential subordinations. *Mathematics* **2019**, *7*, 724. [CrossRef]

21. Kumar, S.S.; Kumar, V.; Ravichandran, V.; Cho, N.E. Sufficient conditions for starlike functions associated with the lemniscate of Bernoulli. *J. Inequal. Appl.* **2013**, *2013*, 176. [CrossRef]
22. Kumar, S.; Ravichandran, V. Subordinations for functions with positive real part. *Complex Anal. Oper. Theory* **2018**, *12*, 1179–1191. [CrossRef]
23. Paprocki, E.; Sokół, J. The extremal problems in some subclass of strongly starlike functions. *Zeszyty Nauk. Politech. Rzeszowskiej Mat* **1996**, *20*, 89–94.
24. Raza, M.; Sokół, J.; Mushtaq, S. Differential subordinations for analytic functions. *Iran. J. Sci. Technol. Trans. A Sci.* **2019**, *43*, 883–890. [CrossRef]
25. Sharma, K.; Ravichandran, V. Applications of subordination theory to starlike functions. *Bull. Iran. Math. Soc.* **2016**, *42*, 61–777.
26. Tuneski, N. Some simple sufficient conditions for starlikeness and convexity. *Appl. Math. Lett.* **2009**, *22*, 693–697. [CrossRef]
27. Jack, I.S. Functions starlike and convex of order alpha. *J. Lond. Math. Soc.* **1971**, *2*, 469–474. [CrossRef]
28. Libera, R.J. Some classes of regular univalent functions. *Proc. Am. Math. Soc.* **1965**, *16*, 755–758. [CrossRef]
29. Bernardi, S.D. Convex and starlike univalent functions. *Trans. Am. Math. Soc.* **1969**, *135*, 429–446. [CrossRef]
30. Noor, K.I.; Arif, M. Mapping properties of an integral operator. *Appl. Math. Lett.* **2012**, *25*, 1826–1829. [CrossRef]

© 2020 by the authors. Licensee MDPI, Basel, Switzerland. This article is an open access article distributed under the terms and conditions of the Creative Commons Attribution (CC BY) license (http://creativecommons.org/licenses/by/4.0/).

Systems of Simultaneous Differential Inclusions Implying Function Containment

José A. Antonino [1] and Sanford S. Miller [2,*]

[1] Departamento de Matemática Aplicada, ETSICCP, Universidad Politécnica de Valencia, 46071 Valencia, Spain; jantonin@mat.upv.es
[2] Department of Mathematics, SUNY Brockport, Brockport, NY 14420, USA
* Correspondence: smiller@brockport.edu

Abstract: An important problem in complex analysis is to determine properties of the image of an analytic function p defined on the unit disc $\mathbf{U}$ from an inclusion or containment relation involving several of the derivatives of p. Results dealing with differential inclusions have led to the development of the field of Differential Subordinations, while results dealing with differential containments have led to the development of the field of Differential Superordinations. In this article, the authors consider a mixed problem consisting of special differential inclusions implying a corresponding containment of the form $D[p](\mathbf{U}) \subset \Omega \Rightarrow \Delta \subset p(\mathbf{U})$, where Ω and Δ are sets in $\mathbb{C}$, and D is a differential operator such that $D[p]$ is an analytic function defined on $\mathbf{U}$. We carry out this research by considering the more general case involving a system of two simultaneous differential operators in two unknown functions.

Keywords: differential inclusions; differential containments; differential inequalities; differential subordinations; univalent functions

MSC: primary 34A40; 34A60; secondary 30C80

1. Introduction

We begin by introducing the important classes of functions considered in this article. Let $\mathcal{H} = \mathcal{H}[\mathbf{U}]$ denote the class of functions analytic in the unit disk $\mathbf{U}$, and let

$$\mathcal{H}[a,n] = \{f \in \mathcal{H} : f(z) = a + a_n z^n + \cdots \}.$$

A common problem in complex analysis is to determine the range of a function $p \in \mathcal{H}[a,n]$ from a differential inclusion or containment relation involving several of the derivatives of p. Let Ω and Δ be sets in $\mathbb{C}$, and D be a differential operator such that $D[p]$ is an analytic function defined on $\mathbf{U}$. A natural question is to ask what conditions on D, Ω and Δ are needed so that

$$D[p](\mathbf{U}) \subset \Omega \Rightarrow p(\mathbf{U}) \subset \Delta. \tag{1}$$

In this case, we have a *differential inclusion* ⇒ *function inclusion*. There are many papers of this type that deal with special differential inclusions implying an inclusion for the image of the function p. Similarly, there are many papers that deal with special differential containments and corresponding containments for the image of the function p of the form

$$\Omega \subset D[p](\mathbf{U}) \Rightarrow \Delta \subset p(\mathbf{U}). \tag{2}$$

In this case, we have a *differential containment* ⇒ *function containment*. Both sets of papers have resulted in many applications in complex analysis. See the monographs [1,2] for many results, applications and extensive bibliographies of results such as (1) and (2).

An open question to consider is to combine the two concepts in (1) and (2) and determine conditions on D, Ω and Δ so that the mixed problem of differential inclusions implies a function containment of the form

$$D[p](\mathbf{U}) \subset \Omega \Rightarrow \Delta \subset p(\mathbf{U}). \tag{3}$$

In this case, we have a *differential inclusion* $\Rightarrow$ *function containment*.

In a recent article [3] the authors have extended results described in (1) to systems of two simultaneous second-order differential operators in two complex-valued functions. It is our intention to do the same with (3).

2. Definitions

We first indicate the forms of the two simultaneous second-order analytic differential operators that we will consider in this article.

Definition 1. *Let $D_i : \mathbb{C}^7 \to \mathbb{C}$ and let $\lambda_i(z)$ be analytic in $\mathbf{U}$ for $i = 1, 2$. For $p \in \mathcal{H}[a, n]$ and $q \in \mathcal{H}[b, n]$ we define the second-order differential operators $D_i[p, q, \lambda_i]$, for $i = 1, 2$, by*

$$D_i[p, q, \lambda_i](z) \equiv D_i[p(z), zp'(z), z^2 p''(z), q(z), zq'(z), z^2 q''(z), \lambda_i(z)]. \tag{4}$$

Throughout this article we will assume that $D_i[p, q, \lambda_i]$ is analytic in $\mathbf{U}$.

Let Ω_i and Δ_i be sets in $\mathbb{C}$ and $D_i[p, q, \lambda_i]$ be the second-order differential operators defined in (4), for $i = 1, 2$. The analogue of (3) that we will consider in this article deals with *two simultaneous differential inclusions implying function containments* of the following form

$$\begin{cases} D_1[p, q, \lambda_1](\mathbf{U}) \subset \Omega_1 \\ D_2[p, q, \lambda_2](\mathbf{U}) \subset \Omega_2 \end{cases} \Rightarrow \begin{cases} \Delta_1 \subset p(\mathbf{U}) \\ \Delta_2 \subset q(\mathbf{U}) \end{cases}. \tag{5}$$

In many cases, the *containments* on the right-sides of (5) can be written in terms of superordinations. We recall those definitions. Let f and F be members of $\mathcal{H}$. The function f is said to be subordinate to F (or F is superordinate to f), written $f \prec F$, if there exists a function w analytic in $\mathbf{U}$ with $w(0) = 0$ and $|w(z)| < 1$, such that $f(z) = F(w(z))$. If, in addition, F is univalent, then $f \prec F$ if and only if $f(0) = F(0)$ and $f(\mathbf{U}) \subset F(\mathbf{U})$.

If p and q in (5) are univalent, and Δ_1 and Δ_2 are simply connected domains, then it is possible to rephrase the right-side of (5) in terms of superordination. If Δ_1 is a simply connected domain containing the point $p(0) = a$ and $\Delta_1 \neq \mathbb{C}$, then there is a conformal mapping g_1 of $\mathbf{U}$ onto Δ_1 such that $g_1(0) = a$, and if Δ_2 is a simply connected domain containing the point $q(0) = b$ and $\Delta_2 \neq \mathbb{C}$, then there is a conformal mapping g_2 of $\mathbf{U}$ onto Δ_2 such that $g_2(0) = b$. In this case, (5) can be rewritten as

$$\begin{cases} D_1[p, q, \lambda_1](\mathbf{U}) \subset \Omega_1 \\ D_2[p, q, \lambda_2](\mathbf{U}) \subset \Omega_2 \end{cases} \Rightarrow \begin{cases} g_1(z) \prec p(z) \\ g_2(z) \prec q(z) \end{cases}. \tag{6}$$

We shall refer to the left sides of (5) and (6) as a *System of Simultaneous Differential Inclusions (SSDI)*.

There are three basic pairs of elements in (5) and (6): the differential operators D_i, the sets Ω_i, and the sets Δ_i (or functions g_i). If two of these elements are given, one would hope to find conditions on the third.

Our aim in this article is to solve a system of such simultaneous differential inclusions— analogous to solving a system of simultaneous differential equations in the real-plane. We restrict our development to systems consisting of two second-order differential inclusions in two unknown functions. The results presented here can be extended in a natural way to their corresponding third-order cases. We begin by introducing some important definitions.

Definition 2. *Let Ω_i be sets in $\mathbb{C}$ and let $D_i[p,q,\lambda_i]$ be the analytic differential operators defined in (4) for $i = 1, 2$. If $p \in \mathcal{H}[a,n]$ and $q \in \mathcal{H}[b,n]$ satisfy the SSDI*

$$\begin{cases} D_1[p,q,\lambda_1](U) \subset \Omega_1 \\ D_2[p,q,\lambda_2](U) \subset \Omega_2 \end{cases} \tag{7}$$

*then p and q are called **Solutions of the SSDI**.*

We will show that certain SSDI's have solutions, and that these solutions have particular properties such as those given on the right-sides of (5) and (6).

Example 1. *Let $p \in \mathcal{H}[0,1]$ and $q \in \mathcal{H}[0,1]$ and consider the SSDI given by*

$$\begin{cases} \{-zp'(z) + 2q(z) : z \in U\} \subset U \\ \{2p(z) - zq'(z) : z \in U\} \subset U \end{cases} \tag{8}$$

It is easy to check that the univalent functions $p(z) = q(z) = z + z^2/2$ are Solutions of the SSDI given in (8).

Example 2. *Let $\Omega_i = \{z : \operatorname{Re} z > 0\}$, the right half plane for $i = 1, 2$. Let $p \in \mathcal{H}[0,1]$ and $q \in \mathcal{H}[0,1]$ and consider the SSDI given by*

$$\begin{cases} \{-zp'(z) + 2zq'(z) : z \in U\} \subset \Omega_1 \\ \{z^2 p''(z) - 5zq'(z) : z \in U\} \subset \Omega_2 \end{cases}$$

It is clear that this SSDI has no solutions since there are no analytic functions p and q that can satisfy this system at $z = 0$.

Definition 3. *The set of analytic functions $\{g_1, g_2\}$ as given in (6) is called a **set of subordinants of the Solutions of the SSDI** (6) or more simply a **set of subordinants** if $g_1 \prec p$ and $g_2 \prec q$ for all p and q satisfying the left-side of (6). A set of subordinants $\{\tilde{g}_1, \tilde{g}_2\}$ that satisfies $g_1 \prec \tilde{g}_1$ and $g_2 \prec \tilde{g}_2$ for all subordinants $\{g_1, g_2\}$ of (6) is called a **set of best subordinants** of (6). Please note that the set of best subordinants is unique up to a rotation of U.*

It is our intent to show that for certain types of SSDI we can obtain corresponding sets of subordinants and best subordinants $\{\tilde{g}_1, \tilde{g}_2\}$ of the system.

The analogue of the **best subordinants** in Definition 3 for the SSDI (5) would be finding the *largest inclusion sets* $\tilde{\Delta}_1$ and $\tilde{\Delta}_2$ such that

$$\begin{cases} D_1[p,q,\lambda_1](U) \subset \Omega_1 \Rightarrow \tilde{\Delta}_1 \subset p(U) \\ D_2[p,q,\lambda_2](U) \subset \Omega_2 \Rightarrow \tilde{\Delta}_2 \subset q(U) \end{cases}$$

3. Admissibility and a Fundamental Theorem

For the development of the theory we need to the consider the following class of univalent functions defined on the closed unit disc.

Definition 4. *Let Q denote the set of functions g that are analytic and univalent on the set $\overline{U} \setminus E(g)$, where*

$$E(g) = \left\{ \zeta \in \partial U : \lim_{z \to \zeta} g(z) = \infty \right\},$$

and are such that $\operatorname{Min}|g'(\zeta)| = \rho > 0$ for $\zeta \in \partial U \setminus E(g)$. The subclass of Q for which $g(0) = a$ is denoted by $Q(a)$.

As a simple example of a member of the class $Q(1)$, consider the function $g(z) = (1+z)/(1-z)$. For this function we have $E(g) = \{1\}$, and $\operatorname{Min}|g'(\zeta)| = 1/2 > 0$ for $\zeta \in \partial U \setminus \{1\}$ and hence $g \in Q(1)$.

The following lemma [1] (p. 22) and [4] has played a key role in many results involving the theory of differential subordinations and will also play a key role in this article.

Lemma 1 (Miller/Mocanu Lemma.). *Let $q \in \mathcal{H}[a,n]$ with $q(z) \not\equiv a$ and $n \geq 1$, and let $p \in Q(a)$. If there exist points $z_0 = r_0 e^{i\theta_0} \in U$ and $\zeta_0 \in \partial U \setminus E(p)$ such that $q(z_0) = p(\zeta_0)$, and $q(U_{r_0}) \subset p(U)$, then there exists an m, where $m \geq n \geq 1$ such that*

$$\zeta_0 p'(\zeta_0) = z_0 q'(z_0)/m \text{ and } \operatorname{Re} \frac{\zeta_0 p''(\zeta_0)}{p'(\zeta_0)} + 1 \leq \frac{1}{m}\left[\operatorname{Re} \frac{z_0 q''(z_0)}{q'(z_0)} + 1\right].$$

We first define a special class of differential operators needed to solve a SSDI.

Definition 5. *Let λ_i be analytic in $\overline{U}$, and $g_i \in Q$ with corresponding sets $E(g_i)$ as given in Definition 4 for $i = 1, 2$. Let (Ω_1, Ω_2) be a subset of $\mathbb{C} \times \mathbb{C}$ and let n_1 and n_2 be positive integers. The **Set of Admissible Differential Operators** $\Psi_{(n_1, n_2)}[(\lambda_1, \lambda_2), (\Omega_1, \Omega_2), (g_1, g_2)]$ consists of those pairs of differential operators (D_1, D_2), with $D_i : \mathbb{C}^7 \to \mathbb{C}$ as given in Definition 1, for $i = 1, 2$, which satisfy the two **admissibility conditions***

$$D_1[r, s, t, g_2(\zeta), \zeta g_2'(\zeta), \zeta^2 g_2''(\zeta), \lambda_1(\zeta)] \notin \overline{\Omega}_1 \tag{9}$$

when $r = g_1(z)$, $s = zg_1'(z)/m_1$, $\operatorname{Re} \frac{t}{s} + 1 \leq \frac{1}{m_1} \operatorname{Re}\left[\frac{zg_1''(z)}{g_1'(z)} + 1\right]$,

$z \in U$, $\zeta \in \partial U \setminus E(g_2)$ and $m_1 \geq n_1 \geq 1$.

$$D_2[g_1(\eta), \eta g_1'(\eta), \eta^2 g_1''(\eta), \rho, \sigma, \tau, \lambda_2(\eta)] \notin \overline{\Omega}_2 \tag{10}$$

when $\rho = g_2(z)$, $\sigma = zg_2'(z)/m_2$, $\operatorname{Re} \frac{\tau}{\sigma} + 1 \leq \frac{1}{m_2} \operatorname{Re}\left[\frac{zg_2''(z)}{g_2'(z)} + 1\right]$,

$z \in U$, $\eta \in \partial U \setminus E(g_1)$ and $m_2 \geq n_2 \geq 1$.

In the special case when $n_1 = n_2 = 1$, we denote the set of operators $\Psi_{(1,1)}[(\lambda_1, \lambda_2), (\Omega_1, \Omega_2), (g_1, g_2)]$ by $\Psi[(\lambda_1, \lambda_2), (\Omega_1, \Omega_2), (g_1, g_2)]$. In the special case when $\Omega_1 \neq \mathbb{C}$ and $\Omega_2 \neq \mathbb{C}$ are simply connected domains and h_1 and h_2 are conformal maps of U onto Ω_1 and Ω_2 respectively, we denote the set $\Psi_{(n_1, n_2)}[(\lambda_1, \lambda_2), (h_1(U), h_2(U)), (g_1, g_2)]$ by $\Psi_{(n_1, n_1)}[(\lambda_1, \lambda_2), (h_1, h_2), (g_1, g_2)]$.

In the case of first-order differential operators the admissibility conditions (9) and (10), with $D_i : \mathbb{C}^5 \to \mathbb{C}$ for $i = 1, 2$ simplify to

$$D_1[g_1(z), zg_1'(z)/m_1, g_2(\zeta), \zeta g_2'(\zeta), \lambda_1(\zeta)] \notin \overline{\Omega}_1 \tag{11}$$

when $z \in U$, $\zeta \in \partial U \setminus E(g_2)$ and $m_1 \geq n_1 \geq 1$.

$$D_2[g_1(\eta), \eta g_1'(\eta), g_2(z), zg_2'(z)/m_2, \lambda_2(\eta)] \notin \overline{\Omega}_2 \tag{12}$$

when $z \in U$, $\eta \in \partial U \setminus E(g_1)$ and $m_2 \geq n_2 \geq 1$.

A closer look at conditions (9) and (10) [or (11) and (12)] indicate that there are different conditions on each of the operators D_1 and D_2 in the pair (D_1, D_2). An operator pair (D_α, D_β) may not be in the Set of Admissible Operators as given by Definition 5, but the pair (D_β, D_α) may be in the Set of Admissible Operators. We will see a case of this in Examples 3 and 4. In Example 3 we show that the pair (D_α, D_β) is not in the Set of Admissible Operators, while in Example 4 we show that the pair (D_β, D_α) is in the Set of Admissible Operators.

Example 3. Let $\Omega = \Omega_1 = \Omega_2 = \{\operatorname{Re} z > 0 : z \in U\}$, the right-half complex plane, and let $p \in \mathcal{H}[a, n_1] \cap Q(a)$ and $q \in \mathcal{H}[b, n_2] \cap Q(b)$ satisfy the SSDI

$$\begin{cases} D_1[p, q, \lambda_1](U) \equiv \{2p(z) + zp'(z) - q(z) : z \in U\} \subset \Omega \\ D_2[p, q, \lambda_2](U) \equiv \{q(z) + zq'(z) : z \in U\} \subset \Omega \end{cases} \quad (13)$$

We will show that this pair (D_1, D_2) with the functions $g_1(z) = g_2(z) = (1+z)/(1-z)$ is not in the Set of Admissible Operators. Writing (13) in standard form we see that the functions D_1 and D_2 are of the form

$$\begin{cases} D_1[r, s, t, \rho, \sigma, \tau, \lambda_1(z)] = 2r + s - \rho \\ D_2[r, s, t, \rho, \sigma, \tau, \lambda_2(z)] = \rho + \sigma \end{cases}$$

We need to show that this pair of operators does not satisfy

$$(D_1, D_2) \in \Psi_{(n_1, n_1)}[(0, 0), (\Omega_1, \Omega_2), (g_1, g_2)].$$

In order for this last statement to be true, according to condition (9) of the first part of Definition 5, requires showing that

$$2g_1(z) + zg_1'(z)/m_1 - \frac{1+\zeta}{1-\zeta} \notin \overline{\Omega},$$

when $z \in U$, $\zeta \in \partial U \setminus E(g_2)$ and $m_1 \geq n_1 \geq 1$. This condition is equivalent to requiring that

$$\operatorname{Re}\left[2\frac{1+z}{1-z} + \frac{2z}{(1-z)^2 m_1} - \frac{1+\zeta}{1-\zeta}\right] < 0. \quad (14)$$

Since this is not satisfied when $z = 0$, condition (14) cannot be satisfied and the pair of differential operators given in (13) is not in the Set of Admissible Operators.

We next interchange the differential operators in Example 3 to obtain an appropriate pair of operators.

Example 4. Let $\Omega = \Omega_1 = \Omega_2 = \{\operatorname{Re} z > 0 : z \in U\}$, the right-half complex plane, and let $p \in \mathcal{H}[a, n_1] \cap Q(a)$ and $q \in \mathcal{H}[b, n_2] \cap Q(b)$ satisfy the SSDI

$$\begin{cases} D_1[p, q, \lambda_1](U) \equiv \{q(z) + zq'(z) : z \in U\} \subset \Omega \\ D_2[p, q, \lambda_2](U) \equiv \{2p(z) + zp'(z) - q(z) : z \in U\} \subset \Omega \end{cases} \quad (15)$$

We will show that this pair (D_1, D_2), with the functions $g_1(z) = g_2(z) = (1+z)/(1-z)$, is in the Set of Admissible Operators. Writing (15) in standard form we see that the functions D_1 and D_2 are of the form

$$\begin{cases} D_1[r, s, t, \rho, \sigma, \tau, \lambda_1(z)] = \rho + \sigma \\ D_2[r, s, t, \rho, \sigma, \tau, \lambda_2(z)] = 2r + s - \rho \end{cases}$$

We need to show that $(D_1, D_2) \in \Psi_{(n_1, n_1)}[(0, 0), (\Omega_1, \Omega_2), (g_1, g_2)]$. According to Definition 5, we need to show that

$$g_2(\zeta) + \zeta g_2'(\zeta) \notin \overline{\Omega}$$
$$2g_1(\eta) + \eta g_1'(\eta) - g_2(z) \notin \overline{\Omega}$$

when $z \in U$, $\zeta \in \partial U \setminus E(g_2)$ and $\eta \in \partial U \setminus E(g_1)$. This follows since

$$\operatorname{Re}[g_2(\zeta) + \zeta g_2'(\zeta)] = \operatorname{Re}\left[\frac{1+\zeta}{1-\zeta} + \frac{2\zeta}{(1-\zeta)^2}\right] < 0 \text{ and}$$

$$\operatorname{Re}[2g_1(\eta) + \eta g_1'(\eta) - g_2(z)] = \operatorname{Re}\left[2\frac{1+\eta}{1-\eta} + \frac{2\eta}{(1-\eta)^2} - \frac{1+z}{1-z}\right] < 0.$$

Hence (D_1, D_2) is in the Set of Admissible Operators.
The following theorem is a foundation result for the theory of Second-Order SSDI.

Theorem 1. *Let Ω_1 and Ω_2 be sets in $\mathbb{C}$, let λ_1 and λ_2 be analytic in $\overline{U}$, let $g_1 \in Q(a)$, $g_2 \in Q(b)$, and let $(D_1, D_2) \in \Psi_{(n_1,n_2)}[(\lambda_1, \lambda_2), (\Omega_1, \Omega_2), (g_1, g_2)]$. If $p \in \mathcal{H}[a, n_1] \cap Q(a)$ and $q \in \mathcal{H}[b, n_2] \cap Q(b)$ satisfy the SSDI*

$$\begin{cases} D_1[p, g_2, \lambda_1](U) \subset \Omega_1 \\ D_2[g_1, q, \lambda_2](U) \subset \Omega_2' \end{cases} \quad (16)$$

then $g_1 \prec p$ and $g_2 \prec q$, and $\{g_1, g_2\}$ are a set of subordinants of (16).

Proof. (a) For the first implication, if we assume $g_1 \not\prec p$, then by Lemma 1 there exist points $z_0 = r_0 e^{i\theta_0} \in U$, $\zeta_0 \in \partial U \setminus E(g_1)$ and $m_1 \geq n_1 \geq 1$ such that $g_1(z_0) = p(\zeta_0)$, $g_1(U_{r_0}) \subset p(U)$,

$$\zeta_0 p'(\zeta_0) = z_0 g_1'(z_0)/m_1 \quad \text{and} \quad \operatorname{Re}\frac{\zeta_0 p''(\zeta_0)}{p'(\zeta_0)} + 1 \leq \frac{1}{m_1}\left[\operatorname{Re}\frac{z_0 g_1''(z_0)}{g_1'(z_0)} + 1\right].$$

Using these results in (9) of Definition 5 we conclude

$$D_1[p(\zeta_0), \zeta_0 p'(\zeta_0), \zeta_0^2 p''(\zeta_0), g_2(\zeta_0), \zeta_0 g_2'(\zeta_0), \zeta_0^2 g_2''(\zeta_0), \lambda_1(\zeta_0)] \notin \overline{\Omega}_1.$$

Since this contradicts the first part of (16) we must have $g_1 \prec p$.

(b) For the second implication, if we assume $g_2 \not\prec q$, then by Lemma 1 there exist points $z_0 = r_0 e^{i\theta_0} \in U$, $\eta_0 \in \partial U \setminus E(g_2)$ and $m_2 \geq n_2 \geq 1$ such that $g_2(z_0) = q(\eta_0)$, $g_2(U_{r_0}) \subset q(U)$,

$$\eta_0 q'(\eta_0) = z_0 g_2'(z_0)/m_2 \quad \text{and} \quad \operatorname{Re}\frac{\eta_0 q''(\eta_0)}{q'(\eta_0)} + 1 \leq \frac{1}{m_2}\left[\operatorname{Re}\frac{z_0 g_2''(z_0)}{g_2'(z_0)} + 1\right].$$

Using these results in (10) of Definition 5 we obtain

$$D_2[g_1(\eta_0), \eta_0 g_1'(\eta_0), \eta_0^2 g_1''(\eta_0), q(\eta_0), \eta_0 q'(\eta_0), \eta_0^2 q''(\eta_0), \lambda_2(\eta_0)] \notin \overline{\Omega}_2.$$

Since this contradicts the second part of (16) we must have $g_2 \prec q$. □

As a result of the above theorem we can obtain subordinants of a SSDI of the form (16) by merely checking that the operators D_1 and D_2 satisfy the admissibility conditions (9) and (10) [or (11) and (12)] of Definition 5. This simple algebraic check yields subordinants of various SSDI that would be very difficult to obtain directly.

In the following two examples we use Theorem 1 to find subordinants of a SSDI.

Example 5. *Let $U_r = \{z : |z| < r\}$, $p \in \mathcal{H}[0, 1] \cap Q(0)$, $q \in \mathcal{H}[0, 1] \cap Q(0)$ and suppose*

$$\begin{cases} \{-zp'(z) + 3g_2(z) : z \in U\} \subset U_2 \\ \{2g_1(z) - zq'(z) : z \in U\} \subset U_1' \end{cases} \quad (17)$$

for $g_1(z) = g_2(z) = z$. It is our intention to prove that

$$\begin{cases} \{-zp'(z) + 3g_2(z) : z \in U\} \subset U_2 \\ \{2g_1(z) - zq'(z) : z \in U\} \subset U_1 \end{cases} \implies \begin{cases} z \prec p(z) \\ z \prec q(z) \end{cases}. \tag{18}$$

The differential operators in (17) are of the form

$$\begin{cases} D_1[p, g_2, \lambda_1](z) = -zp'(z) + 3g_2(z) \\ D_2[g_1, q, \lambda_2](z) = 2g_1(z) - zq'(z) \end{cases},$$

with

$$\begin{cases} D_1[r, s, t, \rho, \sigma, \tau, \lambda_1(z)] = -s + 3\rho \\ D_2[r, s, t, \rho, \sigma, \tau, \lambda_2(z)] = 2r - \sigma \end{cases}. \tag{19}$$

We will use Theorem 1 to prove (18) with $\{g_1, g_2\} = \{z, z\}$. We only need to show that the pair of operators (D_1, D_2), as given in (19), satisfy the admissibility conditions of Definition 5, namely that $(D_1, D_2) \in \Psi_{(n_1, n_2)}[(0, 0), (U, U), (g_1, g_2)]$. According to Definition 5 and (19) this requires showing that

$$\begin{cases} -zg_1'(z)/m_1 + 3g_2(\zeta) \notin \overline{U}_2 \\ 2g_1(\eta) - zg_2'(z)/m_2 \notin \overline{U}_1 \end{cases}$$

when $z \in U, \zeta \in \partial U, \eta \in \partial U, m_1 \geq 1$ and $m_2 \geq 1$. This simplifies to the conditions that

$$\begin{cases} -z/m_1 + 3\zeta \notin \overline{U}_2 \\ 2\eta - z/m_2 \notin \overline{U}_1 \end{cases}$$

which are true because of the conditions on the four variables. Hence by Theorem 1 we conclude that

$$\begin{cases} \{-zp'(z) + 3g_2(z) : z \in U\} \subset U_2 \\ \{2g_1(z) - zq'(z) : z \in U\} \subset U_1 \end{cases} \implies \begin{cases} g_1(z) \prec p(z) \\ g_2(z) \prec q(z) \end{cases},$$

which proves (18).

Example 6. *Let $\Omega_1 = \Omega_2 = \{z : \mathrm{Re}\, z > 0\}$ and $\lambda_1(z)$ be analytic in $\overline{U}$, with $\mathrm{Re}\, \lambda_1(z) > 0$. Let $p \in \mathcal{H}[1, n_1] \cap Q(1), q \in \mathcal{H}[1, n_2] \cap Q(1)$ and suppose*

$$\begin{cases} \{-p(z) + g_2(z) + \lambda_1(z) \cdot zg_2'(z) : z \in U\} \subset \Omega_1 \\ \{2g_1(z) + zg_1'(z) + z^2 g_1''(z) - q(z) : z \in U\} \subset \Omega_2 \end{cases} \tag{20}$$

for $g_1(z) = g_2(z) = (1+z)/(1-z)$. It is our intention to prove that

$$\begin{cases} \{-p(z) + g_2(z) + \lambda_1(z) \cdot zg_2'(z) : z \in U\} \subset \Omega_1 \\ \{2g_1(z) + zg_1'(z) + z^2 g_1''(z) - q(z) : z \in U\} \subset \Omega_2 \end{cases} \implies \begin{cases} (1+z)/(1-z) \prec p(z) \\ (1+z)/(1-z) \prec q(z) \end{cases}.$$

The differential operators in (20) are of the form

$$\begin{cases} D_1[p, g_2, \lambda_1](z) = -p(z) + g_2(z) + \lambda_1(z) \cdot zg_2'(z) \\ D_2[g_1, q, \lambda_2](z) = 2g_1(z) + zg_1'(z) + z^2 g_1''(z) - q(z) \end{cases}$$

with

$$\begin{cases} D_1[r,s,t,\rho,\sigma,\tau,\lambda_1(z)] = -r+\rho+\lambda_1(\zeta)\cdot\sigma \\ D_2[r,s,t,\rho,\sigma,\tau,\lambda_2(z)] = 2r+s+t-\rho \end{cases} \quad (21)$$

We will use Theorem 1 to prove that if p and q satisfy (20) then they have subordinants g_1 and g_2 respectively given by $g_1(z) = g_2(z) = (1+z)/(1-z)$. We need to show that the pair (D_1, D_2) as given in (19) is in the Set of Admissible Operators, i.e., that $(D_1, D_2) \in \Psi_{(n_1,n_2)}[(0,0),(h_1,h_2),(g_1,g_2)]$. We need to show that

$$\begin{cases} -g_1(z) + g_2(\zeta) + \lambda_1(\zeta) \cdot \zeta g_2'(\zeta) \notin \overline{\Omega}_1 \\ 2g_1(\eta) + \eta g_1'(\eta) + \eta^2 g_1''(\eta) - g_2(z) \notin \overline{\Omega}_2 \end{cases}$$

when $\zeta = e^{i\theta} \in \partial U \setminus E(g_2)$, $\eta = e^{i\phi} \in \partial U \setminus E(g_1)$ and $z \in U$. This follows since

$$\operatorname{Re}\left[-g_1(z) + g_2(e^{i\theta}) + \lambda_1(e^{i\theta}) \cdot g_2'(e^{i\theta})\right]$$

$$= \operatorname{Re}\left[-\frac{1+z}{1-z} + \frac{1+e^{i\theta}}{1-e^{i\theta}} - \lambda(e^{i\theta})\frac{1}{1-\cos\theta}\right] < -\frac{\operatorname{Re}\lambda(e^{i\theta})}{1-\cos\theta} < 0, \text{ and}$$

$$\operatorname{Re}\left[2\frac{1+e^{i\phi}}{1-e^{i\phi}} + e^{i\phi}g_1'(e^{i\phi}) + e^{2i\phi}g_1''(e^{i\phi}) - \frac{1+z}{1-z}\right] = \operatorname{Re}\left[\frac{2e^{i\phi}}{(1-e^{i\phi})^2} + \frac{4e^{2i\phi}}{(1-e^{i\phi})^3} - \frac{1+z}{1-z}\right]$$

$$= \operatorname{Re}\left[\frac{(2e^{i\phi} + 2e^{2i\phi})(1-e^{-i\phi})^3}{(1-e^{i\phi})^3(1-e^{-i\phi})^3} - \frac{1+z}{1-z}\right] = \operatorname{Re}\left[\frac{4(\sin 2\phi - 2\sin\phi)i}{|1-e^{i\phi}|^6} - \frac{1+z}{1-z}\right] < 0.$$

Hence by Theorem 1 we conclude that

$$\begin{cases} \{-p(z) + g_2(z) + \lambda_1(z) \cdot zg_2'(z) : z \in U\} \subset \Omega_1 \\ \{2g_1(z) + zg_1'(z) + z^2g_1''(z) - q(z) : z \in U\} \subset \Omega_2 \end{cases} \implies \begin{cases} (1+z)/(1-z) \prec p(z) \\ (1+z)/(1-z) \prec q(z) \end{cases}$$

if p and q satisfy (20), then $(1+z)/(1-z) \prec p(z)$ and $(1+z)/(1-z) \prec q(z)$.

The definition of the pair of operators $(D_1, D_2) \in \Psi_{(n_1,n_2)}[(\lambda_1,\lambda_2),(\Omega_1,\Omega_2),(g_1,g_2)]$, and their dependency on the conditions that $g_1 \in Q(a)$ and $g_2 \in Q(b)$ indicates that Theorem 1 depends very heavily on the functions g_1 and g_2 behaving very nicely on the boundary of U. If this is not the case or if their behavior on the boundary is unknown, it may still be possible to obtain a variant of the theorem by the following limiting process.

Theorem 2. *Let λ_1 and λ_2 be analytic in U, let (Ω_1, Ω_2) be a subset of $\mathbb{C} \times \mathbb{C}$ and let g_1 and g_2 be univalent on U, with $g_1(0) = a$ and $g_2(0) = b$. Let $g_{i\rho}(z) = g_i(\rho z)$ and $\lambda_{i\rho}(z) = \lambda_i(\rho z)$ for $i = 1, 2$. Let $D_i : \mathbb{C}^7 \to \mathbb{C}$ for $i = 1, 2$ and suppose there exists $\rho_0 \in (0,1)$ such that $(D_1, D_2) \in \Psi_{(n_1,n_2)}[(\lambda_{1\rho}, \lambda_{2\rho}),(\Omega_1, \Omega_2),(g_{1\rho}, g_{2\rho})]$ for all $\rho \in (\rho_0, 1)$. If $p \in \mathcal{H}[a, n_1] \cap Q(a)$ and $q \in \mathcal{H}[b, n_2] \cap Q(b)$ have the properties that $D_1[p,q,\lambda_1]$ and $D_2[p,q,\lambda_2]$ are analytic in U and*

$$\begin{cases} D_1[p, g_2, \lambda_1](U) \subset \Omega_1 \\ D_2[g_1, q, \lambda_2](U) \subset \Omega_2 \end{cases}$$

then $g_1(z) \prec p(z)$ and $g_2(z) \prec q(z)$.

Proof. If we replace z by ρz in $p(z)$, $q(z)$, $g_1(z)$, $g_2(z)$, $\lambda_1(z)$ and $\lambda_2(z)$ we obtain

$$\begin{cases} D_1[p(\rho z), \rho z p'(\rho z), \rho^2 z^2 p''(\rho z), g_2(\rho z), \rho z g_2'(\rho z), \rho^2 z^2 g_2''(\rho z), \lambda_1(\rho z)] \subset \Omega_1 \\ D_2[g_1(\rho z), \rho z g_1'(\rho z), \rho^2 z^2 g_1''(\rho z), q(\rho z), \rho z q'(\rho z), \rho^2 z^2 q''(\rho z), \lambda_2(\rho z)] \subset \Omega_2 \end{cases}$$

for $z \in U$. If we set $p_\rho(z) = p(\rho z)$ and $q_\rho(z) = q(\rho z)$ we obtain

$$\begin{cases} D_1[p_\rho(z), zp'_\rho(z), z^2 p''_\rho(z), g_{2\rho}(z), zg'_{2\rho}(z), z^2 g''_{2\rho}(z), \lambda_{1\rho}(z)] \subset \Omega_1 \\ D_2[g_{1\rho}(z), zg'_{1\rho}(z), z^2 g''_{1\rho}(z), q_\rho(z), zq'_\rho(z), z^2 q''_\rho(z), \lambda_{2\rho}(z)] \subset \Omega_2 \end{cases},$$

for $z \in \mathbf{U}$. Since $(D_1, D_2) \in \Psi_{(n_1, n_2)}[(\lambda_{1\rho}, \lambda_{2\rho}), (\Omega_1, \Omega_2), (g_{1\rho}, g_{2\rho})]$ we can apply Theorem 1 to conclude that $g_1(\rho z) \prec p_\rho(z) = p(\rho z)$ and $g_2(\rho z) \prec q_\rho(z) = q(\rho z)$ for $\rho \in (\rho_0, 1)$. If we now let $\rho \to 1^-$, we obtain the results $g_1(z) \prec p(z)$ and $g_2(z) \prec q(z)$. □

4. Best Subordinants

In the previous sections we have discussed the problem of finding a set of subordinants for a SSDI. In this section, we discuss a technique for improving that result by finding a set of best subordinants of a SSDI.

Theorem 3. *Let h_1 and h_2 be analytic in $\mathbf{U}$, λ_1 and λ_2 be analytic in $\overline{\mathbf{U}}$, and suppose that the system of simultaneous differential equations*

$$\begin{cases} D_1[u, v, \lambda_1](z) = h_1(z) \\ D_2[u, v, \lambda_2](z) = h_2(z) \end{cases} \tag{22}$$

has solutions $u = g_1 \in Q(a)$ and $v = g_2 \in Q(b)$.

Let $p \in \mathcal{H}[a, n_1] \cap Q(a)$, $q \in \mathcal{H}[b, n_2] \cap Q(b)$ and $(D_1, D_2) \in \Psi_{(n_1, n_2)}[(\lambda_1, \lambda_2), (h_1, h_2), (g_1, g_2)]$. If

$$\begin{cases} \{D_1[p, g_2, \lambda_1](z) : z \in \mathbf{U}\} \subset h_1(\mathbf{U}) \\ \{D_2[g_1, q, \lambda_2](z) : z \in \mathbf{U}\} \subset h_2(\mathbf{U}) \end{cases} \tag{23}$$

then $g_1(z) \prec p(z)$ and $g_2(z) \prec q(z)$, and the set of functions $\{g_1, g_2\}$ is a set of best subordinants of (23).

Proof. Since $(D_1, D_2) \in \Psi_{(n_1, n_2)}[(\lambda_1, \lambda_2), (h_1, h_2), (g_1, g_2)]$, from (23) and Theorem 1 we see that the set of functions $\{g_1, g_2\}$ form a set of subordinants of SSDI (21). Thus, $g_1(z) \prec p(z)$ and $g_2(z) \prec q(z)$ for all p and q satisfying (23). On the other hand, the functions g_1 and g_2, which are solutions of the System of Simultaneous Differential Equations (20), also satisfy the SSDI (23). Thus, they must be dominant to all subordinants of the system and hence $\{g_1, g_2\}$ is a set of best subordinants of system (23). In conclusion, we have the sharp results $g_1(z) \prec p(z)$ and $g_2(z) \prec q(z)$. □

5. Open Problems

This article dealt with describing and defining the key terms and elements for finding subordinants of a System of Simultaneous Second-Order Differential Inclusions. We found conditions for finding subordinants for some special cases of such systems. In particular, if p and q are analytic functions satisfying a differential inclusion system of the form

$$\begin{cases} D_1[p, g_2, \lambda_1](\mathbf{U}) \subset \Omega_1 \\ D_2[g_1, q, \lambda_2](\mathbf{U}) \subset \Omega_2 \end{cases},$$

then we found conditions on the special operators D_1 and D_2 so that

$$\begin{cases} D_1[p, g_2, \lambda_1](\mathbf{U}) \subset \Omega_1 \\ D_2[g_1, q, \lambda_2](\mathbf{U}) \subset \Omega_2 \end{cases} \implies \begin{cases} g_1(z) \prec p(z) \\ g_2(z) \prec q(z) \end{cases}.$$

The general problem of determining conditions on the operators D_1 and D_2, the sets Ω_1 and Ω_2, and the functions g_1 and g_2 so that analytic functions p and q satisfy

$$\begin{cases} D_1[p,q,\lambda_1](\mathbf{U}) \subset \Omega_1 \\ D_2[p,q,\lambda_2](\mathbf{U}) \subset \Omega_2 \end{cases} \implies \begin{cases} g_1(z) \prec p(z) \\ g_2(z) \prec q(z) \end{cases},$$

remains an interesting open problem. In addition, the problem of finding the corresponding set of best subordinants of such systems remains an open question.

Author Contributions: Writing—original draft, J.A.A. and S.S.M. Both authors have read and agreed to the published version of the manuscript.

Funding: This research received no external funding.

Institutional Review Board Statement: Not applicable.

Informed Consent Statement: Not applicable.

Data Availability Statement: Not applicable.

Conflicts of Interest: The authors declare no conflict of interest.

References

1. Miller, S.S.; Mocanu, P.T. Differential Subordinations and Univalent Functions. *Michigan Math. J.* **1981**, *28*, 157–171. [CrossRef]
2. Bulboaca, T. *Differential Subordinations and Superordinations. Recent Results*; Casa Cartii De Stinta: Cluj-Napoca, Romania, 2005.
3. Antonino, J.A.; Miller, S.S. Systems of Simultaneous Differential Inequalities, Inclusions and Subordinations in the Complex Plane. *Anal. Math. Phys.* **2020**, *10*, 32. [CrossRef]
4. Miller, S.S.; Mocanu, P.T. *Differential Subordinations, Theory and Applications*; Marcel Dekker Inc.: New York, NY, USA; Basel, Switzerland, 2000.

Article

Coefficient Estimates for Bi-Univalent Functions in Connection with Symmetric Conjugate Points Related to Horadam Polynomial

S. Melike Aydoğan [1,*] and Zeliha Karahüseyin [2]

1 Department of Mathematics, Istanbul Technical University, 34467 Istanbul, Turkey
2 Department of Mathematics and Computer Science, Istanbul Kültür University, 34158 Istanbul, Turkey; karahuseyinzeliha@gmail.com
* Correspondence: aydogansm@itu.edu.tr

Received: 8 June 2020; Accepted: 30 September 2020; Published: 31 October 2020

Abstract: In the current study, we construct a new subclass of bi-univalent functions with respect to symmetric conjugate points in the open disc E, described by Horadam polynomials. For this subclass, initial Maclaurin coefficient bounds are acquired. The Fekete–Szegö problem of this subclass is also acquired. Further, some special cases of our results are designated.

Keywords: bi-univalent functions; symmetric conjugate points; horadam polynomial; Fekete–Szegö problem

MSC: 30C45

1. Introduction

Let $\mathcal{A}$ represent the class of all functions which are analytic and given by the following form

$$s(z) = z + \sum_{n=2}^{\infty} a_n z^n \tag{1}$$

in the open unit disc $E = \{z : z \in \mathbb{C}, |z| < 1\}$. Let S be class of all functions belonging to $\mathcal{A}$ which are univalent and hold the conditions of normalized $s(0) = s'(0) - 1 = 0$ in E.

For the functions s and r in E analytic, it is known that the function s is subordinate to r in E given by $s(z) \prec r(z)$, $(z \in E)$, if there is an analytic Schwarz function $w(z)$ given in E with the conditions

$$w(0) = 0 \quad \text{and} \quad |w(z)| < 1 \quad \text{for all} \quad z \in E,$$

such that $s(z) = r(w(z))$ for all $z \in E$.

Moreover, it is given by

$$s(z) \prec r(z) \quad (z \in E) \Leftrightarrow s(0) = r(0) \quad \text{and} \quad s(E) \subset r(E)$$

when r is univalent. By the Koebe one-quarter theorem, we know that the range of every function which belongs to S contains the disc $\{w : |w| < \frac{1}{4}\}$ [1]. Therefore, it is obvious that every univalent function s has an inverse s^{-1}, introduced by

$$s(s^{-1}(z)) = z \quad (z \in E),$$

and
$$s(s^{-1}(w)) = w \quad \left(|w| < r_0(s);\ r_0(s) \geq \frac{1}{4}\right),$$

where
$$s^{-1}(w) = w - a_2 w^2 + (2a_2^2 - a_3)w^3 - (5a_2^3 - 5a_2 a_3 + a_4)w^4 + \cdots. \tag{2}$$

A function $s \in \mathcal{A}$ is said to be bi-univalent in E if both $s(z)$ and $s^{-1}(z)$ are univalent in E. The class of all functions $s \in \mathcal{A}$, such that s and $s^{-1} \in \mathcal{A}$ are both univalent in E, will be denoted by σ.

In 1967, the class σ of bi-univalent functions was first enquired by Lewin [2] and it was derived that $|a_2| < 1.51$. Brannan and Taha [3] also considered subclasses of bi-univalent functions, and acquired estimates of initial coefficients. In 2010, Srivastava et al. [4] investigated various classes of bi-univalent functions. Moreover, many authors (see [5–9]) have introduced subclasses for bi-univalent functions.

We define the class $S^*(\varphi)$ of starlike functions and the class $K(\varphi)$ of convex functions by

$$S^*(\varphi) = \left\{s : s \in \mathcal{A},\ \frac{zs'(z)}{s(z)} \prec \varphi(z)\right\},\ z \in E,$$

and
$$K(\varphi) = \left\{s : s \in \mathcal{A},\ 1 + \frac{zs''(z)}{s(z)} \prec \varphi(z)\right\},\ z \in E.$$

These classes were described and studied by Ma and Minda [10].

It is especially clear that $K = K(0)$ and $S^* = S^*(0)$.

It is also obvious that if $s(z) \in K$, then $zs'(z) \in S^*$.

El-Ashwah and Thomas [11] presented the class S^*_{sc} of functions known as starlike with respect to symmetric conjugate points. This class consists of the functions $s \in S$, satisfying the inequality

$$\text{Re}\left\{\frac{zs'(z)}{s(z) - \overline{s(-\bar{z})}}\right\} > 0,\quad z \in E.$$

A function $s \in S$ is said to be convex with respect to symmetric conjugate points if

$$\text{Re}\left\{\frac{(zs'(z))'}{(s(z) - \overline{s(-\bar{z})})'}\right\} > 0,\quad z \in E.$$

The class of all convex functions with respect to symmetric conjugate points is denoted by C_{sc}.

The Horadam polynomials $h_n(x)$ are given by the iteration relation (see [12])

$$h_n(x) = kx h_{n-1}(x) + l h_{n-2}(x),\quad (n \in \mathbb{N} \geq 2), \tag{3}$$

with $h_1(x) = c,\ h_2(x) = dx$, and $h_3(x) = kdx^2 + cl$, where c, d, k, l are some real constants.

Some special cases regarding Horadam polynomials can be found in [12]. For further knowledge related to Horadam polynomials, see [13–16].

Remark 1. ([9,12]). Let $\Omega(x, z)$ be the generating function of the Horadam polynomials $h_n(x)$. At that time

$$\Omega(x, z) = \frac{c + (d - ck)xz}{1 - kxz - lz^2} = \sum_{n=1}^{\infty} h_n(x) z^{n-1}. \tag{4}$$

We took our motivation from the paper written by Wanas and Majeed [17]. They obtained coefficient estimates using Chebyshev polynomials, but in our study we used Horadam Polynomials instead.

In the present paper, we introduce a new subclass of bi-univalent functions with respect to symmetric conjugate points by handling the Horadam polynomials $h_n(x)$ and the generating function $\Omega(x,z)$. Moreover, we find the initial coefficients and the problem of Fekete–Szegö for functions in this new subclass. Some special cases related to our results were also acquired.

2. Main Results

Definition 1. For $0 < \alpha \leq 1$, a function $s \in \sigma$ is belong to the class $\mathcal{F}_\sigma^{sc}(\alpha, x)$ if it satisfies the following conditions

$$\frac{2zs'(z)}{s(z) - \overline{s(-\overline{z})}} + \frac{2(zs'(z))'}{(s(z) - \overline{s(-\overline{z})})'} - \frac{2\alpha z^2 s''(z) + 2zs'(z)}{\alpha z(s(z) - \overline{s(-\overline{z})})' + (1-\alpha)(s(z) - \overline{s(-\overline{z})})}$$
$$\prec \Omega(x,z) + 1 - c \tag{5}$$

and

$$\frac{2wr'(w)}{r(w) - \overline{r(-\overline{w})}} + \frac{2(wr'(w))'}{(r(w) - \overline{r(-\overline{w})})'} - \frac{2\alpha w^2 r''(w) + 2wr'(w)}{\alpha w(r(w) - \overline{r(-\overline{w})})' + (1-\alpha)(r(w) - \overline{r(-\overline{w})})}$$
$$\prec \Omega(x,w) + 1 - c \tag{6}$$

where c, d, and l are real constants as in (3), and r is the extension of s^{-1}, presented by (2).

In particular, if we set $\alpha = 0$, we obtain the class $\mathcal{F}_\sigma^{sc}(0, x) = \mathcal{F}_\sigma^{sc}(x)$, which holds the following conditions:

$$\frac{2(zs'(z))'}{(s(z) - \overline{s(-\overline{z})})'} \prec \Omega(x,z) + 1 - c$$

and

$$\frac{2(wr'(w))'}{(r(w) - \overline{r(-\overline{w})})'} \prec \Omega(x,w) + 1 - c,$$

where the function $r = s^{-1}$ is presented by (2).

We prove that our first theorem includes initial coefficients of the class $\mathcal{F}_\sigma^{sc}(\alpha, x)$.

Theorem 1. Let the function $s \in \sigma$ denoted by (1) belong to the class $\mathcal{F}_\sigma^{sc}(\alpha, x)$. Then

$$|a_2| \leq \frac{|dx|\sqrt{|dx|}}{\sqrt{2\left|[(3-2\alpha)d - 2(2-\alpha)^2 k]dx^2 - 2(2-\alpha)^2 cl\right|}} \tag{7}$$

and

$$|a_3| \leq \frac{|dx|}{2(3-2\alpha)} + \frac{(dx)^2}{4(2-\alpha)^2} \tag{8}$$

Proof. Let $s \in \sigma$ be presented by Maclaurin expansion (1). Let us consider the functions Ψ and Φ, which are analytic, and satisfy $\Psi(0) = \Phi(0) = 0$, $|\Psi(w)| < 1$ and $|\Phi(z)| < 1$, $z, w \in E$. Note that if

$$|\Phi(z)| = |p_1 z + p_2 z^2 + p_3 z^3 + \ldots| < 1 \quad (z \in E)$$

and

$$|\Psi(w)| = |q_1 w + q_2 w^2 + q_3 w^3 + \ldots| < 1 \quad (w \in E),$$

then

$$|p_i| \leq 1 \quad \text{and} \quad |q_i| \leq 1 \quad (i \in \mathbb{N}).$$

In light of Definition 1, we have

$$\frac{2zs'(z)}{s(z) - \overline{s(-\overline{z})}} + \frac{2(zs'(z))'}{(s(z) - \overline{s(-\overline{z})})'}$$
$$- \frac{2\alpha z^2 s''(z) + 2zs'(z)}{\alpha z(s(z) - \overline{s(-\overline{z})})' + (1-\alpha)(s(z) - \overline{s(-\overline{z})})}$$
$$= \Omega(x, \Phi(z)) + 1 - c$$

and

$$\frac{2wr'(w)}{r(w) - \overline{r(-\overline{w})}} + \frac{2(wr'(w))'}{(r(w) - \overline{r(-\overline{w})})'}$$
$$- \frac{2\alpha w^2 r''(w) + 2wr'(w)}{\alpha w(r(w) - \overline{r(-\overline{w})})' + (1-\alpha)(r(w) - \overline{r(-\overline{w})})}$$
$$= \Omega(x, \Psi(w)) + 1 - c$$

or equivalently

$$\frac{2zs'(z)}{s(z) - \overline{s(-\overline{z})}} + \frac{2(zs'(z))'}{(s(z) - \overline{s(-\overline{z})})'}$$
$$- \frac{2\alpha z^2 s''(z) + 2zs'(z)}{\alpha z(s(z) - \overline{s(-\overline{z})})' + (1-\alpha)(s(z) - \overline{s(-\overline{z})})}$$
$$= 1 + h_1(x) - c + h_2(x)\Phi(z) + h_3(x)[\Phi(z)]^3 + \cdots \quad (9)$$

and

$$\frac{2wr'(w)}{r(w) - \overline{r(-\overline{w})}} + \frac{2(wr'(w))'}{(r(w) - \overline{r(-\overline{w})})'}$$
$$- \frac{2\alpha w^2 r''(w) + 2wr'(w)}{\alpha w(r(w) - \overline{r(-\overline{w})})' + (1-\alpha)(r(w) - \overline{r(-\overline{w})})}$$
$$= 1 + h_1(x) - c + h_2(x)\Psi(w) + h_3(x)[\Psi(w)]^3 + \cdots \quad (10)$$

If $\Phi(z) = p_1 z + p_2 z^2 + p_3 z^3 + \cdots$ $(z \in E)$ and $\Psi(w) = q_1 w + q_2 w^2 + q_3 w^3 + \cdots$ $(w \in E)$, from the equalities of (9) and (10), we obtain

$$\frac{2zs'(z)}{s(z) - \overline{s(-\overline{z})}} + \frac{2(zs'(z))'}{(s(z) - \overline{s(-\overline{z})})'}$$
$$- \frac{2\alpha z^2 s''(z) + 2zs'(z)}{\alpha z(s(z) - \overline{s(-\overline{z})})' + (1-\alpha)(s(z) - \overline{s(-\overline{z})})}$$
$$= 1 + h_2(x) p_1 z + \left[h_2(x) p_2 + h_3(x) p_1^2 \right] z^2 + \cdots \qquad (11)$$

and

$$\frac{2wr'(w)}{r(w) - \overline{r(-\overline{w})}} + \frac{2(wr'(w))'}{(r(w) - \overline{r(-\overline{w})})'}$$
$$- \frac{2\alpha w^2 r''(w) + 2wr'(w)}{\alpha w(r(w) - \overline{r(-\overline{w})})' + (1-\alpha)(r(w) - \overline{r(-\overline{w})})}$$
$$= 1 + h_2(x) q_1 w + \left[h_2(x) q_2 + h_3(x) q_1^2 \right] w^2 + \cdots \qquad (12)$$

Thus, upon equating the coincident coefficients in (11) and (12), after some basic calculations, we acquired

$$2(2-\alpha) a_2 = h_2(x) p_1 \qquad (13)$$
$$2(3-2\alpha) a_3 = h_2(x) p_2 + h_3(x) p_1^2 \qquad (14)$$
$$-2(2-\alpha) a_2 = h_2(x) q_1 \qquad (15)$$
$$2(3-2\alpha)(2a_2^2 - a_3) = h_2(x) q_2 + h_3(x) q_1^2 \qquad (16)$$

From (13) and (15), we obtain that

$$p_1 = -q_1 \qquad (17)$$

and

$$8(2-\alpha)^2 a_2^2 = h_2^2(x)(p_1^2 + q_1^2). \qquad (18)$$

Furthermore, by using (16) and (14), we obtain

$$4(3 - 2\alpha) a_2^2 = h_2(x)(p_2 + q_2) + h_3(x)(p_1^2 + q_1^2) \qquad (19)$$

By using (18) in (19), we get

$$\left[4(3-2\alpha) - h_3(x) \frac{8(2-\alpha)^2}{h_2^2(x)} \right] a_2^2 = h_2(x)(p_2 + q_2). \qquad (20)$$

From (3) and (20), we acquired the result which is desired in (7).
Later, in order to derive the coefficient bound on $|a_3|$, by subtracting (16) from (14)

$$-4(3-2\alpha)(a_2^2 - a_3) = h_2(x)(p_2 - q_2) + h_3(x)(p_1^2 - q_1^2)$$

and using (17) and (18), we have

$$\frac{-4(3-2\alpha)h_2^2(x)(p_1^2+q_1^2)}{8(2-\alpha)^2} + 4(3-2\alpha)a_3 = h_2(x)(p_2-q_2)$$

$$a_3 = \frac{h_2(x)(p_2-q_2)}{4(3-2\alpha)} + \frac{h_2^2(x)(p_1^2+q_1^2)}{8(2-\alpha)^2}. \tag{21}$$

Hence, using (17) and applying (3), we obtain the desired result in (8). □

For $\alpha = 0$ the class $\mathcal{F}_\sigma^{sc}(\alpha, x)$ reduced to the class $\mathcal{F}_\sigma^{sc}(x)$. The following corollary belongs to reduced class $\mathcal{F}_\sigma^{sc}(x)$.

Corollary 1. *Let the function $s \in \sigma$, presented by (1), belong to the class $\mathcal{F}_\sigma^{sc}(x)$. Then*

$$|a_2| \leq \frac{|dx|\sqrt{|dx|}}{\sqrt{2|(3d-8k)dx^2-8cl|}} \tag{22}$$

$$|a_3| \leq \frac{|dx|}{6} + \frac{(dx)^2}{16}. \tag{23}$$

3. Fekete–Szegö Problem

For $s \in S$, $|a_3 - \xi a_2^2|$ is the Fekete–Szegö functional, well-known for its productive history in the area of GFT. It started from the disproof by Fekete and Szegö [18] conjecture of Littlewood and Paley, suggesting that the coefficients of odd univalent functions are restricted by unity.

Theorem 2. *For $0 < \alpha \leq 1$ and $\xi \in \mathbb{R}$, let s, given by (1), be in the class $\mathcal{F}_\sigma^{sc}(\alpha, x)$. Then*

$$\left|a_3 - \xi a_2^2\right| \leq \begin{cases} \frac{|dx|}{2(3-2\alpha)} & ; \text{for } |\xi-1| \leq 1 - \frac{2(2-\alpha)^2(kdx^2+cl)}{(3-2\alpha)(dx)^2} \\ \frac{|dx|^3|1-\xi|}{|2(3-2\alpha)(dx)^2-4(2-\alpha)^2(kdx^2+cl)|} & ; \text{for } |\xi-1| \geq 1 - \frac{2(2-\alpha)^2(kdx^2+cl)}{(3-2\alpha)(dx)^2}. \end{cases}$$

Proof. It follows from (20) and (21) that

$$a_3 - \xi a_2^2 = \frac{[h_2(x)]^3(1-\xi)(p_2+q_2)}{4(3-2\alpha)h_2^2(x) - 8(2-\alpha)^2 h_3(x)} + \frac{h_2(x)(p_2-q_2)}{4(3-2\alpha)}$$

$$= h_2(x)\left[\left(\Theta(\xi,x) + \frac{1}{4(3-2\alpha)}\right)p_2 + \left(\Theta(\xi,x) - \frac{1}{4(3-2\alpha)}\right)q_2\right],$$

where

$$\Theta(\xi, x) = \frac{[h_2(x)]^2(1-\xi)}{4(3-2\alpha)h_2^2(x) - 8(2-\alpha)^2 h_3(x)}.$$

Thus, we conclude that

$$\left|a_3 - \xi a_2^2\right| \leq \begin{cases} \frac{|h_2(x)|}{2(3-2\alpha)} & , |\Theta(\xi,x)| \leq \frac{1}{4(3-2\alpha)} \\ 2|h_2(x)||\Theta(\xi,x)| & , |\Theta(\xi,x)| \geq \frac{1}{4(3-2\alpha)}. \end{cases}$$

In this way, the proof of Theorem 2 is completed. □

For $\alpha = 0$ the class $\mathcal{F}_\sigma^{sc}(\alpha, x)$ reduced to the class $\mathcal{F}_\sigma^{sc}(x)$. The following corollary belongs to reduced class $\mathcal{F}_\sigma^{sc}(x)$.

Corollary 2. *For $\xi \in \mathbb{R}$, let s, presented by (1), belong to the class $\mathcal{F}_\sigma^{sc}(x)$. Then*

$$\left|a_3 - \xi a_2^2\right| \leq \begin{cases} \frac{|dx|}{6} & ; \text{for } |\xi - 1| \leq 1 - \frac{8(kdx^2 + cl)}{3(dx)^2} \\ \frac{|dx|^3 |1-\xi|}{|6(dx)^2 - 16(kdx^2 + cl)|} & ; \text{for } |\xi - 1| \geq 1 - \frac{8(kdx^2 + cl)}{3(dx)^2}. \end{cases}$$

Upon taking $\xi = 1$ in Theorem 2, we easily acquire the corollary given below

Corollary 3. *For $0 < \alpha \leq 1$, let s, presented by (1), belong to the class $\mathcal{F}_\sigma^{sc}(\alpha, x)$. Then*

$$\left|a_3 - a_2^2\right| \leq \frac{|dx|}{2(3 - 2\alpha)}.$$

Remark 2. *Different subclasses and results were obtained for some special cases of parameters in our results, such as corollaries. Furthermore, when we take $d = 2, k = 2, c = -1, l = 1$, in our results, it can be seen that these results enhance the study by Wanas and Majeed [17].*

Author Contributions: Data curation, S.M.A. and Z.K.; Funding acquisition, S.M.A.; Methodology, Z.K.; Resources, S.M.A.; Software, Z.K; Supervision, S.M.A. All authors have read and agreed to the published version of the manuscript.

Funding: This research received no external funding.

Conflicts of Interest: The authors declare no conflict of interest.

References

1. Duren, P.L. *Univalent Functions*; Springer: New York, NY, USA, 1983.
2. Lewin, M. On a coefficient problem for bi-univalent functions. *Proc. Am. Math. Soc.* **1967**, *18*, 63–68. [CrossRef]
3. Brannan, D.A.; Taha, T.S. On some classes of bi-univalent functions. In *Mathematical Analysis and Its Applications*; Mazhar, S.M., Hamoul, A., Faour, N.S., Eds.; Elsevier Science Limited: Oxford, UK, 1988; pp. 53–60.
4. Srivastava, H.M.; Mishra, A.K.; Gochhayat, P. Certain subclasses of analytic and bi-univalent function. *Appl. Math. Lett.* **2010**, *23*, 1188–1192. [CrossRef]
5. Alamoush, A.G.; Darus, M. Coefficient bounds for new subclasses of bi-univalent functions using Hadamard products. *Acta Univ. Apulensis* **2014**, *38*, 153–161.
6. Akgül, A.; Sakar, F.M. A certain subclass of bi-univalent analytic functions introduced by means of the q-analogue of Noor integral operator and Horadam polynomials. *Turk. J. Math.* **2019**, *43*, 2275–2286. [CrossRef]
7. Altınkaya, Ş.; Yalçın, S. Coefficient estimates for two new subclasses of bi-univalent functions with respect to symmetric points. *J. Funct. Spaces* **2015**, *2015*, 145242.
8. Sakar, F.M. Estimating Coefficients for Certain Subclasses of Meromorphic and Bi-univalent Functions. *J. Inequal. Appl.* **2018**, *2018*, 283. [CrossRef] [PubMed]
9. Sakar, F.M.; Aydoğan, S.M. Initial Bounds for Certain Subclasses of Generalized Salagean Type Bi-univalent Functions Associated with the Horadam Polynomials. *Qual. Meas. Anal.* **2019**, *15*, 89–100.
10. Ma, W.C.; Minda, D. A unified treatment of some special classes of univalent functions. In Proceedings of the Conference on Complex Analysis (Nankai Institute of Mathematics), Tianjin, China, 19–23 June 1992; pp. 157–169.

11. El-Ashwah, R.M.; Thomas, D.K. Some subclasses of close-to-convex functions. *J. Ramanujan Math. Soc.* **1987**, *2*, 85–100.
12. Horcum, T.; Kocer, E.G. On some properties of Horadam Polynomials. *Internat. Math. Forum* **2009**, *4*, 1243–1252.
13. Horadam, A.F. Jacobsthal Representation Polynomials. *Fibonacci Q.* **1997**, *35*, 137–148.
14. Horadam, A.F.; Mahon, J.M. Pell and Pell-Lucas Polynomials. *Fibonacci Q.* **1985**, *23*, 7–20.
15. Koshy, T. *Fibonacci and Lucas Numbers with Applications*; A Wiley-Interscience Publication: New York, NY, USA, 2001.
16. Lupas, A. A Guide of Fibonacci and Lucas Polynomials. *Octagon Math. Mag.* **1999**, *7*, 2–12.
17. Wanas, A.K.; Majeed, A.H. Chebyshev polynomial bounded for analytic and bi-univalent functions with respect to symmetric conjugate points. *Appl. Math. E-Notes* **2019**, *19*, 14–21.
18. Fekete, M.; Szegö, G. Eine bemerkung uber ungerade schlichte funktionen. *J. Lond. Math. Soc.* **1933**, *8*, 85–89. [CrossRef]

Publisher's Note: MDPI stays neutral with regard to jurisdictional claims in published maps and institutional affiliations.

© 2020 by the authors. Licensee MDPI, Basel, Switzerland. This article is an open access article distributed under the terms and conditions of the Creative Commons Attribution (CC BY) license (http://creativecommons.org/licenses/by/4.0/).

Article

q-Generalized Linear Operator on Bounded Functions of Complex Order

Rizwan Salim Badar * and Khalida Inayat Noor

Department of Mathematics, COMSATS University Islamabad, Islamabad 44000, Pakistan; khalidainayat@comsats.edu.pk
* Correspondence: rizwansbadar@gmail.com

Received: 16 June 2020; Accepted: 9 July 2020; Published: 14 July 2020

Abstract: This article presents a q-generalized linear operator in Geometric Function Theory (GFT) and investigates its application to classes of analytic bounded functions of complex order $S_q(c; M)$ and $C_q(c; M)$ where $0 < q < 1$, $0 \neq c \in \mathbb{C}$, and $M > \frac{1}{2}$. Integral inclusion of the classes related to the q-Bernardi operator is also proven.

Keywords: q-difference operator; subordinating factor sequence; bounded analytic functions of complex order; q-generalized linear operator

MSC: Primary 30C45; Secondary 30C50; 30H05

1. Introduction

Quantum calculus or q-calculus is attributed to the great mathematicians L.Euler and C. Jacobi, but it became popular when Albert Einstein used it in quantum mechanics in his paper [1] published in 1905. F.H. Jackson [2,3] introduced and studied the q-derivative and q-integral in a proper way. Later, quantum groups gave the geometrical aspects to q-calculus. It is pertinent to mention that q-calculus can be considered an extension of classical calculus discovered by I. Newton and G.W. Leibniz. In fact, the operators defined as:

$$d_h f(z) = \frac{f(z+h) - f(z)}{h}$$

and:

$$d_q f(z) = \frac{f(z) - f(qz)}{(1-q)z}, \; 0 < q < 1,$$

where $z \in \mathbb{C}$ and $h > 0$ are the h-derivative and q-derivative, respectively, where h is Planck's constant, are related as: $q = e^{ih} = e^{2\pi i \bar{h}}$ where $\bar{h} = h/2\pi$. Srivastava [4] applied the concepts of q-calculus by using the basic (or q-) hypergeometric functions in Geometric Function Theory (GFT). Ismail [5] and Agarwal [6] introduced the class of q-starlike functions by using the q-derivative. The q-close-to-convex functions were defined in [7], and Sahoo and Sharma [8] obtained several interesting results for q-close-to-convex functions. Several convolution and fractional calculus q-operators were defined by the researchers, which were reposited by Srivastava in [9]. Darus [10] defined a new differential operator called the q-generalized operator by using q-hypergeometric functions. Let A be the class of functions of the form:

$$f(z) = z + \sum_{k=2}^{\infty} a_k z^k, \tag{1}$$

analytic in the open unit disc $E = \{z : |z| < 1\}$.

Let $f(z)$ be given by (1) and $g(z)$ defined as:

$$g(z) = z + \sum_{k=2}^{\infty} b_k z^k.$$

The Hadamard product (or convolution) of f and g is defined by:

$$(f * g)(z) = z + \sum_{k=2}^{\infty} a_k b_k z^k.$$

Let f, h be analytic functions. Then, f is subordinate to h, written as $f \prec h$ or $f(z) \prec h(z), z \in E$, if there exists a Schwartz function $w(z)$ analytic in E with $w(0) = 0$ and $|w(z)| < 1$ for $z \in E$, such that $f(z) = h(w(z))$. If h is univalent in E, then $f \prec h$, if and only if $f(0) = h(0)$ and $f(E) \subset h(E)$.

A sequence $\{b_k\}_{k=1}^{\infty}$ of complex numbers is a subordinating factor if, whenever $f(z) = \sum_{k=1}^{\infty} a_k z^k, a_1 = 1$ is regular, univalent, and convex in E, we have $\sum_{n=1}^{\infty} b_n a_n z^n \prec f(z), z \in E$ [11].

We recall some basic concepts from q-calculus that are used in our discussion and refer to [2,3,12] for more details.

A subset $B \subset \mathbb{C}$ is called q-geometric if $zq \in B$ whenever $z \in B$, and it contains all the geometric sequences $\{zq^k\}_0^{\infty}$. In GFT, the q-derivative of $f(z)$ is defined as:

$$d_q f(z) = \frac{f(z) - f(qz)}{(1-q)z}, \quad q \in (0,1), \quad (z \in B \setminus \{0\}),$$

and $d_q f(0) = f'(0)$. For a function $g(z) = z^k$, the q-derivative is:

$$d_q g(z) = [k] z^{k-1},$$

where $[k] = \frac{1-q^k}{1-q} = 1 + q + q^2 + \ldots + q^{k-1}$.

We note that as $q \to 1^-$, $d_q f(z) \to f'(z)$, which is the ordinary derivative. From (1), we deduce that:

$$d_q f(z) = 1 + \sum_{k=2}^{\infty} [k] a_k z^k.$$

Let $f(z)$ and $g(z)$ be defined on a q-geometric set B. Then, for complex numbers a, b, we have:

$$d_q(af(z) \pm bg(z)) = a d_q f(z) \pm b d_q g(z).$$

$$d_q(f(z)g(z)) = f(qz) d_q g(z) + g(z) d_q f(z).$$

$$d_q\left(\frac{f(z)}{g(z)}\right) = \frac{g(z) d_q f(z) - f(z) d_q g(z)}{g(z)g(qz)}, \quad g(z)g(qz) \neq 0.$$

$$d_q(\log f(z)) = \frac{\ln q^{-1}}{1-q} \frac{d_q f(z)}{f(z)}.$$

Jackson [2] introduced the q-integral of a function f, given by:

$$\int_0^z f(t) d_q t = z(1-q) \sum_{k=0}^{\infty} q^k f(q^k z),$$

provided that the series converges.

For any non-negative integer n, the q-number shift factorial is defined as:

$$[n]! = \begin{cases} [1][2]\ldots[n] & \text{if } n \neq 0, \\ 1 & \text{if } n = 0. \end{cases}$$

Let $\lambda \in \mathbb{R}$ and $n \in \mathbb{N}$; the q-generalized Pochhammer symbol is defined as:

$$[\lambda]_n = [\lambda][\lambda+1][\lambda+2]\ldots[\lambda+n-1].$$

The q-Gamma function is defined for $\lambda > 0$ as:

$$\Gamma_q(\lambda+1) = [\lambda]\Gamma_q(\lambda) \quad \text{and} \quad \Gamma_q(1) = 1.$$

For complex parameters a_i $(1 \leq i \leq l), b_j \neq 0, -1, -2, \ldots (1 \leq j \leq m)$ with $l \leq m+1$, the basic q-hypergeometric function is defined as,

$$_l F_m(a_1, \ldots, a_l; b_1, \ldots, b_m, z) = \sum_{k=0}^{\infty} \frac{(a_1)_k \ldots (a_l)_k}{(q)_k (b_1)_k \ldots (b_m)_k} \left[(-1)^n q^{\binom{n}{2}}\right]^{1+m-l} z^k. \tag{2}$$

with $\binom{n}{2} = \frac{n(n-1)}{2}$ and $l, m \in \mathbb{N}_0 = \mathbb{N} \cup \{0\}$. Here, the q-shifted factorial is defined for $a \in \mathbb{C}$ as:

$$(a)_k = \begin{cases} (1-a)(1-aq)\ldots(1-aq^{k-1}) & \text{if } k \in \mathbb{N}, \\ 1 & \text{if } k = 0. \end{cases}$$

Let $l = m+1$, $a_1 = q^{\lambda+1}(\lambda > -1)$, $a_i = q$ $(\forall\, 2 \leq i \leq l)$, and $b_j = q$ $(\forall\, 1 \leq j \leq m)$, and by using the property $(q^a)_k = \Gamma_q(a+k)(1-q)^k / \Gamma_q(a)$, from (2), we get the function,

$$F_{q,\lambda+1}(z) = z + \sum_{k=2}^{\infty} \frac{\Gamma_q(\lambda+k)}{[k-1]!\Gamma_q(\lambda+1)} z^k = z + \sum_{k=2}^{\infty} \frac{[\lambda+1]_{k-1}}{[k-1]!} z^k, \; z \in E.$$

In [13], the q-Srivastava–Attiya convolution operator is defined as:

$$G_{q,a}^s(z) = z + \sum_{k=2}^{\infty} \left(\frac{[1+a]}{[k+a]}\right)^s z^k, z \in E,$$

$(a \in \mathbb{C} \setminus \mathbb{Z}_0^-; s \in \mathbb{C}$ when $|z| < 1$; $\operatorname{Re}(s) > 1$ when $|z| = 1)$.
Using convolution, the operator $D_{q,a,\lambda}^s$ for $\lambda > -1$ is defined as:

$$D_{q,a,\lambda}^s f(z) = J_{q,a,\lambda}^s(z) * f(z)$$

$$= z + \sum_{k=2}^{\infty} \left(\frac{[k+a]}{[1+a]}\right)^s \frac{[\lambda+1]_{k-1}}{[k-1]!} a_k z^k, z \in E,$$

where:

$$J_{q,a,\lambda}^s(z) = \left(G_{q,a}^s(z)\right)^{-1} * F_{q,\lambda+1}(z) = z + \sum_{k=2}^{\infty} \left(\frac{[k+a]}{[1+a]}\right)^s \frac{[\lambda+1]_{k-1}}{[k-1]!} z^k.$$

It is a convergent series with a radius of convergence of one. We observe that $D_{q,a,0}^0 f(z) = f(z)$ and $D_{q,0,0}^1 f(z) = zd_q f(z)$. The operator $D_{q,a,\lambda}^s$ reduces to known linear operators for different values of parameters $a, s,$ and λ as:

(i) If $q \to 1^-$, it reduces to the operator $D_{a,\lambda}^s$ discussed by Noor et al. in [14].
(ii) For $s = 0$, it is a q-Ruscheweyh differential operator [15].
(iii) If $s = -1$, $\lambda = 0$, and $q \to 1^-$, it is an Owa–Srivastava integral operator [16].
(iv) If $s \in \mathbb{N}_0$, $a = 1$, $\lambda = 0$, and $q \to 1^-$, it reduces to the generalized Srivastava–Attiya integral operator [17].
(v) If $s \in \mathbb{N}_0$, $a = 0, \lambda = 0$, it is a q-Salagean differential operator [18].
(vi) For $s, \lambda \in \mathbb{N}_0$, and $a = 0$, it is the operator defined in [19].

The following identities hold for the operator $D^s_{q,a,\lambda}f(z)$,

$$zd_q\left(D^s_{q,a,\lambda}f(z)\right) = \left(\frac{[1+a]}{q^a}\right)D^{s+1}_{q,a,\lambda}f(z) - \frac{[a]}{q^a}D^s_{q,a,\lambda}f(z) \qquad (3)$$

$$zd_q(D^s_{q,a,\lambda}f(z)) = \left(\frac{[1+\lambda]}{q^\lambda}\right)D^s_{q,a,\lambda+1}f(z) - \frac{[\lambda]}{q^\lambda}D^s_{q,a,\lambda}f(z). \qquad (4)$$

Let $P(q)$ be the class of functions of the form $p(z) = 1 + c_1 z + c_2 z^2 + \ldots$, analytic in E, and satisfying:

$$\left|p(z) - \frac{1}{1-q}\right| \leq \frac{1}{1-q}, \qquad (z \in E, q \in (0,1)).$$

It is known from [20] that $p \in P(q)$ implies $p(z) \prec \frac{1+z}{1-qz}$. It follows immediately that $\operatorname{Re} p(z) > 0$, $z \in E$.

The classes of bounded q-starlike functions $S_q(c, M)$ and bounded q-convex functions $C_q(c, M)$ of complex order c were defined in [21], respectively, as:

$$S_q(c, M) = \left\{f \in A : \left|\frac{c - 1 + \frac{zd_q f(z)}{f(z)}}{c} - M\right| < M\right\},$$

$$\left(c \in \mathbb{C}^*; M > \frac{1}{2}, z \in E\right),$$

or equivalently,

$$S_q(c, M) = \left\{f \in A : \frac{zd_q f(z)}{f(z)} \prec \frac{1 + \{c(1+m) - m\}z}{1 - mz}\right\},$$

$$\left(c \in \mathbb{C}^*; m = 1 - \frac{1}{M}; M > \frac{1}{2}\right).$$

The class of bounded q-convex functions $C_q(c, M)$ of complex order c is defined as:

$$C_q(c, M) = \left\{f \in A : \left|\frac{c - 1 + \frac{d_q(zd_q f(z))}{d_q f(z)}}{c} - M\right| < M\right\},$$

$$\left(c \in \mathbb{C}^*; M > \frac{1}{2}, z \in E\right),$$

or equivalently,

$$C_q(c, M) = \left\{f \in A : \frac{d_q(zd_q f(z))}{d_q f(z)} \prec \frac{1 + \{c(1+m) - m\}z}{1 - mz}\right\}$$

$$\left(c \in *; m = 1 - \frac{1}{M}; M > \frac{1}{2}\right).$$

Using the operator $D^s_{q,a,\lambda}f(z)$, we now define the following new classes $S_{q,a,s,\lambda}(c, M)$ and $C_{q,a,s,\lambda}(c, M)$ as:

$$S_{q,a,s,\lambda}(c, M) = \left\{ f \in A : \frac{z(d_q D^s_{q,a,\lambda}(f(z)))}{D^s_{q,a,\lambda}(f(z))} \prec \frac{1 + \{c(1 + m) - m\}z}{1 - mz}, z \in E \right\},$$

$$\left(0 < q < 1, c \in \mathbb{C}^*; m = 1 - \frac{1}{M}; M > \frac{1}{2} \right).$$

Special cases:

(i) If $c = 1$, $m = 1$, and $q \to 1^-$, then $S_{q,a,s,\lambda}(c, M)$ reduces to class $S^s(a, \lambda)$ discussed in [22].
(ii) If $c = 1$, $s = 0$, $\lambda = 0$, $m = -q$, then $S_{q,a,s,\lambda}(c, M)$ reduces to class S^*_q introduced by Noor et al. [23].
(iii) If $s = 0$, $c = \frac{m}{1+m}$ $(-1 < m < 0)$, $m = -q$, then $S_{q,a,s,\lambda}(c, M)$ reduces to class ST_q studied by Noor [24].
(iv) If $s = 0, \lambda = 0, c = ae^{-i\beta} \cos \beta$ $(a \in \mathbb{C}^*, |\beta| < \frac{\pi}{2})$, and $q \to 1^-$, then $S_{q,a,s,\lambda}(c, M)$ becomes special cases of Janowski β-spiral like functions of complex order $S^\beta(A, B, a)$ discussed in [25].
(v) If $s \in \mathbb{N}_0$, $\lambda = 0, a = 0$, and $q \to 1^-$, then $S_{q,a,s,\lambda}(c, M)$ reduces to class $H_n(c, M)$ discussed by Aouf et al. in [26].
(vi) If $0 < c \leq 1$, $-1 < m < 0$, and $q \to 1^-$, then $S_{q,a,s,\lambda}(c, M)$ becomes a special case of the class $S^s_{a,\lambda}(\eta, A, B)$ with $\eta = 0$ discussed in [19].

A function $f \in A$ is in the class $S_{q,a,s,\lambda}(c, M)$ if and only if:

$$\left| \frac{\frac{zd_q(D^s_{q,a,\lambda} f(z))}{D^s_{q,a,\lambda} f(z)} - 1}{A - B \left\{ \frac{zd_q(D^s_{q,a,\lambda} f(z))}{D^s_{q,a,\lambda} f(z)} \right\}} \right| < 1, \quad (5)$$

where $A = c(1 + m) - m$ and $B = -m$.

The class $C_{q,a,s,\lambda}(c, M)$ is defined as:

$$C_{q,a,s,\lambda}(c, M) = \left\{ f \in A : \frac{d_q(zd_q(D^s_{q,a,\lambda} f(z)))}{d_q(D^s_{q,a,\lambda} f(z))} \prec \frac{1 + \{c(1 + m) - m\}z}{1 - mz}, z \in E \right\},$$

$$\left(0 < q < 1, c \in \mathbb{C}^*; m = 1 - \frac{1}{M}; M > \frac{1}{2} \right).$$

It is easy to see that $f \in C_{q,a,s,\lambda}(c, M) \Leftrightarrow zd_q f \in S_{q,a,s,\lambda}(c, M)$. In order to develop results for the classes $S_{q,a,s,\lambda}(c, M)$ and $C_{q,a,s,\lambda}(c, M)$, we need the following:

Lemma 1 ([27]). *Let β and γ be complex numbers with $\beta \neq 0$, and let $h(z)$ be regular in E with $h(0) = 1$ and $\text{Re}[\beta h(z) + \gamma] > 0$. If $p(z) = 1 + p_1 z + p_2 z^2 + ...$ is analytic in E, then $p(z) + \frac{zd_q p(z)}{\beta p(z) + \gamma} \prec h(z) \Rightarrow p(z) \prec h(z)$.*

Lemma 2 ([11]). *The sequence $\{b_n\}_{n=1}^{\infty}$ is a subordinating factor sequence if and only if:*

$$\text{Re} \left\{ 1 + 2 \sum_{k=1}^{\infty} b_k z^k \right\} > 0, \; z \in E.$$

2. Properties of Classes $S_{q,a,s,\lambda}(c, M)$ and $C_{q,a,s,\lambda}(c, M)$

We start the section with the necessary and sufficient condition for a function to be in the class $S_{q,a,s,\lambda}(c, M)$.

Theorem 1. Let $f \in A$. Then, $f \in S_{q,a,s,\lambda}(c, M)$ if and only if:

$$\sum_{k=2}^{\infty} \{[k] - 1 + |c(1+m) + m([k]-1)|\} \frac{[\lambda+1]_{k-1}}{[k-1]!} \left|\left(\frac{[k+a]}{[1+a]}\right)^s\right| |a_k| < |c(1+m)|, \quad (6)$$

where $m = 1 - \frac{1}{M}$, $(M > \frac{1}{2})$.

Proof. Let us assume first that Inequality (6) holds. To show $f \in S_{q,a,s,\lambda}(c, M)$, we need to prove Inequality (5).

$$\left| \frac{\frac{z(d_q(D^s_{q,a,\lambda}f(z)))}{D^s_{q,a,\lambda}f(z)} - 1}{A - B \left\{ \frac{zd_q(D^s_{q,a,\lambda}f(z))}{D^s_{q,a,\lambda}f(z)} \right\}} \right| = \left| \frac{\sum_{k=2}^{\infty} \left(\frac{[k+a]}{[1+a]}\right)^s \cdot \frac{[\lambda+1]_{k-1}}{[k-1]!}([k]-1)a_k z^k}{(A-B)z + \sum_{k=2}^{\infty}(A - B[k])\left(\frac{[1+a]}{[k+a]}\right)^s \cdot \frac{[\lambda+1]_{k-1}}{[k-1]!} a_k z^k} \right|$$

$$\leq \frac{\sum_{k=2}^{\infty} \left|\left(\frac{[k+a]}{[1+a]}\right)^s\right| \cdot \frac{[\lambda+1]_{k-1}}{[k-1]!}([k]-1)|a_k|}{|A - B| - \left|\sum_{k=2}^{\infty}(A - B[k])\left(\frac{[k+a]}{[1+a]}\right)^s \cdot \frac{[\lambda+1]_{k-1}}{[k-1]!} a_k\right|}$$

$$\leq \frac{\sum_{k=2}^{\infty} \left|\left(\frac{[k+a]}{[1+a]}\right)^s\right| \cdot \frac{[\lambda+1]_{k-1}}{[k-1]!}([k]-1)|a_k|}{|c(1+m)| - \sum_{k=2}^{\infty}|c(1+m) + m([k]-1)| \frac{[\lambda+1]_{k-1}}{[k-1]!} \left|\left(\frac{[k+a]}{[1+a]}\right)^s\right| |a_k|}$$

$$< 1.$$

Hence, $f \in S_{q,a,s,\lambda}(c, M)$ by using Inequality (6). Conversely, let $f \in S_{q,a,s,\lambda}(c, M)$ be of the form (1), then:

$$\left| \frac{\frac{z(d_q(D^s_{q,a,\lambda}f(z)))}{D^s_{q,a,\lambda}(f(z))} - 1}{A - B \left\{ \frac{zd_q(D^s_{q,a,\lambda}f(z))}{D^s_{q,a,\lambda}f(z)} \right\}} \right| = \left| \frac{\sum_{k=2}^{\infty} \left(\frac{[k+a]}{[1+a]}\right)^s \cdot \frac{[\lambda+1]_{k-1}}{[k-1]!}([k]-1)a_k z^k}{(A-B)z + \sum_{k=2}^{\infty}(A - B[k])\left(\frac{[k+a]}{[1+a]}\right)^s \cdot \frac{[\lambda+1]_{k-1}}{[k-1]!} a_k z^k} \right|.$$

Since $|\operatorname{Re} z| \leq |z|$, we have:

$$\operatorname{Re} \left\{ \frac{\sum_{k=2}^{\infty} \left(\frac{[k+a]}{[1+a]}\right)^s \cdot \frac{[\lambda+1]_{k-1}}{[k-1]!}([k]-1)a_k z^k}{(A-B)z + \sum_{k=2}^{\infty}(A - B[k])\left(\frac{[k+a]}{[1+a]}\right)^s \cdot \frac{[\lambda+1]_{k-1}}{[k-1]!} a_k z^k} \right\} < 1.$$

Now, we choose values of z on the real axis such that $zd_q(D^s_{q,a,\lambda}f(z))/D^s_{q,a,\lambda}f(z)$ is real. Letting $z \to 1^-$ through real values, after some calculations, we obtain Inequality (6). □

Remark 1. (i) If $q \to 1^-$, $s \in \mathbb{N}_0$, $a = 0$, and $\lambda = 0$, the above result reduces to the sufficient condition for $f(z)$ to be in class $H_n(c, M)$ ($c \in \mathbb{C}^*$, $M > \frac{1}{2}$) discussed in [26]. (ii) If $c = 1 - \alpha$ ($\alpha \in [0, 1)$), $m = 0$, $\lambda = 0$, and $q \to 1^-$, the above result reduces to the sufficient condition for $f(z)$ to be in class $S^*_{s,a}(\alpha)$ discussed in [28].

Theorem 2. Let $f_i \in S_{q,a,s,\lambda}(c, M)$ having the form:

$$f_i(z) = z + \sum_{k=2}^{\infty} a_{k,i} z^k, \qquad \text{for } i = 1, 2, 3, .., l.$$

Then, $F \in S_{q,a,s,\lambda}(c, M)$, where $F(z) = \sum_{i=1}^{l} c_i f_i(z)$ with $\sum_{i=1}^{l} c_i = 1$.

Proof. From Theorem 1, we can write:

$$\sum_{k=2}^{\infty} \left\{ \frac{\{[k]-1+|b(1+m)+m([k]-1)|\} \frac{[\lambda+1]_{k-1}}{[k-1]!} \left|\left(\frac{[k+a]}{[1+a]}\right)^s\right|}{|b(1+m)|} \right\} a_{k,i} < 1. \quad (7)$$

Therefore:

$$F(z) = \sum_{i=1}^{l} c_i \left(z + \sum_{k=2}^{\infty} a_{k,i} z^k \right)$$

$$= z + \sum_{k=2}^{\infty} \left(\sum_{i=1}^{l} c_i a_{k,i} \right) z^k;$$

where however due to (7), we have:

$$\sum_{k=2}^{\infty} \frac{\{[k]-1+|b(1+m)+m([k]-1)|\} \frac{[\lambda+1]_{k-1}}{[k-1]!} \left|\left(\frac{[k+a]}{[1+a]}\right)^s\right|}{|b(1+m)|} \left(\sum_{i=2}^{l} c_i a_{k,i} \right)$$

$$= \sum_{i=2}^{l} \left[\frac{\{[k]-1+|b(1+m)+m([k]-1)|\} \frac{[\lambda+1]_{k-1}}{[k-1]!} \left|\left(\frac{[k+a]}{[1+a]}\right)^s\right|}{|b(1+m)|} \right] c_i \leq 1;$$

Therefore, $F \in S_{q,a,s,\lambda}(c, M)$. □

Theorem 3. *Let f_i with $i = 1, 2, ..., v$ belong to the class $S_{q,a,s,\lambda}(c, M)$. The arithmetic mean h of f_i is given by:*

$$h(z) = \frac{1}{v} \sum_{i=1}^{v} f_i(z) \quad (8)$$

belonging to class $S_{q,a,s,\lambda}(c, M)$.

Proof. From (8), we can write:

$$h(z) = \frac{1}{v} \sum_{i=1}^{v} \left(z + \sum_{k=2}^{\infty} a_{k,i} z^k \right) = z + \sum_{k=2}^{\infty} \left(\frac{1}{v} \sum_{i=1}^{v} a_{k,i} \right) z^k. \quad (9)$$

Since $f_i \in S_{q,a,s,\lambda}(c, M)$ for every $i = 1, 2, ..., v$, using (6) and (9), we have:

$$\sum_{k=2}^{\infty} \{[k]-1+|b(1+m)+m([k]-1)|\} \frac{[\lambda+1]_{k-1}}{[k-1]!} \left|\left(\frac{[k+a]}{[1+a]}\right)^s\right| \left(\frac{1}{v} \sum_{i=1}^{v} a_{k,i} \right)$$

$$= \frac{1}{v} \sum_{i=1}^{v} \left(\sum_{k=2}^{\infty} \{[k]-1+|b(1+m)+m([k]-1)|\} \frac{[\lambda+1]_{k-1}}{[k-1]!} \left|\left(\frac{[k+a]}{[1+a]}\right)^s\right| a_{k,i} \right)$$

$$\leq \frac{1}{v} \sum_{i=1}^{v} (|b(1+m)|) = |b(1+m)|,$$

and this completes the proof. □

Now, we give the subordination relation for the functions in class $S_{q,a,s,\lambda}(c, M)$ by using the subordination theorem.

Theorem 4. *Let $m = 1 - \frac{1}{M}$ ($M > \frac{1}{2}$). Furthermore, $c \neq 0$ with $\text{Re}(c) > \frac{-m}{2(1+m)}$ when $m > 0$ and $\text{Re}(c) < \frac{-m}{2(1+m)}$ when $m < 0$ and $\lambda \geq 0$. If $f \in S_{q,a,s,\lambda}(c, M)$, then:*

$$\frac{\{q+|c(1+m)+mq|\}C_{\lambda,2}B_{s,a}(2)}{2[\{q+|c(1+m)+mq|\}C_{\lambda,2}B_{s,a}(2)+|c(1+m)|]}(f*g)(z) \prec g(z) \qquad (10)$$

where $g(z)$ is a convex function in E, $C_{\lambda,k} = \frac{[\lambda+1]_{k-1}}{[k-1]!}$, $B_{s,a}(k) = \left|\left(\frac{[k+a]}{[1+a]}\right)^s\right|$, and:

$$\mathrm{Re}\, f(z) > -1 - \frac{(1+m)|c|}{\{q+|c(1+m)+mq|\}C_{\lambda,2}B_{s,a}(2)}. \qquad (11)$$

The constant $\frac{\{q+|c(1+m)+mq|\}C_{\lambda,2}B_{s,a}(2)}{2[\{q+|c(1+m)+mq|\}C_{\lambda,2}B_{s,a}(2)+|c(1+m)|]}$ is the best estimate.

Proof. Let $f(z) \in S_{q,a,s,\lambda}(c,M)$ and $g(z) = z + \sum_{k=2}^{\infty} c_k z^k$. Then:

$$\frac{\{q+|c(1+m)+mq|\}C_{\lambda,2}B_{s,a}(2)}{2[\{q+|c(1+m)+mq|\}C_{\lambda,2}B_{s,a}(2)+|c(1+m)|]}(f*g)(z)$$

$$= \frac{\{q+|c(1+m)+mq|\}C_{\lambda,2}B_{s,a}(2)}{2[\{q+|c(1+m)+mq|\}C_{\lambda,2}B_{s,a}(2)+|c(1+m)|]}\left(z + \sum_{k=2}^{\infty} a_k c_k z^k\right). \qquad (12)$$

Thus, (10) holds true if:

$$\left\{\frac{\{q+|c(1+m)+mq|\}C_{\lambda,2}B_{s,a}(2)}{2[\{q+|c(1+m)+mq|\}C_{\lambda,2}B_{s,a}(2)+|c(1+m)|]}a_k\right\}_{k=1}^{\infty} \qquad (13)$$

is a subordinating factor sequence with $a_1 = 1$. From Lemma 2, it suffices to show:

$$\mathrm{Re}\left\{1 + \sum_{k=1}^{\infty} \frac{\{q+|c(1+m)+mq|\}C_{\lambda,2}B_{s,a}(2)}{[\{q+|c(1+m)+mq|\}C_{\lambda,2}B_{s,a}(2)+|c(1+m)|]}a_k z^k\right\} > 0. \qquad (14)$$

Now, as $\{[k] - 1 + |c(1+m) + m([k]-1)|\}C_{\lambda,k}B_{s,a}(k)$ is an increasing function of k ($k \geq 2$), we have:

$$\mathrm{Re}\left\{1 + \sum_{k=1}^{\infty} \frac{\{q+|c(1+m)+mq|\}C_{\lambda,2}B_{s,a}(2)}{[\{q+|c(1+m)+mq|\}C_{\lambda,2}B_{s,a}(2)+|c(1+m)|]}a_k z^k\right\}$$

$$= \mathrm{Re}\left\{1 + \frac{\{q+|c(1+m)+mq|\}C_{\lambda,2}B_{s,a}(2)}{[\{q+|c(1+m)+mq|\}C_{\lambda,2}B_{s,a}(2)+|c(1+m)|]}z + \frac{\sum_{k=2}^{\infty}\{q+|c(1+m)+mq|\}C_{\lambda,2}B_{s,a}(2)a_k z^k}{[\{q+|c(1+m)+mq|\}C_{\lambda,2}B_{s,a}(2)+|c(1+m)|]}\right\}$$

$$\geq 1 - \frac{\{q+|c(1+m)+mq|\}C_{\lambda,2}B_{s,a}(2)}{[\{q+|c(1+m)+mq|\}C_{\lambda,2}B_{s,a}(2)+|c(1+m)|]}r -$$

$$\frac{\sum_{k=2}^{\infty}\{q+|c(1+m)+mq|\}C_{\lambda,2}B_{s,a}(2)|a_k|r^k}{[\{q+|c(1+m)+mq|\}C_{\lambda,2}B_{s,a}(2)+|c(1+m)|]}$$

$$> 1 - \frac{\{q+|c(1+m)+mq|\}C_{\lambda,2}B_{s,a}(2)}{[\{q+|c(1+m)+mq|\}C_{\lambda,2}B_{s,a}(2)+|c(1+m)|]}r -$$

$$\frac{(1+m)|c|}{[\{q+|c(1+m)+mq|\}C_{\lambda,2}B_{s,a}(2)+|c(1+m)|]}r$$

$$> 0. \qquad (|z| = r < 1)$$

Hence, (14) holds true in E, and the subordination result (10) is affirmed by Theorem 4. The inequality (11) follows by taking $g(z) = \frac{z}{1-z} = \sum_{k=1}^{\infty} z^k$ in (10).

Let us consider the function:

$$\phi(z) = z - \frac{|c(1+m)|}{[\{q+|c(1+m)+mq|\}C_{\lambda,2}B_{s,a}(2)+|c(1+m)|]}z^2 \quad (z \in E)$$

which is a member of $S_{q,a,s,\lambda}(c,M)$. Then, by using (10), we have:

$$\frac{\{q+|c(1+m)+mq|\}C_{\lambda,2}B_{s,a}(2)}{2[\{q+|c(1+m)+mq|\}C_{\lambda,2}B_{s,a}(2)+|c(1+m)|]}\phi(z) \prec \frac{z}{1-z}.$$

It is easily verified that:

$$\min \text{Re}\left\{\frac{\{q+|c(1+m)+mq|\}C_{\lambda,2}B_{s,a}(2)}{2[\{q+|c(1+m)+mq|\}C_{\lambda,2}B_{s,a}(2)+|c(1+m)|]}\phi(z)\right\} = -\frac{1}{2} \quad (z \in E),$$

then the constant $\frac{\{q+|c(1+m)+mq|\}C_{\lambda,2}B_{s,a}(2)}{2[\{q+|c(1+m)+mq|\}C_{\lambda,2}B_{s,a}(2)+|c(1+m)|]}$ cannot be replaced by a larger one. □

Remark 2. *If $s \in \mathbb{N}_0$, $a = 0$, $\lambda = 0$, and $q \to 1^-$, Theorem 4 reduces to the subordination result proven in [29].*

Now, we discuss the inclusion results pertaining to classes $S_{q,a,s,\lambda}(c,M)$ and $C_{q,a,s,\lambda}(c,M)$ in reference to parameters s and λ.

Theorem 5. *For any complex number s, $S_{q,a,s+1,\lambda}(c,M) \subset S_{q,a,s,\lambda}(c,M)$ if $\text{Re}(\frac{1+\{c(1+m)-m\}z}{1-mz}) > \frac{1}{q^{a_1}(1-q)}\{1-\cos(a_2 \ln q)\}$ where $a = a_1 + ia_2$.*

Proof. Let $f \in S_{q,a,s+1,\lambda}(c,M)$, then:

$$\frac{zd_q(D_{q,a,\lambda}^{s+1}f(z))}{D_{q,a,\lambda}^{s+1}f(z)} \prec \frac{1+\{c(1+m)-m\}z}{1-mz}, \tag{15}$$

Let:

$$h(z) = \frac{1+\{c(1+m)-m\}z}{1-mz}$$

and:

$$r(z) - \frac{zd_q(D_{q,a,\lambda}^s f(z))}{D_{q,a,\lambda}^s f(z)}.$$

We will show:

$$r(z) \prec h(z),$$

which would prove $S_{q,a,s,\lambda}(c,M) \subset S_{q,a,s+1,\lambda}(c,M)$. From the identity relation (3), after a few calculations, we have:

$$\frac{zd_q(D_{q,a,\lambda}^s f(z))}{D_{q,a,\lambda}^s f(z)} = \frac{[1+a]}{q^a} \cdot \frac{D_{q,a,\lambda}^{s+1}f(z)}{D_{q,a,\lambda}^s f(z)} - \frac{[a]}{q^a}.$$

After some calculations, we have:

$$\frac{D_{q,a,\lambda}^{s+1}f(z)}{D_{q,a,\lambda}^{s}f(z)} = \frac{1}{[1+a]}\left\{\frac{q^a z d_q(D_{q,a,\lambda}^{s}f(z))}{D_{q,a,\lambda}^{s}f(z)} + [a]\right\}$$

$$= \frac{1}{[1+a]}\{q^a r(z) + [a]\}.$$

Applying logarithmic q-differentiation, we have:

$$\frac{z d_q(D_{q,a,\lambda}^{s+1}f(z))}{D_{q,a,\lambda}^{s+1}f(z)} = r(z) + \frac{z d_q r(z)}{r(z) + q^{-a}[a]}. \tag{16}$$

From (15) and (16), we have:

$$r(z) + \frac{z[d_q r(z)]}{r(z) + q^{-a}[a]} \prec \frac{1 + \{c(1+m) - m\}z}{1 - mz}.$$

If $\mathrm{Re}(h(z)) > \frac{1}{q^{a_1}(1-q)}\{1 - \cos(a_2 \ln q)\}$, then from Lemma 1, it implies:

$$r(z) \prec h(z),$$

which implies $f(z) \in S_{q,a,s,\lambda}(c, M)$. Therefore, $S_{q,a,s,\lambda}(c, M) \subset S_{q,a,s+1,\lambda}(c, M)$. □

Theorem 6. *For any complex number s, $C_{q,a,s+1,\lambda}(c, M) \subset C_{q,a,s,\lambda}(c, M)$ if $\mathrm{Re}(\frac{1+\{c(1+m)-m\}z}{1-mz}) > \frac{1}{q^{a_1}(1-q)}\{1 - \cos(a_2 \ln q)\}$ where $a = a_1 + ia_2$.*

Proof. It is obvious from the fact $f \in C_{q,a,s,\lambda}(c, M) \Leftrightarrow z d_q f \in S_{q,a,s,\lambda}(c, M)$. □

Theorem 7. *For any complex number s, $S_{q,a,s,\lambda+1}(c, M) \subset S_{q,a,s,\lambda}(c, M)$ if $\mathrm{Re}(\frac{1+\{c(1+m)-m\}z}{1-mz}) > \frac{1-q^{-\lambda}}{1-q}$, $\lambda > -1$.*

Proof. Let $f \in S_{q,a,s,\lambda+1}(c, M)$, then:

$$\frac{z d_q(D_{q,a,\lambda+1}^{s}f(z))}{D_{q,a,\lambda+1}^{s}f(z)} \prec \frac{1 + \{c(1+m) - m\}z}{1 - mz}. \tag{17}$$

Consider:

$$h(z) = \frac{1 + \{c(1+m) - m\}z}{1 - mz}$$

and:

$$q(z) = \frac{z d_q(D_{q,a,\lambda}^{s}f(z))}{D_{q,a,\lambda}^{s}f(z)}.$$

We will show:

$$q(z) \prec h(z),$$

which would conveniently prove $S_{q,a,s,\lambda+1}(c, M) \subset S_{q,a,s,\lambda}(c, M)$. From the identity relation (4), after a few calculations, we have:

$$\frac{z d_q(D_{q,a,\lambda}^{s}f(z))}{D_{q,a,\lambda}^{s}f(z)} = \frac{[1+\lambda]}{q^\lambda} \frac{D_{q,a,\lambda+1}^{s}f(z)}{D_{q,a,\lambda}^{s}f(z)} - \frac{[\lambda]}{q^\lambda}.$$

After some calculations, we have:

$$\frac{D^s_{q,a,\lambda+1}f(z)}{D^s_{q,a,\lambda}f(z)} = \frac{1}{[1+\lambda]}\left\{\frac{q^a.zd_q(D^s_{q,a,\lambda}f(z))}{D^s_{q,a,\lambda}f(z)} + [\lambda]\right\}$$

$$= \frac{1}{[1+\lambda]}\left\{q^\lambda q(z) + [\lambda]\right\}.$$

Applying logarithmic q-differentiation, we have:

$$\frac{zd_q(D^s_{q,a,\lambda+1}f(z))}{D^s_{q,a,\lambda+1}f(z)} = q(z) + \frac{zd_q q(z)}{q(z) + q^{-\lambda}[\lambda]} \quad (18)$$

From (17) and (18), we have:

$$q(z) + \frac{z[d_q q(z)]}{q(z) + q^{-\lambda}[\lambda]} \prec \frac{1 + \{c(1+m) - m\}z}{1 - mz}.$$

If $\text{Re}(h(z)) > \frac{1-q^{-\lambda}}{1-q}$ for any value of $\lambda > -1$, so by Lemma 1, we have $q(z) \prec h(z)$, which implies $f(z) \in S_{q,a,s,\lambda}(c,M)$. Therefore, $S_{q,a,s,\lambda+1}(c,M) \subset S_{q,a,s,\lambda}(c,M)$. □

Remark 3. *If we consider $q \to 1^-$ with $\text{Re}\, a \geq 0$, $c = 1, m = 1$ in Theorem 5 and $\lambda \geq 0, c = 1, m = 1$ in Theorem 7, we obtain the special cases of the inclusion results, Theorems 2.4 and 2.5 in [19].*

In [30], the q-Bernardi integral operator $L_b f(z)$ is defined as:

$$L_b f(z) = \frac{[1+b]}{z^b}\int_0^z t^{b-1}f(t)d_q t$$

$$= z + \sum_{k=2}^\infty \left(\frac{[1+b]}{[k+b]}\right) a_k z^k, \quad b = 1,2,3,....$$

Now, we apply the generalized operator $D^s_{q,a,\lambda}$ on $L_b f(z)$ as:

$$D^s_{q,a,\lambda}(L_b f(z)) = z + \sum_{k=2}^\infty \left(\frac{[k+a]}{[1+a]}\right)^s \frac{[\lambda+1]_{k-1}}{[k-1]!}\left(\frac{[1+b]}{[k+b]}\right)a_k z^k.$$

The identity relation of $D^s_{q,a,\lambda}(L_b f(z))$ is given as:

$$zd_q\left[D^s_{q,a,\lambda}\{L_b f(z)\}\right] = \left(\frac{[1+b]}{q^b}\right)D^s_{q,a,\lambda}f(z) - \frac{[b]}{q^b}D^s_{q,a,\lambda}\{L_b f(z)\}. \quad (19)$$

The following theorems are the integral inclusions of the classes $S_{q,a,s,\lambda}(c,M)$ and $C_{q,a,s,\lambda}(c,M)$ with respect to the q-Bernardi integral operator.

Theorem 8. *If $f(z) \in S_{q,a,s,\lambda}(c,M)$ then $L_b f(z) \in S_{q,a,s,\lambda}(c,M)$ if $\text{Re}(\frac{1+\{c(1+m)-m\}z}{1-mz}) > \frac{1-q^{-b}}{1-q}$ for any complex number s.*

Proof. Let $g(z) \in S_{q,a,s,\lambda}(c,M)$, then:

$$\frac{zd_q(D^s_{q,a,\lambda}g(z))}{D^s_{q,a,\lambda}g(z)} \prec \frac{1 + \{c(1+m) - m\}z}{1 - mz}. \quad (20)$$

Consider:

and:
$$h(z) = \frac{1 + \{c(1+m) - m\}z}{1 - mz}$$

$$u(z) = \frac{zd_q(D^s_{q,a,\lambda} L_b g(z))}{D^s_{q,a,\lambda} L_b g(z)}.$$

We will show:
$$u(z) \prec h(z),$$

which would prove $L_b g(z) \in S_{q,a,s,\lambda}(c, M)$. From the identity relation (19), after some calculations, we have:

$$\frac{zd_q(D^s_{q,a,\lambda} L_b g(z))}{D^s_{q,a,\lambda} L_b g(z)} = \left(\frac{[1+b]}{q^b}\right) \frac{D^s_{q,a,\lambda} g(z)}{(D^s_{q,a,\lambda} L_b g(z))} - \frac{[b]}{q^b}.$$

After some calculations, we have:

$$\frac{D^s_{q,a,\lambda} g(z)}{D^s_{q,a,\lambda} L_b g(z)} = \frac{1}{[1+b]} \left[\frac{q^b \cdot zd_q(D^s_{q,a,\lambda} L_b g(z))}{D^s_{q,a,\lambda} L_b g(z)} + [b]\right].$$

Applying logarithmic q-differentiation, we have:

$$\frac{zd_q(D^s_{q,a,\lambda} g(z))}{D^s_{q,a,\lambda} g(z)} = u(z) + \frac{z[d_q u(z)]}{u(z) + q^{-b}[b]} \tag{21}$$

From (20) and (21), we have:

$$u(z) + \frac{z[d_q u(z)]}{u(z) + q^{-b}[b]} \prec \frac{1 + \{c(1+m) - m\}z}{1 - mz}$$

If $\text{Re}(h(z)) > \frac{1-q^{-b}}{1-q}$, so by Lemma 1, we have $u(z) \prec h(z)$, which implies $L_b g(z) \in S_{q,a,s,\lambda}(c, M)$. □

Theorem 9. *If $f(z) \in C_{q,a,s,\lambda}(c, M)$, then $L_b f(z) \in C_{q,a,s,\lambda}(c, M)$ for any complex number s.*

Proof. It is an immediate consequence of the fact $C_{q,a,s,\lambda}(c, M) \Leftrightarrow zd_q f \in S_{q,a,s,\lambda}(c, M)$. □

Author Contributions: Conceptualization: K.I.N.; formal analysis: R.S.B. and K.I.N.; investigation: R.S.B. and K.I.N.; methodology: R.S.B. and K.I.N.; supervision: K.I.N.; validation: R.S.B.; writing, original draft: R.S.B. All authors read and agreed to the published version of the manuscript.

Funding: The authors received no funding for this research.

Conflicts of Interest: The authors declare no conflict of interest.

References

1. Einstein, A. Concerning on heuristic point of view toward the emission and transformation of light. *Ann. Phys.* **1905**, *17*, 132–148. [CrossRef]
2. Jackson, F.H. On q-definite integrals. *Q. J. Pure Appl. Math.* **1910**, *41*, 193–203.
3. Jackson, F.H. q-difference equations. *Am. J. Math.* **1910**, *32*, 305–314. [CrossRef]

4. Srivastava, H.M. Univalent functions, fractional calculus and associated generalized hypergeometric functions. In *Univalent Functions, Fractional Calculus and Their Applications*; Srivastava, H.M., Owa, S., Eds.; Halsted Press: Chichester, UK; John Wiley and Sons: New York, NY, USA; Chichester, UK; Brisbane, Australia; Toronto, ON, Canada, 1989; pp. 329–354.
5. Ismail, M.E.H.; Markes, E.; Styer, D. A generalization of starlike functions. *Complex Var.* **1990**, *14*, 77–84. [CrossRef]
6. Agrawal, S.; Sahoo SK. A generalization of starlike functions of order alpha. *Hokkaido Math. J.* **2017**, *46*, 15–27. [CrossRef]
7. Purohit, S.D.; Raina, R.K. Certain subclasses of analytic functions associated with fractional q-calculus operator. *Math. Scand.* **2007**, *109*, 55–70. [CrossRef]
8. Sahoo, S.K.; Sharma, N.L. On a generalization of close-to-convex functions. *Ann. Polon. Math.* **2015**, *113*, 93–108. [CrossRef]
9. Srivastava, H.M. Operators of basic (or q-) calculus and fractional q-calculus and their applications in geometric function theory. *Iran. J. Sci. Technol. Trans. Sci.* **2020**, *44*, 327–344. [CrossRef]
10. Mohammad, A.; Darus, M. A generalized operator involving the q-hypergeometric function. *Mat. Vesnik* **2013**, *65*, 454–465.
11. Wilf, H.S. Subordinating factor sequence for convex maps of the unit circle. *Proc. Am. Math. Soc.* **1962**, *12*, 689–693. [CrossRef]
12. Ernst, T. *A Comprehensive Treatment of q-Calculus*; Springer: Basel, Switzerland; Heidelberg, Germany; New York, NY, USA; Dordrecht, The Netherlands; London, UK, 2012.
13. Ali, S.; Noor, K.I. Study on the q-analogue of a certain family of linear operators. *Turkish J. Math.* **2019**, *43*, 2707–2714.
14. Noor, K.I.; Bukhari, S.Z.H. Some subclasses of analytic and spiral-like functions of complex order involving the Srivastava-Attiya integral operator. *Integral Transforms Spec. Funct.* **2010**, *21*, 907–916. [CrossRef]
15. Kanas, S.; Răducanu, D. Some classes of analytical functions related to conic domains. *Math. Slovaca* **2014**, *64*, 1183–1196. [CrossRef]
16. Owa, S.; Srivastava, H.M. Some applications of generalized Libera integral operator. *Proc. Japan Acad. Ser. A. Math. Sci.* **1986**, *62*, 125–128. [CrossRef]
17. Wang, Z.G.; Li, Q.G.; Jiang, Y.P. Certain subclasses of multivalent analytic functions involving generalized Srivastava-Attiya operator. *Integral Transforms Spec. Funct.* **2010**, *21*, 221–234. [CrossRef]
18. Govindaraj, M.; Sivasubramanian, S. On a class of analytic functions related to conic domains involving q-calculus. *Anal. Math.* **2017**, *43*, 475–487. [CrossRef]
19. Al-Shaqsi, K.; Darus, M. A multiplier transformation defined by convolution involving nth order polylogarithm functions. *Int. Math. Forum* **2009**, *4*, 1823–1837.
20. Çetinkaya, A.; Polatoğlu, Y. q-Harmonic mappings for which analytic part is q-convex functions of complex order. *Hacet. J. Math. Stat.* **2018**, *47*, 813–820. [CrossRef]
21. Aouf, M.K.; Seoudy, T.M. Convolution properties for classes of bounded analytic functions with complex order defined by q-derivative. *RACSAM* **2019**, *113*, 1279–1288. [CrossRef]
22. Bukhari, S.Z.H.; Noor, K.I.; Malik, B. Some applications of generalized Srivastava-Attiya integral operator. *Iran. J. Sci. Technol Trans. A Sci.* **2018**, *42*, 2251–2257. [CrossRef]
23. Noor, K.I.; Riaz, S. Generalized q-starlike functions. *Studia Sci. Math. Hungar.* **2017**, *54*, 509–522. [CrossRef]
24. Noor, K.I. Some classes analytic functions associated with q-Ruscheweyh differential operator. *Facta Univ. Ser. Math. Inform.* **2018**, *33*, 531–538.
25. Polatoğlu, Y.; Şen, A. Some results on subclasses of Janowski -spiral like functions of complex order. *Gen. Math.* **2007**, *15*, 88–97.
26. Aouf, M.K.; Darwish, H.E.; Attiya, AA. On a class of certain analytic functions of complex order. *Indian J. Pure Appl. Math.* **2001**, *32*, 1443–1452.
27. Shamsan, H.; Latha, S. On generalized bounded Mocanu variation related to q-derivative and conic regions. *Ann. Pure Appl. Math.* **2018**, *17*, 67–83. [CrossRef]
28. Răducanu, D.; Srivastava, H.M. A new class of analytic functions defined by means of a convolution operator involving the Hurwitz-Lerch Zeta function. *Integral Transforms Spec. Funct.* **2007**, *18*, 933–943. [CrossRef]

29. Güney, H.O.; Attiya, A.A. A subordination result with Salagean-type certain analytic functions of complex order. *Bull. Belg. Math. Soc.* **2011**, *18*, 253-258. [CrossRef]
30. Noor, K.I.; Riaz, S.; Noor, M.A. On *q*-Bernardi linear operator. *TMWS J. Pure Appl. Math.* **2017**, *8*, 3–11.

© 2020 by the authors. Licensee MDPI, Basel, Switzerland. This article is an open access article distributed under the terms and conditions of the Creative Commons Attribution (CC BY) license (http://creativecommons.org/licenses/by/4.0/).

Article

New Criteria for Meromorphic Starlikeness and Close-to-Convexity

Ali Ebadian [1], Nak Eun Cho [2,*], Ebrahim Analouei Adegani [1] and Sibel Yalçın [3]

1. Department of Mathematics, Faculty of Science, Urmia University, Urmia 5756151818, Iran; a.ebadian@urmia.ac.ir (A.E.); analoey.ebrahim@gmail.com (E.A.A.)
2. Department of Applied Mathematics, College of Natural Sciences, Pukyong National University, Busan 608-737, Korea
3. Department of Mathematics, Faculty of Arts and Sciences, Bursa Uludag University, Bursa 16059, Turkey; syalcin@uludag.edu.tr
* Correspondence: necho@pknu.ac.kr

Received: 23 March 2020; Accepted: 20 May 2020; Published: 23 May 2020

Abstract: The main purpose of current paper is to obtain some new criteria for meromorphic strongly starlike functions of order α and strongly close-to-convexity of order α. Furthermore, the main results presented here are compared with the previous outcomes obtained in this area.

Keywords: differential subordination; strongly close-to-convex functions; starlike functions; meromorphic strongly starlike functions

MSC: Primary 30C45; Secondary 30C80

---

1. Introduction and Preliminaries

For two analytic functions f and F in $\mathbb{U} := \{z : z \in \mathbb{C} \text{ and } |z| < 1\}$, it is stated that the function f is subordinate to the function F in $\mathbb{U}$, written as $f(z) \prec F(z)$, if there exists a Schwarz function ϖ, which is analytic in $\mathbb{U}$ with

$$\varpi(0) = 0 \quad \text{and} \quad |\varpi(z)| < 1 \quad (z \in \mathbb{U}),$$

such that $f(z) = F(\varpi(z))$ for all $z \in \mathbb{U}$. In particular, if F be a univalent function in $\mathbb{U}$, then we have below equivalence:

$$f(z) \prec F(z) \iff f(0) = F(0) \quad \text{and} \quad f(\mathbb{U}) \subset F(\mathbb{U}).$$

Let Σ_n denote the category of all functions analytic in the punctured open unit disk $\mathbb{U}^*$ given by

$$\mathbb{U}^* := \{z : z \in \mathbb{C} \text{ and } 0 < |z| < 1\} = \mathbb{U} \setminus \{0\},$$

which have the form

$$f(z) = \frac{1}{z} + \sum_{k=n}^{\infty} a_{k-1} z^{k-1} \quad (n \in \mathbb{N} := \{1, 2, \cdots\}). \tag{1}$$

A function $f \in \Sigma$, where Σ is the union of Σ_n for all positive integers n, is said to be in the class $\widetilde{\mathcal{MS}}^*(\alpha)$ of meromorphic strongly starlike functions of order α if we have the condition

$$\left| \arg\left(-\frac{zf'(z)}{f(z)} \right) \right| < \frac{\alpha \pi}{2} \quad (z \in \mathbb{U}^*; \ 0 < \alpha \leqq 1).$$

In particular, $\mathcal{MS}^* := \widetilde{\mathcal{MS}}^*(1)$ is the class of meromorphic starlike functions in the open unit disk $\mathbb{U}$.

Let $\mathcal{A}_n$ be the category of all functions analytic in $\mathbb{U}$ which have the following form

$$f(z) = z^n + \sum_{k=n+1}^{\infty} a_k z^k \quad (n \in \mathbb{N}). \tag{2}$$

The class $\mathcal{A}_1$ is denoted by $\mathcal{A}$.

Let $\widetilde{\mathcal{S}}^*(\alpha)$ be the subcategory of $\mathcal{A}$ defined as follows

$$\widetilde{\mathcal{S}}^*(\alpha) := \left\{ f : f \in \mathcal{A} \text{ and } \left| \arg\left(\frac{zf'(z)}{f(z)}\right) \right| < \frac{\alpha\pi}{2} \quad (z \in \mathbb{U}; \, 0 < \alpha \leqq 1) \right\}.$$

The classes $\widetilde{\mathcal{S}}^*(\alpha)$ will be called the class of *strongly starlike functions of order* α. In particular, $\mathcal{S}^* := \widetilde{\mathcal{S}}^*(1)$ is the class of *starlike functions* in $\mathbb{U}$.

By means of the principle of subordination between analytic functions, the above definition is equivalent to

$$\widetilde{\mathcal{S}}^*(\alpha) := \left\{ f : f \in \mathcal{A} \text{ and } \frac{zf'(z)}{f(z)} \prec \left(\frac{1+z}{1-z}\right)^{\alpha} \quad (z \in \mathbb{U}; \, 0 < \alpha \leqq 1) \right\}.$$

Furthermore, let $\widetilde{\mathcal{CC}}(\alpha)$ denote the category of all functions in $\mathcal{A}$ which are *strongly close-to-convex of order* α in $\mathbb{U}$ if there exists a function $g \in \mathcal{S}^*$ such that

$$\left| \arg\left(\frac{zf'(z)}{g(z)}\right) \right| < \frac{\alpha\pi}{2} \quad (z \in \mathbb{U}; \, 0 < \alpha \leqq 1).$$

In particular, $\mathcal{CC} := \widetilde{\mathcal{CC}}(1)$ is the class of *close-to-convex functions* in $\mathbb{U}$.

In the year 1978, Miller and Mocanu [1] introduced the method of differential subordinations. Because of the interesting properties and applications possessed by the Briot-Bouquet differential subordination, there have been many attempts to extend these results. Then, in recent years, several authors obtained several applications of the method of differential subordinations in geometric function theory by using differential subordination associated with starlikeness, convexity, close-to-convexity and so on (see, for example, [2–13]). Furthermore, based on the generalized Jack lemma, the well-known lemma of Nunokawa and so on, certain sufficient conditions were derived in [14–16] considering concept of arg, real part and imaginary part for function to be p-valently starlike and convex one in the unit disk.

The aim of the current paper is to obtain some new criteria for univalence, strongly starlikeness and strongly close-to-convexity of functions in the normalized analytic function class $\mathcal{A}_n$ in the open unit disk $\mathbb{U}$ and meromorphic strongly starlikeness in the punctured open unit disk $\mathbb{U}^*$ by using a lemma given by Nunokawa (see [17,18]). Further, the current results are compared with the previous outcomes obtained in this area.

In order to prove our main results, we require the following lemma.

Lemma 1 (see [17,18]). *Let the function $p(z)$ given by*

$$p(z) = 1 + \sum_{n=m}^{\infty} c_n z^n \quad (c_m \neq 0; \, m \in \mathbb{N})$$

be analytic in $\mathbb{U}$ with

$$p(0) = 1 \quad \text{and} \quad p(z) \neq 0 \quad (z \in \mathbb{U}).$$

If there exists a point z_0 (with $|z_0| < 1$) such that

$$|\arg(p(z))| < \frac{\gamma\pi}{2} \quad (|z| < |z_0|)$$

and

$$|\arg(p(z_0))| = \frac{\gamma\pi}{2}$$

for some $\gamma > 0$, then

$$\frac{z_0 p'(z_0)}{p(z_0)} = ik\gamma \quad (i = \sqrt{-1}),$$

where

$$k \geq \frac{m(a + a^{-1})}{2} \geq m \quad \text{when} \quad \arg(p(z_0)) = \frac{\gamma\pi}{2} \tag{3}$$

and

$$k \leq -\frac{m(a + a^{-1})}{2} \leq -m \quad \text{when} \quad \arg(p(z_0)) = -\frac{\gamma\pi}{2}, \tag{4}$$

where

$$[p(z_0)]^{1/\gamma} = \pm ia \quad \text{and} \quad a > 0.$$

2. Main Results

Theorem 1. *Let p be an analytic function in $\mathbb{U}$, given by*

$$p(z) = 1 + \sum_{n=m}^{\infty} c_n z^n \quad (c_m \neq 0;\ m \geq 2)$$

and $p(z) \neq 0$ for $z \in \mathbb{U}$. Let α_0 is the only root of the equation

$$\arctan(2m\alpha) - \pi\alpha = 0.$$

If

$$\left|\arg\left(p^2(z) - 2p(z)zp'(z)\right)\right| < \frac{\pi}{2}\left[\frac{2}{\pi}\arctan(2m\alpha) - 2\alpha\right], \tag{5}$$

where $0 < \alpha < \alpha_0$, then

$$|\arg(p(z))| < \frac{\alpha\pi}{2} \quad (z \in \mathbb{U}).$$

Proof. To prove our result we suppose that there exists a point $z_0 \in \mathbb{U}$ so that

$$|\arg(p(z))| < \frac{\alpha\pi}{2} \quad \text{for} \quad |z| < |z_0|$$

and

$$|\arg(p(z_0))| = \frac{\alpha\pi}{2}.$$

Then, Lemma 1, gives us that

$$\frac{z p'(z_0)}{p(z_0)} = ik\alpha,$$

where $[p(z_0)]^{\frac{1}{\alpha}} = \pm ia$ $(a > 0)$ and k is given by (3) or (4).
For the case $\arg(p(z_0)) = \frac{\alpha\pi}{2}$ when

$$[p(z_0)]^{\frac{1}{\alpha}} = ia \quad (a > 0),$$

with $k \geqq m$, we have

$$\arg\left(p^2(z_0) - 2p(z_0)z_0 p'(z_0)\right) = \arg\left(p^2(z_0)\left(1 - 2\frac{z_0 p'(z_0)}{p(z_0)}\right)\right)$$

$$= \arg(p^2(z_0)) + \arg\left(1 - 2\frac{z_0 p'(z_0)}{p(z_0)}\right)$$

$$= 2\arg(p(z_0)) + \arg(1 - i2k\alpha)$$

$$\leqq \alpha\pi - \arctan(2m\alpha)$$

$$= \frac{-\pi}{2}\left(\frac{2}{\pi}\arctan(2m\alpha) - 2\alpha\right),$$

which contradicts with condition (5).

Next, for the case $\arg(p(z_0)) = -\frac{\alpha\pi}{2}$ when

$$[p(z_0)]^{\frac{1}{\alpha}} = -ia \quad (a > 0),$$

with $k \leqq -m$, applying the similar method as the above, we can get

$$\arg\left(p^2(z_0) - 2p(z_0)z_0 p'(z_0)\right) \leqq -\alpha\pi + \arctan(2m\alpha)$$

$$= \frac{\pi}{2}\left(\frac{2}{\pi}\arctan(2m\alpha) - 2\alpha\right),$$

which is a contradiction to (5).

Therefore, from the two mentioned contradictions, we obtain

$$|\arg(p(z))| < \frac{\alpha\pi}{2} \quad (z \in \mathbb{U}).$$

This completes our proof. □

Let $\psi(r,s,t;z) : \mathbb{C}^3 \times \mathbb{U} \to \mathbb{C}$ and let h be univalent in $\mathbb{U}$. If p is analytic in $\mathbb{U}$ and satisfies the (second order) differential subordination

$$\psi(p(z), zp'(z), z^2 p''(z); z) \prec h(z), \tag{6}$$

then p is called a solution of the differential subordination. The univalent function q is called a dominant of the solution of the differential subordination or more simply a dominant, if $p \prec q$ for all p satisfying (6). A dominant $\tilde{q}$ satisfying $\tilde{q} \prec q$ for all dominants q of (6) is said to be the best dominant of (6). The best dominant is unique up to a rotation of $\mathbb{U}$. If $p(z) = 1 + a_n z^n + a_{n+1} z^{n+1} + \cdots$ be analytic in $\mathbb{U}$, then p will be called a $(1,n)$-solution, q a $(1,n)$-dominant, and $\tilde{q}$ the best $(1,n)$-dominant.

The following result, which is one of the types of differential subordinations was expressed in [1].

Theorem 2 ([19], Theorem 3.1e, p. 77). *Let h be convex in $\mathbb{U}$, with $h(0) = 1$ and $\operatorname{Re} h(z) > 0$. Let also $p(z) = 1 + a_n z^n + a_{n+1} z^{n+1} + \cdots$ be analytic in $\mathbb{U}$. If p satisfies*

$$p^2(z) + 2p(z)zp'(z) \prec h(z), \tag{7}$$

then

$$p(z) \prec q(z) = \sqrt{Q(z)},$$

where
$$Q(z) = \frac{1}{nz^{\frac{1}{n}}} \int_0^z h(t) t^{\frac{1}{n}-1} dt$$

and the function q is the best $(1,n)$-dominant.

Remark 1. *The form (5) cannot be used to obtain in inequality (7). Therefore, Theorem 1 is a small extension of Theorem 2.*

For $m = 2$ in Theorem 1 we have
$$\sigma_2(\alpha) =: \frac{2}{\pi} \arctan(4\alpha) - 2\alpha > 0 \qquad (8)$$

for $\alpha \in (0, \alpha_0)$ which $\alpha_0 = 1/4$ is the smallest positive root of the equation $\sigma_2(\alpha)$. So we have the following results

Remark 2. *Suppose that $f \in \Sigma_1$ with*
$$p(z) := -\frac{zf'(z)}{f(z)} \neq 0,$$

and $0 < \alpha < 1/4$ satisfy the following inequality
$$\left| \arg \left(\left(\frac{zf'(z)}{f(z)} \right)^2 \left(1 - 2 \left[1 + \frac{zf''(z)}{f'(z)} - \frac{zf'(z)}{f(z)} \right] \right) \right) \right| < \frac{\sigma_2(\alpha)\pi}{2},$$

where $\sigma_2(\alpha)$ is given by (8). Then f is meromorphic strongly starlike function of order α.

Remark 3. *Suppose that $f \in \mathcal{A}_2$ with*
$$p(z) := \sqrt{\frac{f(z)}{z}} \neq 0,$$

and $0 < \alpha < \frac{1}{2}$ satisfy the following inequality
$$\left| \arg \left(\frac{2f(z)}{z} - f'(z) \right) \right| < \frac{\sigma_2(\alpha)\pi}{2},$$

where $\sigma_2(\alpha)$ is given by (8). Then
$$\left| \arg \sqrt{\frac{f(z)}{z}} \right| < \frac{\alpha\pi}{2} \qquad (z \in \mathbb{U}).$$

Since $\sigma_2(\alpha)$ given by (8) takes its maximum value at $\alpha = \sqrt{(4-\pi)/16\pi}$, we obtain the following result.

Corollary 1. *Let p be an analytic function in $\mathbb{U}$, given by*
$$p(z) = 1 + \sum_{n=2}^{\infty} c_n z^n$$

and $p(z) \neq 0$ for $z \in \mathbb{U}$. Let

$$\left| \arg \left(p^2(z) - 2p(z)zp'(z) \right) \right| < \frac{\sigma_2 \left(\sqrt{\frac{4-\pi}{16\pi}} \right) \pi}{2} \simeq 0.071125,$$

then

$$\left| \arg \left(p(z) \right) \right| < \frac{\sqrt{\frac{4-\pi}{16\pi}} \pi}{2} \simeq 0.20528 \quad (z \in \mathbb{U}).$$

Theorem 3. *Let p be an analytic function in $\mathbb{U}$, given by*

$$p(z) = 1 + \sum_{n=2}^{\infty} c_n z^n$$

and $p(z) \neq 0$ for $z \in \mathbb{U}$. Let α_0 be the smallest positive root of the equation

$$\frac{2}{\pi} \arctan \left(\frac{2\alpha \left(\frac{1-2\alpha}{1+2\alpha} \right)^{(1+2\alpha)/2} \cos(\pi\alpha)}{1 - 2\alpha - 2\alpha \left(\frac{1-2\alpha}{1+2\alpha} \right)^{(1+2\alpha)/2} \sin(\pi\alpha)} \right) - \alpha = 0. \tag{9}$$

Suppose that

$$\left| \arg \left(p(z) - \frac{zp'(z)}{[p(z)]^2} \right) \right| < \frac{\delta(\alpha)\pi}{2}, \tag{10}$$

where

$$\delta(\alpha) = \frac{2}{\pi} \arctan \left(\frac{2\alpha \left(\frac{1-2\alpha}{1+2\alpha} \right)^{(1+2\alpha)/2} \cos(\pi\alpha)}{1 - 2\alpha - 2\alpha \left(\frac{1-2\alpha}{1+2\alpha} \right)^{(1+2\alpha)/2} \sin(\pi\alpha)} \right) - \alpha \tag{11}$$

and $0 < \alpha < \alpha_0$. Then

$$\left| \arg \left(p(z) \right) \right| < \frac{\alpha\pi}{2} \quad (z \in \mathbb{U}).$$

Proof. First, let us define

$$\delta(\alpha) = \frac{2}{\pi} \arctan \left(\frac{n(\alpha)}{m(\alpha)} \right) - \alpha$$

where

$$n(\alpha) = 2\alpha \left(\frac{1-2\alpha}{1+2\alpha} \right)^{(1+2\alpha)/2} \cos(\pi\alpha) \quad \text{and} \quad m(\alpha) = 1 - 2\alpha - 2\alpha \left(\frac{1-2\alpha}{1+2\alpha} \right)^{(1+2\alpha)/2} \sin(\pi\alpha),$$

then we have $\delta(0) = 0$, $\delta(\alpha)_{\alpha \to 1/2} = -1/2$, and $\delta'(0) > 0$. Therefore, there exists in $\left(0, 1/2 \right)$ the smallest positive root α_0 of the equality (9), so that $\delta(\alpha) > 0$ for $\alpha \in (0, \alpha_0)$.

Now we suppose that there exists a point $z_0 \in \mathbb{U}$ such that

$$\left| \arg \left(p(z) \right) \right| < \frac{\alpha\pi}{2} \quad \text{for} \quad |z| < |z_0|$$

and

$$\left| \arg \left(p(z_0) \right) \right| = \frac{\alpha\pi}{2}.$$

Then, from Lemma 1, it follows that

$$\frac{zp'(z_0)}{p(z_0)} = ik\alpha,$$

where $[p(z_0)]^{\frac{1}{\alpha}} = \pm ia \ (a > 0)$ and k is given by (3) or (4) for $m = 2$.
For the case $\arg(p(z_0)) = \frac{\alpha\pi}{2}$ when

$$[p(z_0)]^{\frac{1}{\alpha}} = ia \quad (a > 0),$$

we have

$$\arg\left(p(z_0) - \frac{zp'(z_0)}{[p(z_0)]^2}\right) = \arg\left(p(z_0)\left(1 - \frac{zp'(z_0)}{p(z_0)}\frac{1}{[p(z_0)]^2}\right)\right)$$

$$= \arg(p(z_0)) + \arg\left(1 - ik\alpha\frac{1}{(ia)^{2\alpha}}\right)$$

$$= \arg(p(z_0)) + \arg\left(1 + \frac{k\alpha}{a^{2\alpha}}e^{-i\pi(1+2\alpha)/2}\right).$$

Since

$$\frac{k\alpha}{a^{2\alpha}} \geq \alpha(a^{1-2\alpha} + a^{-1-2\alpha}),$$

we now define a real function g by

$$g(a) = a^{1-2\alpha} + a^{-1-2\alpha} \quad (a > 0).$$

Then this function takes on the minimum value for a given by

$$a = \sqrt{\frac{1+2\alpha}{1-2\alpha}}.$$

Therefore, from the above inequality we obtain

$$\frac{k\alpha}{a^{2\alpha}} \geq \alpha\left(\left(\frac{1+2\alpha}{1-2\alpha}\right)^{(1-2\alpha)/2} + \left(\frac{1+2\alpha}{1-2\alpha}\right)^{(-1-2\alpha)/2}\right) = \frac{2\alpha}{1-2\alpha}\left(\frac{1-2\alpha}{1+2\alpha}\right)^{(1+2\alpha)/2} =: l(\alpha).$$

Therefore

$$\arg\left(p(z_0) - \frac{zp'(z_0)}{[p(z_0)]^2}\right) \leq \frac{\alpha\pi}{2} + \arctan\left(\frac{-l(\alpha)\cos(\pi\alpha)}{1 - l(\alpha)\sin(\pi\alpha)}\right)$$

$$= \frac{\alpha\pi}{2} - \arctan\left(\frac{l(\alpha)\cos(\pi\alpha)}{1 - l(\alpha)\sin(\pi\alpha)}\right)$$

$$= \frac{\delta(\alpha)\pi}{2},$$

which is contradict with condition (10).
Next, for the case $\arg(p(z_0)) = -\frac{\alpha\pi}{2}$ when

$$[p(z_0)]^{\frac{1}{\alpha}} = -ia \quad (a > 0),$$

with

$$\frac{k\alpha}{a^{2\alpha}} \leq -\alpha(a^{1-2\alpha} + a^{-1-2\alpha}),$$

applying the similar method as the above, we can get

$$\arg\left(p(z_0) - \frac{zp'(z_0)}{[p(z_0)]^2}\right) = \arg(p(z_0)) + \arg\left(1 - ik\alpha \frac{1}{(-ia)^{2\alpha}}\right)$$

$$= \arg(p(z_0)) + \arg\left(1 - \frac{k\alpha}{a^{2\alpha}} e^{i\pi(1+2\alpha)/2}\right)$$

$$\geq -\frac{\alpha\pi}{2} + \arctan\left(\frac{l(\alpha)\cos(\pi\alpha)}{1 - l(\alpha)\sin(\pi\alpha)}\right)$$

$$= \frac{\delta(\alpha)\pi}{2},$$

which is a contradiction to condition (10).

Therefore, from the two mentioned contradictions, we obtain

$$|\arg(p(z))| < \frac{\alpha\pi}{2} \quad (z \in \mathbb{U}).$$

This completes the proof of Theorem 3. □

Theorem 4 ([19], Corollary 3.4a.3, p. 124). *Let β and γ be complex numbers with $\beta \neq 0$ and let p and h be analytic in $\mathbb{U}$ with $p(0) = h(0)$. If $P(z) = \beta h(z) + \gamma$ satisfies*
(i) $\operatorname{Re} P^2(z) > 0$
(ii) P or P^{-1} *is convex, then*

$$p(z) + zp'(z) \cdot [\beta p(z) + \gamma]^{-2} \prec h(z), \tag{12}$$

implies $p(z) \prec h(z)$.

The condition (10) can be written as a generalized Briot-Bouquet differential subordination. However, It is remarkable that the condition (12) among the outcomes on the generalized Briot-Bouquet differential subordination collected in ([19], Ch. 3) is not taken into account the case $\gamma = 0$, $\beta = i$ which we have in (10).

Corollary 2. *Let $f \in \Sigma_2$ with*

$$p(z) := -\frac{zf'(z)}{f(z)} \neq 0$$

and $0 < \alpha < \alpha_0$ satisfy the following inequality

$$\left|\arg\left(\frac{f(z)}{zf'(z)}\left(1 + \frac{zf''(z)}{f'(z)}\right) - \frac{zf'(z)}{f(z)} - 1\right)\right| < \frac{\delta(\alpha)\pi}{2},$$

where $\delta(\alpha)$ is given by (11). Then f is meromorphic strongly starlike function of order α.

Theorem 5. *Let p be an analytic function in $\mathbb{U}$, given by*

$$p(z) = 1 + \sum_{n=m}^{\infty} c_n z^n \quad (c_m \neq 0;\ m \in \mathbb{N})$$

and $p(z) \neq 0$ for $z \in \mathbb{U}$. Let $\alpha > 0$ and $\beta > 0$ satisfy the inequality

$$\arctan(m\alpha) > \frac{\pi\alpha}{2\beta}.$$

Suppose that
$$\left|\arg\left(p(z)\left[1-\frac{zp'(z)}{p(z)}\right]^{\beta}\right)\right| < \frac{\pi}{2}\left[\frac{2\beta}{\pi}\arctan(m\alpha) - \alpha\right]. \qquad (13)$$

Then
$$|\arg(p(z))| < \frac{\alpha\pi}{2} \qquad (z \in \mathbb{U}).$$

Proof. Suppose that there exists a point $z_0 \in \mathbb{U}$ such that
$$|\arg(p(z))| < \frac{\alpha\pi}{2} \quad \text{for} \quad |z| < |z_0|$$

and
$$|\arg(p(z_0))| = \frac{\alpha\pi}{2}.$$

Then, from Lemma 1, it follows that
$$\frac{zp'(z_0)}{p(z_0)} = ik\alpha,$$

where $[p(z_0)]^{\frac{1}{\alpha}} = \pm ia$ ($a > 0$) and k is given by (3) or (4).

For the case $\arg(p(z_0)) = \frac{\alpha\pi}{2}$ when
$$[p(z_0)]^{\frac{1}{\alpha}} = ia \qquad (a > 0),$$

with $k \geqq m$, we have
$$\arg\left(p(z_0)\left[1 - \frac{z_0 p'(z_0)}{p(z_0)}\right]^{\beta}\right) = \arg(p(z_0)) + \beta \arg\left(1 - \frac{z_0 p'(z_0)}{p(z_0)}\right)$$
$$= \arg(p(z_0)) + \beta \arg(1 - ik\alpha)$$
$$\leqq \frac{\alpha\pi}{2} - \beta \arctan(m\alpha)$$
$$= \frac{-\pi}{2}\left(\frac{2\beta}{\pi}\arctan(m\alpha) - \alpha\right),$$

which contradicts our hypothesis in (13).

Next, for the case $\arg(p(z_0)) = -\frac{\alpha\pi}{2}$ when
$$[p(z_0)]^{\frac{1}{\alpha}} = -ia \qquad (a > 0),$$

with $k \leqq -m$, applying the similar method as the above, we can get
$$\arg\left(p(z_0)\left[1 - \frac{z_0 p'(z_0)}{p(z_0)}\right]^{\beta}\right) \geqq -\frac{\alpha\pi}{2} + \beta \arctan(m\alpha)$$
$$= \frac{\pi}{2}\left(\frac{2\beta}{\pi}\arctan(m\alpha) - \alpha\right),$$

which is a contradiction to (13).

Therefore, from the two mentioned contradictions, we obtain
$$|\arg(p(z))| < \frac{\alpha\pi}{2} \qquad (z \in \mathbb{U}).$$

This completes the proof of Theorem 5. □

Remark 4. *By choosing $m = 2$ and $\beta = 1$ in Theorem 5, we have the result obtained by Nunokawa and Sokół in ([11], Theorem 2.4).*

By choosing
$$p(z) := -\frac{zf'(z)}{f(z)} \neq 0,$$
in Theorem 6, we obtain a sufficient condition for strongly meromorphic starlikeness as follows.

Corollary 3. *Let $f \in \Sigma_2$ with*
$$p(z) := -\frac{zf'(z)}{f(z)} \neq 0.$$
Let $\alpha > 0$ and $\beta > 0$ satisfy the inequality
$$\arctan(2\alpha) > \frac{\pi\alpha}{2\beta}.$$
Suppose that
$$\left| \arg\left(-\frac{zf'(z)}{f(z)}\left[1 + \frac{zf'(z)}{f(z)} - \left(1 + \frac{zf''(z)}{f'(z)}\right)\right]^{\beta} \right) \right| < \frac{\pi}{2}\left[\frac{2\beta}{\pi}\arctan(2\alpha) - \alpha\right]. \tag{14}$$
Then f is meromorphic strongly starlike function of order α.

Theorem 6. *Let p be an analytic function in $\mathbb{U}$ with $p(0) = 1$, $p'(0) \neq 0$ and $p(z) \neq 0$ for $z \in \mathbb{U}$ that satisfies the following inequality*
$$\left| \arg\left(\frac{p(z)(p(z) + zp'(z))}{p(z) - \beta zp'(z)} \right) \right| < \frac{\xi(\alpha)\pi}{2},$$
where
$$\xi(\alpha) = \alpha + \frac{2}{\pi}(\arctan(\alpha) + \arctan(\beta\alpha)) \quad (\alpha > 0; \beta \geq 0). \tag{15}$$
Then
$$|\arg(p(z))| < \frac{\alpha\pi}{2} \quad (z \in \mathbb{U}).$$

Proof. To prove the result asserted by Theorem 6, we suppose that there exists a point $z_0 \in \mathbb{U}$ such that
$$|\arg(p(z))| < \frac{\alpha\pi}{2} \quad \text{for} \quad |z| < |z_0|$$
and
$$|\arg(p(z_0))| = \frac{\alpha\pi}{2}.$$
Then, from Lemma 1, it follows that
$$\frac{zp'(z_0)}{p(z_0)} = ik\alpha,$$
where $[p(z_0)]^{\frac{1}{\alpha}} = \pm ia$ ($a > 0$) and k is given by (3) or (4) for $m = 1$.
For the case
$$\arg(p(z_0)) = \frac{\alpha\pi}{2},$$

where $[p(z_0)]^{\frac{1}{a}} = ia$ $(a > 0)$ and $k \geq 1$, we have

$$\arg\left(\frac{p(z_0)(p(z_0) + z_0 p'(z_0))}{p(z_0) - \beta z_0 p'(z_0)}\right) = \arg\left(p(z_0)\left(\frac{1 + \frac{z p'(z_0)}{p(z_0)}}{1 - \beta \frac{z p'(z_0)}{p(z_0)}}\right)\right)$$

$$= \arg(p(z_0)) + \arg\left(\frac{1 + ik\alpha}{1 - i\beta k\alpha}\right)$$

$$= \arg(p(z_0)) + \arg(1 + ik\alpha) - \arg(1 - i\beta k\alpha)$$

$$= \frac{\alpha \pi}{2} + \arctan(k\alpha) - \arctan(-\beta k\alpha)$$

$$= \frac{\alpha \pi}{2} + \arctan(k\alpha) + \arctan(\beta k\alpha)$$

$$\geq \frac{\alpha \pi}{2} + \arctan(\alpha) + \arctan(\beta \alpha)$$

$$= \frac{\xi(\alpha) \pi}{2},$$

which contradicts our hypothesis in Theorem 6.

Next, for the case

$$\arg(p(z_0)) = -\frac{\alpha \pi}{2},$$

where $[p(z_0)]^{\frac{1}{a}} = -ia$ $(a > 0)$ and $k \leq -1$, applying the similar method as the above, we can get

$$\arg\left(\frac{p(z_0)(p(z_0) + z_0 p'(z_0))}{p(z_0) - \beta z_0 p'(z_0)}\right) = \arg(p(z_0)) + \arg\left(\frac{1 + ik\alpha}{1 - i\beta k\alpha}\right)$$

$$= -\frac{\alpha \pi}{2} + \arctan(k\alpha) + \arctan(\beta k\alpha)$$

$$\leq -\frac{\alpha \pi}{2} - \arctan(\alpha) - \arctan(\beta \alpha)$$

$$= -\frac{\xi(\alpha) \pi}{2},$$

which is a contradiction to the assumption of Theorem 6.

Therefore, from the two mentioned contradictions, we obtain

$$|\arg(p(z))| < \frac{\alpha \pi}{2} \quad (z \in \mathbb{U}).$$

This completes the proof of Theorem 6. □

Remark 5.

(i) If $\beta \alpha^2 < 1$ in Theorem 6, then (15) is equal to

$$\xi = \alpha + \frac{2}{\pi} \arctan\left(\frac{\alpha(1 + \beta)}{1 - \beta \alpha^2}\right).$$

(ii) By setting $\beta = 0$ and $p(z) := f'(z) \neq 0$ in Theorem 6, we have the result obtained by Nunokawa et al. in ([20], Theorem 3).

By setting

$$p(z) := \frac{z f'(z)}{g(z)} \neq 0,$$

in Theorem 6, we obtain a sufficient condition for strongly close-to-convexity as follows.

Corollary 4. For $g \in \mathcal{S}^*$ and $f \in \mathcal{A}$ such that $2f''(0) \neq g''(0)$, suppose that the following inequality

$$\left| \arg \left(\frac{2(f'(z))^2}{f'(z)g'(z) - f''(z)g(z)} - \frac{zf'(z)}{g(z)} \right) \right| < \frac{\xi(\alpha)\pi}{2}$$

is satisfied, where

$$\xi(\alpha) = \alpha + \frac{2}{\pi}(\arctan(\alpha) + \arctan(\beta\alpha)) \quad (\alpha > 0, \ \beta \geq 0). \tag{16}$$

Then

$$\left| \arg \left(\frac{zf'(z)}{g(z)} \right) \right| < \frac{\alpha\pi}{2} \quad (z \in \mathbb{U}).$$

Remark 6. Similar to Corollary 4 by setting

$$p(z) := \frac{zf'(z)}{f(z)} \neq 0,$$

in Theorem 6, (or $g =: f$ in Corollary 4), we can obtain a sufficient condition for strongly starlikeness.

Author Contributions: Investigation: A.E., N.E.C., E.A.A. and S.Y. All authors have read and agreed to the published version of the manuscript.

Funding: The second author was supported by the Basic Science Research Program through the National Research Foundation of Korea (NRF) funded by the Ministry of Education, Science and Technology (No. 2019R1I1A3A01050861).

Conflicts of Interest: The authors declare no conflict of interest.

References

1. Miller, S.S.; Mocanu, P.T. Second-order differential inequalities in the complex plane. *J. Math. Anal. Appl.* **1978**, *65*, 298–305. [CrossRef]
2. Ali, R.M.; Cho, N.E.; Ravichandran, V.; Kumar, S.S. Differential subordination for functions associated with the lemniscate of Bernoulli. *Taiwanese J. Math.* **2012**, *16*, 1017–1026. [CrossRef]
3. Adegani, E.A.; Bulboacă, T.; Motamednezhad, A. Simple sufficient subordination conditions for close-to-convexity. *Mathematics* **2019**, *7*, 241. [CrossRef]
4. Bulboacă, T. On some classes of differential subordinations. *Studia Univ. Babeş-Bolyai Math.* **1986**, *31*, 45–50.
5. Cho, N.E.; Kumar, S.; Kumar, V.; Ravichandran, V.; Srivastava, H.M. Starlike functions related to the Bell numbers. *Symmetry* **2019**, *11*, 219. [CrossRef]
6. Cho, N.E.; Srivastava, H.M.; Adegani, E.A.; Motamednezhad, A. Criteria for a certain class of the Carathéodory functions and their applications. *J. Inequal. Appl.* **2020**, *2020*, 85. [CrossRef]
7. Kargar, R.; Ebadian, A.; Trojnar-Spelina, L. Further results for starlike functions related with Booth lemniscate. *Iran J. Sci. Technol. A* **2019**, *43*, 1235–1238. [CrossRef]
8. Kim, I.H.; Sim, Y.J.; Cho, N.E. New criteria for Carathéodory functions. *J. Inequal. Appl.* **2019**, *2019*, 13. [CrossRef]
9. Masih, V.S.; Ebadian, A.; Najafzadeh, S. On applications of Nunokawa and Sokół theorem for p-valency. *Bull. Iran. Math. Soc.* **2020**, *46*, 471–486. [CrossRef]
10. Nunokawa, M.; Sokół, J. On the order of strong starlikeness and the radii of starlikeness for of some close-to-convex functions. *Anal. Math. Phys.* **2019**, *9*, 2367–2378. [CrossRef]
11. Nunokawa, M.; Sokół, J. On meromorphic and starlike functions. *Complex Var. Elliptic Equ.* **2015**, *60*, 1411–1423. [CrossRef]
12. Nunokawa, M.; Sokół, J. On multivalent starlike functions and Ozaki condition. *Complex Var. Elliptic Equ.* **2019**, *64*, 78–92. [CrossRef]
13. Srivastava, H.M.; Răducanu, D.; Zaprawa, P. A certain subclass of analytic functions defined by means of differential subordination. *Filomat* **2016**, *30*, 3743–3757. [CrossRef]

14. Cang, Y.-L.; Liu, J.-L. Some sufficient conditions for starlikeness and convexity of order α. *J. Appl. Math.* **2013**, *2013*, 869469. [CrossRef]
15. Nunokawa, M.; Sokół, J.; Tuneski, N.; Jolevska-Tuneska, B. On multivalent starlike functions. *Stud. Univ. Babeş-Bolyai Math.* **2019**, *64*, 91–102. [CrossRef]
16. Nunokawa, M.; Sokół, J. Conditions for starlikeness of multivalent functions. *Results Math.* **2017**, *72*, 359–367. [CrossRef]
17. Nunokawa, M. On properties of non-Carathéodory functions. *Proc. Jpn. Acad. Ser. A Math. Sci.* **1992**, *68*, 152–153. [CrossRef]
18. Nunokawa, M. On the order of strongly starlikeness of strongly convex functions. *Proc. Jpn. Acad. Ser. A Math. Sci.* **1993**, *69*, 234–237. [CrossRef]
19. Miller, S.S.; Mocanu, P.T. *Differential Subordinations. Theory and Applications*; Marcel Dekker Inc.: New York, NY, USA, 2000.
20. Nunokawa, M.; Owa, S.; Polatoğlu, Y.; Çağlar, M.; Duman, E.M. Some sufficient conditions for starlikeness and convexity. *Turk. J. Math.* **2010**, *3*, 333–338.

© 2020 by the authors. Licensee MDPI, Basel, Switzerland. This article is an open access article distributed under the terms and conditions of the Creative Commons Attribution (CC BY) license (http://creativecommons.org/licenses/by/4.0/).

Article

Maclaurin Coefficient Estimates of Bi-Univalent Functions Connected with the q-Derivative

Sheza M. El-Deeb [1,†], Teodor Bulboacă [2,*] and Bassant M. El-Matary [1,†]

1. Department of Mathematics, Faculty of Science, Damietta University, New Damietta 34517, Egypt; shezaeldeeb@yahoo.com (S.M.E.-D.); bassantmarof@yahoo.com (B.M.E.-M.)
2. Faculty of Mathematics and Computer Science, Babeş-Bolyai University, 400084 Cluj-Napoca, Romania
* Correspondence: bulboaca@math.ubbcluj.ro
† Current address: Department of Mathematics, College of Science and Arts in Badaya, Al-Qassim University, Al-Badaya, Al-Qassim Province, Saudi Arabia.

Received: 21 February 2020; Accepted: 11 March 2020; Published: 14 March 2020

Abstract: In this paper we introduce a new subclass of the bi-univalent functions defined in the open unit disc and connected with a q-analogue derivative. We find estimates for the first two Taylor-Maclaurin coefficients $|a_2|$ and $|a_3|$ for functions in this subclass, and we obtain an estimation for the Fekete-Szegő problem for this function class.

Keywords: bi-univalent functions; Hadamard (convolution) product; coefficients bounds; q-derivative operator; differential subordination

MSC: Primary 05A30, 30C45; Secondary 11B65, 47B38

---

1. Introduction, Definitions and Preliminaries

Let $\mathcal{A}$ denote the class of normalized analytic functions in the unit disc $\mathbb{U} := \{z \in \mathbb{C} : |z| < 1\}$ of the form

$$f(z) = z + \sum_{k=2}^{\infty} a_k z^k, \ z \in \mathbb{U}, \tag{1}$$

and let $\mathcal{S} \subset \mathcal{A}$ consisting on functions that are univalent in $\mathbb{U}$.

If the function $h \in \mathcal{A}$ is given by

$$h(z) = z + \sum_{k=2}^{\infty} c_k z^k, \ z \in \mathbb{U}, \tag{2}$$

then, the *Hadamard (or convolution) product* of f and h is defined by

$$(f * h)(z) := z + \sum_{k=2}^{\infty} a_k c_k z^k, \ z \in \mathbb{U}.$$

The theory of q-calculus plays an important role in many areas of mathematical, physical, and engineering sciences. Jackson ([1,2]) was the first to have some applications of the q-calculus and introduced the q-analogue of the classical derivative and integral operators [3].

For $0 < q < 1$, the q-derivative operator [2] for $f * g$ is defined by

$$D_q(f*h)(z) := D_q\left(z + \sum_{k=2}^{\infty} a_k c_k z^k\right)$$

$$= \frac{(f*h)(z) - (f*h)(qz)}{z(1-q)} = 1 + \sum_{k=2}^{\infty} [k,q]a_k c_k z^{k-1}, \; z \in \mathbb{U},$$

where

$$[k,q] := \frac{1-q^k}{1-q} = 1 + \sum_{j=1}^{k-1} q^j, \quad [0,q] := 0. \tag{3}$$

Using the definition formula (3) we will define the next two products:

(i) For any non negative integer k, the q-shifted factorial is given by

$$[k,q]! := \begin{cases} 1, & \text{if } k = 0, \\ \prod_{i=1}^{k} [i,q], & \text{if } k \in \mathbb{N} := \{1,2,\dots\}. \end{cases}$$

(ii) For any positive number r, the q-generalized Pochhammer symbol is defined by

$$[r,q]_k := \begin{cases} 1, & \text{if } k = 0, \\ \prod_{i=1}^{k} [r+i-1,q], & \text{if } k \in \mathbb{N}. \end{cases}$$

For $\lambda > -1$ and $0 < q < 1$, we define the linear operator $\mathcal{H}_h^{\lambda,q} : \mathcal{A} \to \mathcal{A}$ by

$$\mathcal{H}_h^{\lambda,q} f(z) * \mathcal{M}_{q,\lambda+1}(z) = z D_q(f*h)(z), \; z \in \mathbb{U},$$

where the function $\mathcal{M}_{q,\lambda+1}$ is given by

$$\mathcal{M}_{q,\lambda+1}(z) := z + \sum_{k=2}^{\infty} \frac{[\lambda+1,q]_{k-1}}{[k-1,q]!} z^k, \; z \in \mathbb{U}.$$

A simple computation shows that

$$\mathcal{H}_h^{\lambda,q} f(z) := z + \sum_{k=2}^{\infty} \phi_{k-1} a_k z^k, \; z \in \mathbb{U} \; (\lambda > -1, \; 0 < q < 1), \tag{4}$$

where

$$\phi_{k-1} := \frac{[k,q]!}{[\lambda+1,q]_{k-1}} c_k, \; k \geq 2. \tag{5}$$

Remark 1. *From the definition relation (4) we can easily verify that the next relations hold for all $f \in \mathcal{A}$:*

(i) $[\lambda+1,q]\mathcal{H}_h^{\lambda,q} f(z) = [\lambda,q]\mathcal{H}_h^{\lambda+1,q} f(z) + q^\lambda z D_q\left(\mathcal{H}_h^{\lambda+1,q} f(z)\right), \; z \in \mathbb{U};$

(ii) $\mathcal{I}_h^\lambda f(z) := \lim_{q \to 1^-} \mathcal{H}_h^{\lambda,q} f(z) = z + \sum_{k=2}^{\infty} \frac{k!}{(\lambda+1)_{k-1}} a_k c_k z^k, \; z \in \mathbb{U}. \tag{6}$

Remark 2. *Taking different particular cases for the coefficients c_k we obtain the next special cases for the operator $\mathcal{H}_h^{\lambda,q}$:*

(i) For $c_k = \dfrac{(-1)^{k-1}\Gamma(v+1)}{4^{k-1}(k-1)!\Gamma(k+v)}$, $v > 0$, we obtain the operator $\mathcal{N}_{v,q}^{\lambda}$ studied by El-Deeb and Bulboacă [4]:

$$\mathcal{N}_{v,q}^{\lambda}f(z) := z + \sum_{k=2}^{\infty} \frac{(-1)^{k-1}\Gamma(v+1)}{4^{k-1}(k-1)!\Gamma(k+v)} \cdot \frac{[k,q]!}{[\lambda+1,q]_{k-1}} a_k z^k$$

$$= z + \sum_{k=2}^{\infty} \frac{[k,q]!}{[\lambda+1,q]_{k-1}} \psi_k a_k z^k, \; z \in \mathbb{U}, \; (v > 0, \; \lambda > -1, \; 0 < q < 1), \tag{7}$$

where

$$\psi_k := \frac{(-1)^{k-1}\Gamma(v+1)}{4^{k-1}(k-1)!\Gamma(k+v)}; \tag{8}$$

(ii) For $c_k = \left(\dfrac{n+1}{n+k}\right)^{\alpha}$, $\alpha > 0$, $n \geq 0$, we obtain the operator $\mathcal{N}_{n,1,q}^{\lambda,\alpha} =: \mathcal{M}_{n,q}^{\lambda,\alpha}$ studied by El-Deeb and Bulboacă [5]:

$$\mathcal{M}_{n,q}^{\lambda,\alpha}f(z) := z + \sum_{k=2}^{\infty} \left(\frac{n+1}{n+k}\right)^{\alpha} \cdot \frac{[k,q]!}{[\lambda+1,q]_{k-1}} a_k z^k, \; z \in \mathbb{U}; \tag{9}$$

(iii) For $c_k = 1$ we obtain the operator $\mathfrak{J}_q^{\lambda}$ studied by Arif et al. [6], defined by

$$\mathfrak{J}_q^{\lambda}f(z) := z + \sum_{k=2}^{\infty} \frac{[k,q]!}{[\lambda+1,q]_{k-1}} a_k z^k, \; z \in \mathbb{U};$$

(iv) For $c_k = \dfrac{m^{k-1}}{(k-1)!}e^{-m}$, $m > 0$, we obtain the q-analogue of Poisson operator defined in [7] by:

$$\mathcal{I}_q^{\lambda,m}f(z) := z + \sum_{k=2}^{\infty} \frac{m^{k-1}}{(k-1)!}e^{-m} \cdot \frac{[k,q]!}{[\lambda+1,q]_{k-1}} a_k z^k, \; z \in \mathbb{U}; \tag{10}$$

(v) For $c_k = \left[\dfrac{1+\ell+\mu(k-1)}{1+\ell}\right]^m$, $m \in \mathbb{Z}$, $\ell \geq 0$, $\mu \geq 0$, we obtain the q-analogue of Prajapat operator defined in [8] by

$$\mathcal{J}_{q,\ell,\mu}^{\lambda,m}f(z) := z + \sum_{k=2}^{\infty} \left[\frac{1+\ell+\mu(k-1)}{1+\ell}\right]^m \cdot \frac{[k,q]!}{[\lambda+1,q]_{k-1}} a_k z^k, \; z \in \mathbb{U}. \tag{11}$$

The Koebe one-quarter theorem ([9]) proves that the image of $\mathbb{U}$ under every univalent function $f \in \mathcal{S}$ contains the disk of radius $\dfrac{1}{4}$. Therefore, every function $f \in \mathcal{S}$ has an inverse f^{-1} that satisfies

$$f(f^{-1}(w)) = w, \; \left(|w| < r_0(f), \; r_0(f) \geq \frac{1}{4}\right),$$

where

$$f^{-1}(w) = w - a_2 w^2 + \left(2a_2^2 - a_3\right)w^3 - \left(5a_2^3 - 5a_2 a_3 + a_4\right)w^4 + \ldots.$$

A function $f \in \mathcal{A}$ is said to be bi-univalent in $\mathbb{U}$ if both f and f^{-1} are univalent in $\mathbb{U}$. Let Σ denote the class of bi-univalent functions in $\mathbb{U}$ given by (1). Note that the functions $f_1(z) = \dfrac{z}{1-z}$, $f_2(z) = \dfrac{1}{2}\log\dfrac{1+z}{1-z}$, $f_3(z) = -\log(1-z)$, with their corresponding inverses $f_1^{-1}(w) = \dfrac{w}{1+w}$, $f_2^{-1}(w) = \dfrac{e^{2w}-1}{e^{2w}+1}$, $f_3^{-1}(w) = \dfrac{e^w-1}{e^w}$, are elements of Σ (see [10]). For a brief history and interesting examples in the class Σ see [11]. Brannan and Taha [12] (see also [10]) introduced certain subclasses of the bi-univalent functions class Σ similar to the familiar subclasses $\mathcal{S}^*(\alpha)$ and $\mathcal{K}(\alpha)$ of starlike and convex functions of order α ($0 \leq \alpha < 1$), respectively (see [11]). Following Brannan and Taha [12], a function $f \in \mathcal{A}$ is said to be in the class $S_{\Sigma}^{*}(\alpha)$ of bi-starlike functions of order α ($0 < \alpha \leq 1$), if each of the following conditions are satisfied:

$$f \in \Sigma, \quad \text{with} \quad \left|\arg \frac{zf'(z)}{f(z)}\right| < \frac{\alpha\pi}{2}, \quad z \in \mathbb{U},$$

and

$$\left|\arg \frac{zg'(w)}{g(w)}\right| < \frac{\alpha\pi}{2}, \quad w \in \mathbb{U},$$

where the function g is the analytic extension of f^{-1} to $\mathbb{U}$, given by

$$g(w) = w - a_2 w^2 + \left(2a_2^2 - a_3\right) w^3 - \left(5a_2^3 - 5a_2 a_3 + a_4\right) w^4 + \ldots, \quad w \in \mathbb{U}. \tag{12}$$

A function $f \in \mathcal{A}$ is said to be in the class $K_{\Sigma}^{*}(\alpha)$ of bi-convex functions of order α ($0 < \alpha \leq 1$), if each of the following conditions are satisfied:

$$f \in \Sigma, \quad \text{with} \quad \left|\arg\left(1 + \frac{zf''(z)}{f'(z)}\right)\right| < \frac{\alpha\pi}{2}, \quad z \in \mathbb{U},$$

and

$$\left|\arg\left(1 + \frac{zg''(w)}{g'(w)}\right)\right| < \frac{\alpha\pi}{2}, \quad w \in \mathbb{U}.$$

The classes $S_{\Sigma}^{*}(\alpha)$ and $K_{\Sigma}(\alpha)$ of bi-starlike functions of order α and bi-convex functions of order α ($0 < \alpha \leq 1$), corresponding to the function classes $S^{*}(\alpha)$ and $K(\alpha)$, were also introduced analogously. For each of the function classes $S_{\Sigma}^{*}(\alpha)$ and $K_{\Sigma}(\alpha)$, they found non-sharp estimates on the first two Taylor-Maclaurin coefficients $|a_2|$ and $|a_3|$ ([10,12]).

The object of the paper is to introduce a new subclass of functions $\mathcal{L}_{\Sigma}^{q,\lambda}(\eta; h; \Phi)$ of the class Σ, that generalize the previous defined classes. This subclass is defined with the aid of a general $\mathcal{H}_h^{\lambda,q}$ linear operator defined by convolution products together with the aid of q-derivative operator. This new class extends and generalizes many previous operators as it was presented in Remark 2, and the main goal of the paper is to find estimates on the coefficients $|a_2|, |a_3|$, and for the Fekete-Szegő functional for functions in these new subclasses.

These classes will be introduced by using the subordination and the results are obtained by employing the techniques used earlier by Srivastava et al. [10]. This last work represents one of the most important study of the bi-univalent functions, and inspired many investigations in this area including the present paper, while many other recent papers deal with problems initiated in this work, like [13–16], and many others. The novelty of our paper consists of the fact that the operator used by defining the new subclass of Σ is a very general operator that generalizes many earlier defined operators, it does not overlap with those studied in the above mentioned papers (that $\Phi'(0) > 0$ and $\Phi(\mathbb{U})$ is symmetric with respect to the real axis), while for the function Φ from Definition 1 we did not assume any restrictions like in many other papers, excepting the fact that $\Phi(0) = 1$ is necessary for the subordinations (13) and (14).

If f and F are analytic functions in $\mathbb{U}$, we say that f is subordinate to F, written $f(z) \prec F(z)$, if there exists a Schwarz function s, which is analytic in $\mathbb{U}$, with $s(0) = 0$, and $|s(z)| < 1$ for all $z \in \mathbb{U}$, such that $f(z) = F(s(z))$, $z \in \mathbb{U}$. Furthermore, if the function F is univalent in $\mathbb{U}$, then we have the following equivalence ([17,18])

$$f(z) \prec F(z) \Leftrightarrow f(0) = F(0) \text{ and } f(\mathbb{U}) \subset F(\mathbb{U}).$$

Throughout this paper we assume that Φ is an analytic function in $\mathbb{U}$ with $\Phi(0) = 1$ of the form

$$\Phi(z) = 1 + B_1 z + B_2 z^2 + B_3 z^3 + \ldots, \quad z \in \mathbb{U}.$$

Now we define the following subclass of bi-univalent functions $\mathcal{L}_{\Sigma}^{q,\lambda}(\eta; h; \Phi)$:

Definition 1. *If the function f has the form (1) and h is given by (2), the function f is said to be in the class $\mathcal{L}_{\Sigma}^{q,\lambda}(\eta; h; \Phi)$ if the following conditions are satisfied:*

$$f \in \Sigma, \quad \text{with} \quad 1 + \frac{1}{\eta}\left(\frac{z D_q\left(\mathcal{H}_h^{\lambda,q} f(z)\right)}{\mathcal{H}_h^{\lambda,q} f(z)} - 1\right) \prec \Phi(z), \tag{13}$$

and

$$1 + \frac{1}{\eta}\left(\frac{w D_q\left(\mathcal{H}_h^{\lambda,q} g(w)\right)}{\mathcal{H}_h^{\lambda,q} g(w)} - 1\right) \prec \Phi(w), \tag{14}$$

with $\lambda > -1, 0 < q < 1$, and $\eta \in \mathbb{C} \setminus \{0\}$, where the function g is the analytic extension of f^{-1} to $\mathbb{U}$, and is given by (12).

Remark 3.

(i) *Putting $q \to 1^-$ we obtain that $\lim_{q\to 1^-} \mathcal{L}_{\Sigma}^{q,\lambda}(\eta; h; \Phi) =: \mathcal{G}_{\Sigma}^{\lambda}(\eta; h; \Phi)$, where $\mathcal{G}_{\Sigma}^{\lambda}(\eta; h; \Phi)$ represents the functions $f \in \Sigma$ that satisfy (13) and (14) for $\mathcal{H}_h^{\lambda,q}$ replaced with $\mathcal{I}_h^{\lambda}$ (6).*

(ii) *Putting $c_k = \frac{(-1)^{k-1}\Gamma(v+1)}{4^{k-1}(k-1)!\Gamma(k+v)}$, $v > 0$, we obtain the class $\mathcal{B}_{\Sigma}^{q,\lambda}(\eta, v; \Phi)$, that represents the functions $f \in \Sigma$ that satisfy (13) and (14) for $\mathcal{H}_h^{\lambda,q}$ replaced with $\mathcal{N}_{v,q}^{\lambda}$ (7).*

(iii) *Putting $c_k = \left(\frac{n+1}{n+k}\right)^{\alpha}$, $\alpha > 0, n \geq 0$, we obtain the class $\mathcal{M}_{\Sigma}^{q,\lambda}(\eta, n, \alpha; \Phi)$, that represents the functions $f \in \Sigma$ that satisfy (13) and (14) for $\mathcal{H}_h^{\lambda,q}$ replaced with $\mathcal{M}_{n,q}^{\lambda,\alpha}$ (9).*

(iv) *Putting $c_k = \frac{m^{k-1}}{(k-1)!}e^{-m}$, $m > 0$, we obtain the class $\mathcal{I}_{\Sigma}^{q,\lambda}(\eta, m; \Phi)$, that represents the functions $f \in \Sigma$ that satisfy (13) and (14) for $\mathcal{H}_h^{\lambda,q}$ replaced with $\mathcal{I}_q^{\lambda,m}$ (10).*

(v) *Putting $c_k = \left[\frac{1+\ell+\mu(k-1)}{1+\ell}\right]^m$, $m \in \mathbb{Z}, \ell \geq 0, \mu \geq 0$, we obtain the class $\mathcal{J}_{\Sigma}^{q,\lambda}(\eta, m, \ell, \mu; \Phi)$, that represents the functions $f \in \Sigma$ that satisfy (13) and (14) for $\mathcal{H}_h^{\lambda,q}$ replaced with $\mathcal{J}_{q,\ell,\mu}^{\lambda,m}$ (11).*

Remark 4. *If the function h_* is given by*

$$h_*(z) = \frac{z}{1-z}, \quad z \in \mathbb{U},$$

then h_* has the form (2) with $c_k = 1, k \geq 2$, and according to (6) we have

$$\mathcal{I}_{h_*}^0 f(z) := \lim_{q \to 1^-} \mathcal{H}_{h_*}^{0,q} f(z) = z + \sum_{k=2}^{\infty} \frac{k!}{(1)_{k-1}} a_k z^k = z + \sum_{k=2}^{\infty} k a_k z^k = z f'(z), z \in \mathbb{U}, \tag{15}$$

for all $f \in \mathcal{A}$ of the form (1). Consider the function $f_*(z) = \frac{z}{1-z} \in \Sigma$, and its inverse analytic extension on $\mathbb{U}$, $g_*(w) = \frac{w}{1+w}$, let $\eta = 1$ and $\Phi_*(z) = \frac{1+z}{1-z}$. Using (15), the relations (13) and (14) become

$$1 + \frac{1}{\eta}\left(\frac{z\left(\mathcal{I}_{h_*}^0 f_*(z)\right)'}{\mathcal{I}_{h_*}^0 f_*(z)} - 1\right) = 1 + \frac{zf_*''(z)}{f_*'(z)} = \Phi_*(z) \prec \Phi_*(z),$$

and

$$1 + \frac{1}{\eta}\left(\frac{w\left(\mathcal{I}_{h_*}^0 g_*(w)\right)'}{\mathcal{I}_{h_*}^0 g_*(w)} - 1\right) = 1 + \frac{wg_*''(w)}{g_*'(w)} = \Phi_*(-w) \prec \Phi_*(w).$$

Hence, using the notation of Remark 3 (i), we have $\mathcal{G}_\Sigma^\lambda(\eta; h; \Phi) \neq \emptyset$ for some values of λ, η, and some special choices of the functions h and Φ.

To prove our main results we need to use the following lemma:

Lemma 1. [19] [p. 172] *If* $s(z) = \sum\limits_{k=1}^{\infty} p_k z^k$ *is a Schwarz function for* $z \in \mathbb{U}$, *then*

$$|p_1| \leq 1, \quad |p_k| \leq 1 - |p_1|^2, \ k \geq 1.$$

2. Coefficient Bounds for the Function Class $\mathcal{L}_\Sigma^{q,\lambda}(\eta; h; \Phi)$

Throughout this paper we are going to assume that $\lambda > -1$ and $0 < q < 1$.

Theorem 1. *If the function f given by (1) belongs to the class* $\mathcal{L}_\Sigma^{q,\lambda}(\eta; h; \Phi)$, *and* $\eta \in \mathbb{C}^* := \mathbb{C} \setminus \{0\}$, *then*

$$|a_2| \leq \frac{|B_1|\sqrt{|B_1|}}{\sqrt{\left|\frac{q}{\eta}\left[(1+q)\phi_2 - \phi_1^2\right]B_1^2 - \frac{q^2}{\eta^2}B_2\phi_1^2\right|}},$$

and

$$|a_3| \leq \frac{|\eta||B_1|}{q(q+1)\phi_2} + \frac{\eta^2|B_1|^2}{q^2\phi_1^2},$$

where ϕ_{k-1}, $k \in \{2,3\}$, are given by (5).

Proof. If $f \in \mathcal{L}_\Sigma^{q,\lambda}(\eta; h; \Phi)$, from (13), (14), and the definition of subordination it follows that there exist two functions U and V analytic in $\mathbb{U}$ with $U(0) = V(0) = 0$ and $|U(z)| < 1$, $|V(w)| < 1$ for all $z, w \in \mathbb{U}$, such that

$$1 + \frac{1}{\eta}\left(\frac{zD_q\left(\mathcal{H}_h^{\lambda,q} f(z)\right)}{\mathcal{H}_h^{\lambda,q} f(z)} - 1\right) = \Phi(U(z)), \tag{16}$$

and

$$1 + \frac{1}{\eta}\left(\frac{wD_q\left(\mathcal{H}_h^{\lambda,q} g(w)\right)}{\mathcal{H}_h^{\lambda,q} g(w)} - 1\right) = \Phi(V(w)). \tag{17}$$

If $U(z) = \sum\limits_{k=1}^{\infty} u_k z^k$ and $V(w) = \sum\limits_{k=1}^{\infty} v_k w^k$, $z, w \in \mathbb{U}$, from Lemma 1 we have

$$|u_k| \leq 1 \text{ and } |v_k| \leq 1, \ k \in \mathbb{N}. \tag{18}$$

Relations (16) and (17) lead to

$$\frac{zD_q\left(\mathcal{H}_h^{\lambda,q} f(z)\right)}{\mathcal{H}_h^{\lambda,q} f(z)} - 1 = \eta\left[\Phi(U(z)) - 1\right], \tag{19}$$

and
$$\frac{w D_q \left(\mathcal{H}_h^{\lambda,q} g(w) \right)}{\mathcal{H}_h^{\lambda,q} g(w)} - 1 = \eta \left[\Phi(V(w)) - 1 \right]. \tag{20}$$

Since
$$\frac{z D_q \left(\mathcal{H}_h^{\lambda,q} f(z) \right)}{\mathcal{H}_h^{\lambda,q} f(z)} - 1 = q\phi_1 a_2 z + \left[q(1+q)\phi_2 a_3 - q\phi_1^2 a_2^2 \right] z^2 + \cdots,$$

$$\frac{w D_q \left(\mathcal{H}_h^{\lambda,q} g(w) \right)}{\mathcal{H}_h^{\lambda,q} g(w)} - 1 = -q\phi_1 a_2 w + \left[q(1+q)\phi_2 \left(2a_2^2 - a_3 \right) - q\phi_1^2 a_2^2 \right] w^2 + \cdots,$$

and
$$\eta \left[\Phi(U(z)) - 1 \right] = \eta B_1 u_1 z + \eta \left(B_1 u_2 + B_2 u_1^2 \right) z^2 + \cdots,$$

$$\eta \left[\Phi(V(w)) - 1 \right] = \eta B_1 v_1 w + \eta \left(B_1 v_2 + B_2 v_1^2 \right) w^2 + \cdots.$$

By equalization according the coefficients of z and w in (19) and (20), it follows that

$$q\phi_1 a_2 = \eta B_1 u_1, \tag{21}$$

$$q(1+q)\phi_2 a_3 - q\phi_1^2 a_2^2 = \eta \left(B_1 u_2 + B_2 u_1^2 \right), \tag{22}$$

$$-q\phi_1 a_2 = \eta B_1 v_1, \tag{23}$$

$$q(1+q)\phi_2 \left(2a_2^2 - a_3 \right) - q\phi_1^2 a_2^2 = \eta \left(B_1 v_2 + B_2 v_1^2 \right). \tag{24}$$

Using (21) and (23) we obtain
$$u_1 = -v_1. \tag{25}$$

Squaring (21) and (23), after adding relations, we get
$$2q^2 a_2^2 \phi_1^2 = \eta^2 B_1^2 \left(u_1^2 + v_1^2 \right). \tag{26}$$

Adding (22) and (24) we have
$$2q \left[(1+q)\phi_2 - \phi_1^2 \right] a_2^2 = \eta \left[B_1 (u_2 + v_2) + B_2 \left(u_1^2 + v_1^2 \right) \right].$$

From (26), replacing $u_1^2 + v_1^2$ in the above equation, we have
$$\left\{ 2q \left[(1+q)\phi_2 - \phi_1^2 \right] \eta B_1^2 - 2q^2 \phi_1^2 B_2 \right\} a_2^2 = \eta^2 B_1^3 (u_2 + v_2),$$

that is
$$a_2^2 = \frac{B_1^3 (u_2 + v_2)}{2 \left\{ \frac{q}{\eta} \left[(1+q)\phi_2 - \phi_1^2 \right] B_1^2 - \frac{q^2}{\eta^2} \phi_1^2 B_2 \right\}}. \tag{27}$$

Taking the absolute value of (27) and using the inequalities (18) we conclude that

$$|a_2| \leq \frac{|B_1| \sqrt{|B_1|}}{\sqrt{\left| \frac{q}{\eta} \left[(1+q)\phi_2 - \phi_1^2 \right] B_1^2 - \frac{q^2}{\eta^2} B_2 \phi_1^2 \right|}},$$

which gives the bound for $|a_2|$ as we asserted in our theorem.

To find the bound for $|a_3|$, by subtracting (24) from (22), we get

$$2q(1+q)\phi_2\left(a_3 - a_2^2\right) = \eta\left[B_1(u_2 - v_2) + B_2\left(u_1^2 - v_1^2\right)\right]. \tag{28}$$

Form (25), (26) and (28), we obtain

$$a_3 = \frac{\eta B_1(u_2 - v_2)}{2q(1+q)\phi_2} + \frac{\eta^2 B_1^2\left(u_1^2 + v_1^2\right)}{2q^2\phi_1^2}. \tag{29}$$

Taking the absolute value of (29) and using the inequalities (18) we obtain

$$|a_3| \le \frac{|\eta||B_1|}{q(q+1)\phi_2} + \frac{\eta^2|B_1|^2}{q^2\phi_1^2}.$$

□

Putting $q \to 1^-$ in Theorem 1 we obtain the following corollary:

Corollary 1. *If the function f given by (1) belongs to the class $\mathcal{G}_\Sigma^\lambda(\eta; h; \Phi)$ for $\eta \neq 0$, then*

$$|a_2| \le \frac{|B_1|\sqrt{|B_1|}}{\sqrt{\left|(2\phi_2 - \phi_1^2)\dfrac{B_1^2}{\eta} - \dfrac{B_2\phi_1^2}{\eta^2}\right|}},$$

and

$$|a_3| \le \frac{|\eta||B_1|}{2\phi_2} + \frac{\eta^2|B_1|^2}{\phi_1^2},$$

where ϕ_{k-1}, $k \in \{2,3\}$, are given by (5).

Taking $c_k = \dfrac{(-1)^{k-1}\Gamma(v+1)}{4^{k-1}(k-1)!\Gamma(k+v)}$, $v > 0$, in Theorem 1 we obtain the following special case:

Corollary 2. *If $f \in \mathcal{B}_\Sigma^{q,\lambda}(\eta, v; \Phi)$ is given by (1) and $\eta \neq 0$, then*

$$|a_2| \le \frac{|B_1|\sqrt{|B_1|}}{\sqrt{\left|\dfrac{q}{\eta}[(1+q)\psi_2 - \psi_1^2]B_1^2 - \dfrac{q^2}{\eta^2}B_2\psi_1^2\right|}},$$

and

$$|a_3| \le \frac{|\eta||B_1|}{q(q+1)\psi_2} + \frac{\eta^2|B_1|^2}{q^2\psi_1^2},$$

where ψ_{k-1}, $k \in \{2,3\}$, are given by (8).

Considering $c_k = \left(\dfrac{n+1}{n+k}\right)^\alpha$, $\alpha > 0$, $n \ge 0$, in Theorem 1 we obtain the following result:

Corollary 3. *If $f \in \mathcal{M}_\Sigma^{q,\lambda}(\eta, n, \alpha; \Phi)$ is given by (1) and $\eta \neq 0$, then*

$$|a_2| \le \frac{|B_1|\sqrt{|B_1|}}{\sqrt{\left|\dfrac{q}{\eta}(1+q)\dfrac{[3,q]!}{[\lambda+1,q]_2}\left(\dfrac{n+1}{n+3}\right)^\alpha - \dfrac{([2,q]!)^2}{([\lambda+1,q])^2}\left(\dfrac{n+1}{n+2}\right)^{2\alpha}\right]B_1^2 - \dfrac{q^2}{\eta^2}B_2\dfrac{([2,q]!)^2}{([\lambda+1,q])^2}\left(\dfrac{n+1}{n+2}\right)^{2\alpha}\right|}},$$

and
$$|a_3| \leq \frac{|\eta| |B_1| [\lambda+1,q]_2 (n+3)^\alpha}{q(q+1)[3,q]!(n+1)^\alpha} + \frac{\eta^2 |B_1|^2 ([\lambda+1,q])^2 (n+2)^{2\alpha}}{q^2 ([2,q]!)^2 (n+1)^{2\alpha}}.$$

Putting $c_k = \frac{m^{k-1}}{(k-1)!} e^{-m}$, $m > 0$, in Theorem 1 we obtain the following special case:

Corollary 4. *If $f \in \mathcal{I}_\Sigma^{q,\lambda}(\eta, m; \Phi)$ is given by (1) and $\eta \neq 0$, then*

$$|a_2| \leq \frac{|B_1|\sqrt{|B_1|}}{\sqrt{\left|\frac{q}{\eta}\left[(1+q)\frac{[3,q]!}{2[\lambda+1,q]_2} m^2 e^{-m} - \frac{([2,q]!)^2}{([\lambda+1,q])^2} m^2 e^{-2m}\right] B_1^2 - \frac{q^2}{\eta^2} B_2 \frac{([2,q]!)^2}{([\lambda+1,q])^2} m^2 e^{-2m}\right|}},$$

and
$$|a_3| \leq \frac{2|\eta| |B_1| [\lambda+1,q]_2}{q(q+1)[3,q]! m^2 e^{-m}} + \frac{\eta^2 |B_1|^2 ([\lambda+1,q])^2}{q^2 ([2,q]!)^2 m^2 e^{-2m}}.$$

3. Fekete-Szegő Problem for the Function Class $\mathcal{L}_\Sigma^{q,\lambda}(\eta; h; \Phi)$

Theorem 2. *If the function f given by (1) belongs to the class $\mathcal{L}_\Sigma^{q,\lambda}(\eta; h; \Phi)$ for $\eta \neq 0$, then*

$$\left|a_3 - \mu a_2^2\right| \leq |\eta||B_1| (|M+N| + |M-N|), \tag{30}$$

with
$$M = \frac{(1-\mu)\eta B_1^2}{2q \left[(1+q)\phi_2 - \phi_1^2\right] \eta B_1^2 - 2q^2 \phi_1^2 B_2}, \quad \text{and} \quad N = \frac{1}{2q(1+q)\phi_2}, \tag{31}$$

where $\mu \in \mathbb{C}$, and ϕ_k, $k \in \{2,3\}$, are given by (5).

Proof. If $f \in \mathcal{L}_\Sigma^{q,\lambda}(\eta; h; \Phi)$, like in the proof of Theorem 1, from (25) and (28) we have

$$a_3 - a_2^2 = \frac{\eta B_1 (u_2 - v_2)}{2q(1+q)\phi_2}. \tag{32}$$

Multiplying (27) by $(1-\mu)$ we get

$$(1-\mu) a_2^2 = \frac{(1-\mu) \eta^2 B_1^3 (u_2 + v_2)}{2q \left[(1+q)\phi_2 - \phi_1^2\right] \eta B_1^2 - 2q^2 \phi_1^2 B_2}. \tag{33}$$

Adding (32) and (33), it follows that

$$a_3 - \mu a_2^2 = \eta B_1 \left[(M+N) u_2 + (M-N) v_2\right], \tag{34}$$

where M and N are given by (31). Taking the absolute value of (34), from (18) we obtain the inequality (30). □

Remark 5. *Algebra shows that the inequality $|M| \leq N$ is equivalent to*

$$|\mu - 1| \leq \left|1 - \left(\eta + \frac{B_2 q}{B_1^2}\right) \frac{\phi_1^2}{\eta(1+q)\phi_2}\right|.$$

From Theorem 2 we get the next:

If the function f given by (1) belongs to the class $\mathcal{L}_\Sigma^{q,\lambda}(\eta;h;\Phi)$ for $\eta \neq 0$, then

$$\left|a_3 - \mu a_2^2\right| \leq \frac{|\eta||B_1|}{q(1+q)\phi_2},$$

where $\mu \in \mathbb{C}$, and

$$|\mu - 1| \leq \left|1 - \left(\eta + \frac{B_2 q}{B_1^2}\right)\frac{\phi_1^2}{\eta(1+q)\phi_2}\right|,$$

with ϕ_k, $k \in \{2,3\}$, are given by (5).

Putting $q \to 1^-$ in Theorem 2 we obtain the following corollary:

Corollary 5. *If the function f given by (1) belongs to the class $\mathcal{G}_\Sigma^\lambda(\eta,h;\Phi)$ for $\eta \neq 0$, then*

$$\left|a_3 - \mu a_2^2\right| \leq |\eta||B_1|(|M+N| + |M-N|),$$

with

$$M = \frac{(1-\mu)\eta B_1^2}{2(2\phi_2 - \phi_1^2)\eta B_1^2 - 2\phi_1^2 B_2}, \quad \text{and} \quad N = \frac{1}{4\phi_2},$$

where $\mu \in \mathbb{C}$, and ϕ_k, $k \in \{2,3\}$, are given by (5).

Taking $c_k = \dfrac{(-1)^{k-1}\Gamma(v+1)}{4^{k-1}(k-1)!\Gamma(k+v)}$, $v > 0$ in Theorem 2, we obtain the following special case:

Corollary 6. *If the function f given by (1) belongs to the class $\mathcal{B}_\Sigma^{q,\lambda}(\eta,v;\Phi)$ for $\eta \neq 0$, then*

$$\left|a_3 - \mu a_2^2\right| \leq |\eta||B_1|(|M+N| + |M-N|),$$

with

$$M = \frac{(1-\mu)\eta B_1^2}{2q\left[(1+q)\psi_2 - \psi_1^2\right]\eta B_1^2 - 2q^2\psi_1^2 B_2}, \quad \text{and} \quad N = \frac{1}{2q(1+q)\psi_2},$$

where $\mu \in \mathbb{C}$, and ϕ_k, $k \in \{2,3\}$, are given by (8).

Considering $c_k = \left(\dfrac{n+1}{n+k}\right)^\alpha$, $\alpha > 0$, $n \geq 0$ in Theorem 2, we obtain the next result:

Corollary 7. *If the function f given by (1) belongs to the class $\mathcal{M}_\Sigma^{q,\lambda}(\eta,n,\alpha;\Phi)$ for $\eta \neq 0$, then*

$$\left|a_3 - \mu a_2^2\right| \leq |\eta||B_1|(|M+N| + |M-N|),$$

with

$$M = \frac{(1-\mu)\eta B_1^2}{2qR_1(n,\alpha,\lambda,q)\eta B_1^2 - 2q^2\left(\frac{[2,q]!}{[\lambda+1,q]}\right)^2\left(\frac{n+1}{n+2}\right)^{2\alpha} B_2},$$

and

$$N = \frac{(n+3)^\alpha [\lambda+1,q]_2}{2q(1+q)(n+1)^\alpha [3,q]!},$$

where $\mu \in \mathbb{C}$, and

$$R_1(n,\alpha,\lambda,q) = (1+q)\frac{[3,q]!}{[\lambda+1,q]_2}\left(\frac{n+1}{n+3}\right)^\alpha - \left(\frac{[2,q]!}{[\lambda+1,q]}\right)^2\left(\frac{n+1}{n+2}\right)^{2\alpha}.$$

If we take $c_k = \dfrac{m^{k-1}}{(k-1)!}e^{-m}$, $m > 0$ in Theorem 2, we get the next special case:

Corollary 8. *If the function f given by (1) belongs to the class $\mathcal{L}_\Sigma^{q,\lambda}(\eta, m; \Phi)$ for $\eta \neq 0$, then*

$$\left|a_3 - \mu a_2^2\right| \leq |\eta||B_1|(|M+N| + |M-N|),$$

with

$$M = \dfrac{(1-\mu)\eta B_1^2}{2qR(m,\lambda,q)\eta B_1^2 - 2q^2\left(\dfrac{[2,q]!}{[\lambda+1,q]}\right)^2 m^2 e^{-2m} B_2},$$

and

$$N = \dfrac{[\lambda+1,q]_2}{q(1+q)m^2 e^{-m}[3,q]!},$$

where $\mu \in \mathbb{C}$, and

$$R(m,\lambda,q) = (1+q)\dfrac{[3,q]! m^2 e^{-m}}{2[\lambda+1,q]_2} - \left(\dfrac{[2,q]!}{[\lambda+1,q]}\right)^2 m^2 e^{-2m}.$$

We will give a few applications of the above results obtained for special choices of the function Φ, as follows.

1. The circular function $\Phi(z) = \dfrac{1+Az}{1+Bz}$ $(-1 < B < A \leq 1)$ is convex in $\mathbb{U}$ and

$$\Phi(\mathbb{U}) = \left\{w \in \mathbb{C} : \left|w - \dfrac{1-AB}{1-B^2}\right| < \dfrac{A-B}{1-B^2}\right\}, \text{ if } -1 < B < A \leq 1,$$

$$\Phi(\mathbb{U}) = \left\{w \in \mathbb{C} : \operatorname{Re} w > \dfrac{1-A}{2}\right\}, \text{ if } -1 = B < A \leq 1.$$

Since $B_1 = A - B$ and $B_2 = B(B-a)$, replacing this function in Theorem 1 and Theorem 2 we obtain the next example:

Example 1. *If $f \in \mathcal{L}_\Sigma^{q,\lambda}\left(\eta; h; \dfrac{1+Az}{1+Bz}\right)$ is given by (1) and $\eta \neq 0$, then*

$$|a_2| \leq \dfrac{|A-B|\sqrt{|A-B|}}{\sqrt{\left|\dfrac{q}{\eta}\left[(1+q)\phi_2 - \phi_1^2\right](A-B)^2 - \dfrac{q^2}{\eta^2}B(B-A)\phi_1^2\right|}},$$

$$|a_3| \leq \dfrac{|\eta||A-B|}{q(q+1)\phi_2} + \dfrac{\eta^2|A-B|^2}{q^2\phi_1^2},$$

and

$$\left|a_3 - \mu a_2^2\right| \leq |\eta||A-B|(|M+N|+|M-N|),$$

with

$$M = \dfrac{(1-\mu)\eta(A-B)^2}{2\eta q\left[(1+q)\phi_2 - \phi_1^2\right](A-B)^2 - 2q^2\phi_1^2 B(B-A)}, \text{ and } N = \dfrac{1}{2q(1+q)\phi_2},$$

where $\phi_{k-1}, k \in \{2, 3\}$, are given by (5).

Remark 6. For the special values $A = 1 - 2\beta$ and $B = -1$ $(0 \leq \beta < 1)$, the above example yields to the next special case: if $f \in \mathcal{L}_{\Sigma}^{q,\lambda}\left(\eta; h; \frac{1+(1-2\beta)z}{1-z}\right)$ is given by (1) and $\eta \neq 0$, then

$$|a_2| \leq \frac{2\sqrt{2}\,(1-\beta)^{\frac{3}{2}}}{\sqrt{\left|\frac{4q}{\eta}\left[(1+q)\phi_2 - \phi_1^2\right](1-\beta)^2 - \frac{2q^2}{\eta^2}(1-\beta)\,\phi_1^2\right|}},$$

$$|a_3| \leq \frac{2\,|\eta|\,(1-\beta)}{q(q+1)\phi_2} + \frac{4\eta^2(1-\beta)^2}{q^2\phi_1^2},$$

and

$$\left|a_3 - \mu a_2^2\right| \leq 2|\eta|\,(1-\beta)\,(|M+N| + |M-N|),$$

with

$$M = \frac{4\eta\,(1-\mu)\,(1-\beta)^2}{8\eta q\left[(1+q)\phi_2 - \phi_1^2\right](1-\beta)^2 - 4q^2\phi_1^2(1-\beta)}, \quad \text{and} \quad N = \frac{1}{2q(1+q)\phi_2},$$

where ϕ_{k-1}, $k \in \{2,3\}$, are given by (5).

2. Let consider the binomial function $\Phi(z) = (1+z)^\alpha$, $z \in \mathbb{U}$, with $\alpha \in \mathbb{C}^*$, where the power is considered at the principal branch, that is $\Phi(0) = 1$. Since

$$\Phi(z) = (1+z)^\alpha = 1 + \sum_{n=1}^{\infty} \frac{\alpha(\alpha-1)\ldots(\alpha-n+1)}{n!} z^n, \quad z \in \mathbb{U},$$

it follows that $B_1 = \alpha$ and $B_2 = \dfrac{\alpha(\alpha-1)}{2}$. Replacing this function in Theorems 1 and 2 we get:

Example 2. If $f \in \mathcal{L}_{\Sigma}^{q,\lambda}\left(\eta; h; (1+z)^\alpha\right)$ is given by (1) and $\eta \neq 0$, then

$$|a_2| \leq \frac{|\alpha|\sqrt{|\alpha|}}{\sqrt{\left|\frac{q}{\eta}\left[(1+q)\phi_2 - \phi_1^2\right]\alpha^2 - \frac{q^2}{2\eta^2}\alpha(\alpha-1)\phi_1^2\right|}},$$

$$|a_3| \leq \frac{|\eta|\,|\alpha|}{q(q+1)\phi_2} + \frac{\eta^2|\alpha|^2}{q^2\phi_1^2},$$

and

$$\left|a_3 - \mu a_2^2\right| \leq |\eta|\,|\alpha|\,(|M+N| + |M-N|),$$

with

$$\frac{(1-\mu)\,\eta\alpha^2}{2q\left[(1+q)\phi_2 - \phi_1^2\right]\eta\alpha^2 - q^2\phi_1^2\alpha(\alpha-1)}, \quad \text{and} \quad N = \frac{1}{2q(1+q)\phi_2},$$

where ϕ_{k-1}, $k \in \{2,3\}$, are given by (5).

3. For the function $\Phi(z) = \left(\dfrac{1+z}{1-z}\right)^\sigma$, $z \in \mathbb{U}$, with $\sigma \in \mathbb{C}^*$, where the power is considered at the principal branch, that is $\Phi(0) = 1$, we have $B_1 = 2\sigma$ and $B_2 = 2\sigma^2$. Therefore, from Theorems 1 and 2 we deduce the following example:

Example 3. If $f \in \mathcal{L}_{\Sigma}^{q,\lambda}(\eta;h;(1+z)^{\alpha})$ is given by (1) and $\eta \neq 0$, then

$$|a_2| \leq \frac{2\sqrt{2}|\sigma|\sqrt{|\sigma|}}{\sqrt{\left|\frac{4q}{\eta}\left[(1+q)\phi_2 - \phi_1^2\right]\sigma^2 - \frac{2q^2}{\eta^2}\sigma^2\phi_1^2\right|}},$$

$$|a_3| \leq \frac{2|\eta||\sigma|}{q(q+1)\phi_2} + \frac{4\eta^2|\sigma|^2}{q^2\phi_1^2},$$

and

$$\left|a_3 - \mu a_2^2\right| \leq 2|\eta||\sigma|\left(|M+N| + |M-N|\right),$$

with

$$\frac{4(1-\mu)\eta\sigma^2}{8q\eta\left[(1+q)\phi_2 - \phi_1^2\right]\sigma^2 - 4q^2\phi_1^2\sigma^2}, \quad \text{and} \quad N = \frac{1}{2q(1+q)\phi_2},$$

where ϕ_{k-1}, $k \in \{2,3\}$, are given by (5).

Remark 7. We mention that all the above estimations for the coefficients $|a_2|$, $|a_3|$, and Fekete-Szegő problem for the function class $\mathcal{L}_{\Sigma}^{q,\lambda}(\eta;h;\Phi)$ are not sharp. To find the sharp upper bounds for the above functionals, it still is an interesting open problem, as well as for $|a_n|$, $n \geq 4$.

Author Contributions: Conceptualization, S.M.E.-D., T.B. and B.M.E.-M.; methodology, S.M.E.-D., T.B. and B.M.E.-M.; investigation, S.M.E.-D., T.B. and B.M.E.-M.; resources, S.M.E.-D., T.B. and B.M.E.-M.; writing–original draft preparation, S.M.E.-D., T.B. and B.M.E.-M.; writing–review and editing, T.B.; supervision, S.M.E.-D. and T.B.; project administration, S.M.E.-D., T.B. and B.M.E.-M. The authors contributed equally to this work. All authors have read and agreed to the published version of the manuscript.

Funding: This research received no external funding.

Acknowledgments: The authors are grateful to the reviewers of this article who gave valuable remarks, comments, and advices, in order to revise and improve the results of the paper.

Conflicts of Interest: The authors declare no conflict of interest.

References

1. Jackson, F.H. On q-functions and a certain difference operator. *Trans. R. Soc. Edinb.* **1909**, *46*, 253–281. [CrossRef]
2. Jackson, F.H. On q-definite integrals. *Quart. J. Pure Appl. Math.* **1910**, *41*, 193–203.
3. Risha, M.H.A.; Annaby, M.H.; Ismail, M.E.H.; Mansour, Z.S. Linear q-difference equations. *Z. Anal. Anwend.* **2007**, *26*, 481–494. [CrossRef]
4. El-Deeb, S.M.; Bulboacă, T. Fekete-Szegő inequalities for certain class of analytic functions connected with q-anlogue of Bessel function. *J. Egypt. Math. Soc.* **2019**, 1–11. [CrossRef]
5. El-Deeb, S.M.; Bulboacă, T. Differential sandwich-type results for symmetric functions connected with a q-analog integral operator. *Mathematics* **2019**, *7*, 1185. [CrossRef]
6. Arif, M.; Haq, M.U.; Liu, J.L. A subfamily of univalent functions associated with q-analogue of Noor integral operator. *J. Funct. Spaces* **2018**, 3818915. [CrossRef]
7. Porwal, S. An application of a Poisson distribution series on certain analytic functions. *J. Complex Anal.* **2014**, 984135. [CrossRef]
8. Prajapat, J.K. Subordination and superordination preserving properties for generalized multiplier transformation operator. *Math. Comput. Model.* **2012**, *55*, 1456–1465. [CrossRef]
9. Duren, P.L. *Univalent Functions, Grundlehren der Mathematischen Wissenschaften, Band 259*; Springer: New York, NY, USA; Berlin/Heidelberg Germany; Tokyo, Japan, 1983.
10. Srivastava, H.M.; Mishra, A.K.; Gochhayat, P. Certain subclasses of analytic and bi-univalent functions. *Appl. Math. Lett.* **2010**, *23*, 1188–1192. [CrossRef]

11. Brannan, D.A.; Clunie, J.; Kirwan, W.E. Coefficient estimates for a class of star-like functions. *Canad. J. Math.* **1970**, *22*, 476–485. [CrossRef]
12. Brannan, D.A.; Taha, T.S. On some classes of bi-univalent functions. In *Mathematical Analysis and Its Applications*; Kuwait, 18–21 February 1985; KFAS Proceedings Series; Smazhar, M., Hamoui, A., Faour, N.S., Eds.; Pergamon Press (Elsevier Science Limited): Oxford, UK, 1988; Volume 3, pp. 53–60; see also *Studia Univ. Babeş-Bolyai Math.* **1986**, *31*, 70–77.
13. Kamble, P.N.; Shrigan, M.G. Coefficient estimates for a subclass of bi-univalent functions defined by Sălăgean type q-calculus operator. *Kyungpook Math. J.* **2018**, *58*, 677–688. [CrossRef]
14. Çaglar, M.; Deniz, E. Initial coefficients for a subclass of bi-univalent functions defined by Sălăgean differential operator. *Commun. Fac. Sci. Univ. Ank. Ser. A1 Math. Stat.* **2017**, *66*, 85–91._ 0000000777. [CrossRef]
15. Elhaddad, S.; Darus, M. Coefficient estimates for a subclass of bi-univalent functions defined by q-derivative operator. *Mathematics* **2020**, *8*, 306. [CrossRef]
16. Aldweby, H.; Darus, M. On a subclass of bi-univalent functions associated with the q-derivative operator. *J. Math. Comput. Sci.* **2019**, *19*, 58–64. [CrossRef]
17. Bulboacă, T. *Differential Subordinations and Superordinations*; Recent Results; House of Scientific Book Publ.: Cluj-Napoca, Romania, 2005.
18. Miller, S.S.; Mocanu, P.T. Differential Subordinations. In *Theory and Applications*; Series on Monographs and Textbooks in Pure and Applied Mathematics; Marcel Dekker Inc.: New York, NY, USA; Basel, Switzerland, 2000; Volume 225.
19. Nehari, Z. *Conformal Mapping*; McGraw-Hill: New York, NY, USA, 1952.

© 2020 by the authors. Licensee MDPI, Basel, Switzerland. This article is an open access article distributed under the terms and conditions of the Creative Commons Attribution (CC BY) license (http://creativecommons.org/licenses/by/4.0/).

Article

Subordination Properties of Meromorphic Kummer Function Correlated with Hurwitz–Lerch Zeta-Function

Firas Ghanim [1,*], Khalifa Al-Shaqsi [2], Maslina Darus [3] and Hiba Fawzi Al-Janaby [4]

1. Department of Mathematics, College of Sciences, University of Sharjah, Sharjah 27272, UAE
2. Department of Information Technology, University of Technology and Applied Science, Nizwa College of Technology, P.O. Box 75, Nizwa 612, Oman; khalifa.alshaqsi@nct.edu.om
3. Department of Mathematical Sciences, Faculty of Science and Technology, Universiti Kebangsaan Malaysia, Bangi 43600, Selangor D. Ehsan, Malaysia; maslina@ukm.edu.my
4. Department of Mathematics, College of Science, University of Baghdad, Baghdad 10081, Iraq; fawzihiba@yahoo.com
* Correspondence: fgahmed@sharjah.ac.ae

Citation: Ghanim, F.; Al-Shaqsi, K.; Darus, M.; Al-Janaby, H.F. Subordination Properties of Meromorphic Kummer Function Correlated with Hurwitz–Lerch Zeta-Function. *Mathematics* **2021**, *9*, 192. https://doi.org/10.3390/math 9020192

Academic Editor: Georgia Irina Oros
Received: 30 October 2020
Accepted: 11 January 2021
Published: 19 January 2021

Publisher's Note: MDPI stays neutral with regard to jurisdictional claims in published maps and institutional affiliations.

Copyright: © 2021 by the authors. Licensee MDPI, Basel, Switzerland. This article is an open access article distributed under the terms and conditions of the Creative Commons Attribution (CC BY) license (https:// creativecommons.org/licenses/by/ 4.0/).

Abstract: Recently, Special Function Theory (SPFT) and Operator Theory (OPT) have acquired a lot of concern due to their considerable applications in disciplines of pure and applied mathematics. The Hurwitz-Lerch Zeta type functions, as a part of Special Function Theory (SPFT), are significant in developing and providing further new studies. In complex domain, the convolution tool is a salutary technique for systematic analytical characterization of geometric functions. The analytic functions in the punctured unit disk are the so-called meromorphic functions. In this present analysis, a new convolution complex operator defined on meromorphic functions related with the Hurwitz-Lerch Zeta type functions and Kummer functions is considered. Certain sufficient stipulations are stated for several formulas of this defining operator to attain subordination. Indeed, these outcomes are an extension of known outcomes of starlikeness, convexity, and close to convexity.

Keywords: meromorphic functions; Hurwitz–Lerch Zeta-function; Riemann zeta function

MSC: 11M35; 30C50

1. Introduction

During the 18th century, complex analysis (complex function theory) had been launched, which has become thereafter one of the major disciplines of mathematics. Prominent complex analysts include Euler, Riemann, and Cauchy. This realm has had a great influence on a variety of research subjects in, for example, engineering, physics, and mathematics, due to its efficient applications to numerous conceptions and problems. Researchers have happened to meet certain unexpected relationships among obviously different research areas. The study of the intriguing and fascinating interplay of geometry and complex analysis has been famed as Geometric Analytic Function Theory (GAFT). In other words, it deals with the structure of analytic functions in the complex domain whose specific geometries are starlike, close-to-starlike, convex, close-to convex, spiral, and so on. In 1851, Riemann contributed to the origin of GAFT by presenting the first significant outcome, namely the Riemann mapping theorem (RIMT). Koebe followed suit in 1907 and proceeded to the study of univalent function. In light of RMT, he initiated the discussion of the merits for univalent analytical functions over the open unit disk rather than in a complex domain. This modified version led to the creation of the Univalent Analytic Function Theory (UAFT). One of the gorgeous problems in UAFT is Bieberbach's conjecture "coefficient conjecture" posed by Bieberbach in 1916. It states the upper bounds of the coefficient of the univalent function in the unit disk [1]. For many years, this conjecture posed a challenge to researchers in the field. Until 1985, De Branges [2] settled all attempts and resolved it.

The difficulty in resolving this conjecture led to several profound and significant contributions in GAFT along with the development of several gadgets. These involve Loewner's parametric technique, Milin's and Fitz Gerald's techniques of exponentiating the Grunsky inequalities, Baernstein's technique of maximal function, and variational techniques in addition to new subclasses of univalent functions imposed by geometric stipulation. Among the subclasses considered are the subclasses of convex functions, starlike functions, close-to-convex function, and quasi-convex functions, consistently. Besides, de Branges employed hypergeometric function, as a sort of the Special Function Theories (SPFT) in order to resolve the Bieberbach problem. From an application point of sight, SPFT are such significant mathematical tools for their interesting merits and remarkable role in the study of the Fractional Calculus (FRC) and Operator Theory (OPT), for instance, Ghanim and Al-Janaby [3]. Accordingly, SPFT plays a giant pivotal role in the development of research in the area of GAFT which includes a lot of new implementations and generalizations. For instance, Noor [4], El-Ashwah and Hassan [5], Xing and Jose Xing, Rassias and Yang ([6,7]), Ghanim and Al-Janaby ([8,9]) and Al-Janaby and Ghanim ([10,11]).

In this context, the term hypergeometric function, first coined by Wallis in the year 1655, also known as the hypergeometric series is in the complex plane $\mathbb{C}$ and the open unit disk $\mathbb{D} = \{z \in \mathbb{C} : |z| < 1\}$. This function was discussed by Euler first, and then systematically investigated by Gauss in 1813. It is formulated as [12]:

$$_2F_1(\varrho, v; \omega; z) = \sum_{\kappa=0}^{\infty} \frac{(\varrho)_\kappa (v)_\kappa}{(\omega)_\kappa} \frac{z^\kappa}{\kappa!}, \quad (\varrho, v \in \mathbb{C}, \omega \in \mathbb{C} \setminus \{0, -1, \ldots\}, |z| < 1).$$

Here $(\omega)_\kappa$ is the Pochhammer (rising) symbol and is defined as:

$$(\omega)_\kappa = \begin{cases} 1 & \kappa = 0, \\ \omega(\omega+1)\cdots(\omega+\kappa-1) & \kappa \in \mathbb{N} = \{1, 2, \ldots\}. \end{cases}$$

Subsequently, in 1837, Kummer presented the Kummer function, namely confluent hypergeometricr function, as a solution of a Kummer differential equation. This function is written as [12]:

$$K(\varrho; \omega, z) = \sum_{\kappa=0}^{\infty} \frac{(\varrho)_\kappa}{(\omega)_\kappa} \frac{z^\kappa}{\kappa!} = {}_1F_1(\varrho; \omega; z), \tag{1}$$

$$(\varrho \in \mathbb{C}, \omega \in \mathbb{C} \setminus \{0, -1, \ldots\}, |z| < 1).$$

Furthermore, the Zeta functions constitute some phenomenal special functions that appear in the study of Analytic Number Theory (ANT). There are a number of generalizations of the Zeta function, such as Euler–Riemann Zeta function, Hurwitz Zeta function, and Lerch Zeta function. The Euler–Riemann Zeta function plays a pioneering role in ANT, due to its advantages in discussing the merits of prime numbers. It also has fruitful implementations in probability theory, applied statistics, and physics. Euler first formulated this function, as a function of a real variable, in the first half of the 18th century. Then, in 1859, Riemann utilized complex analysis to expand on Euler's definition to a complex variable. Symbolized by $S(\varkappa)$, the definition was posed as the Dirichlet series:

$$S(\varkappa) = \sum_{\kappa=1}^{\infty} \frac{1}{\kappa^\varkappa}, \quad \text{for} \quad \Re(\varkappa) > 1.$$

Later, the more general Zeta function, currently called Hurwitz Zeta function, was also propounded by Adolf Hurwitz in 1882, as a general formula of the Riemann Zeta function considered as [13]:

$$S(\mu, \varkappa) = \sum_{\kappa=0}^{\infty} \frac{1}{(\kappa + \mu)^{\varkappa}}, \quad \text{for} \quad \Re(\varkappa) > 1, \Re(\mu) > 1.$$

More generally, the famed Hurwitz–Lerch Zeta function $f(\mu, \varkappa, z)$ is described as [14]:

$$\phi_{\mu,\varkappa}(z) = \sum_{\kappa=0}^{\infty} \frac{z^{\kappa}}{(\kappa + \mu)^{\varkappa}}, \quad \text{for} \quad \Re(\varkappa) > 1, \Re(\mu) > 1. \qquad (2)$$

$(\mu \in \mathbb{C} \setminus \mathbb{Z}_0^-, \varkappa \in \mathbb{C} \text{ when } |z| < 1; \Re(\varkappa) > 1 \text{ when } |z| = 1).$

A generalization of (2) was proposed by Goyal and Laddha [15] in 1997, in the following formula:

$$\psi_{\mu,\varkappa}^{\wp}(z) = \sum_{\kappa=0}^{\infty} \frac{(\wp)_{\kappa}}{\kappa!} \frac{z^{\kappa}}{(\kappa + \mu)^{\varkappa}}, \quad \text{for} \quad \Re(\varkappa) > 1, \Re(\mu) > 1. \qquad (3)$$

$(\mu \in \wp \in \mathbb{C} \ \mathbb{C} \setminus \mathbb{Z}_0^-, \varkappa \in \mathbb{C} \text{ when } |z| < 1; \Re(\varkappa - \wp) > 1 \text{ when } |z| = 1).$

Along with these, there are more remarkable diverse extensions and generalizations that contributed to the rise of new classes of the Hurwitz–Lerch Zeta function in ([16–26]).

In this effort, by utilizing analytic techniques, a new linear (convolution) operator of morphometric functions is investigated and introduced in terms of the generalized Hurwitz–Lerch Zeta functions and Kummer functions. Moreover, sufficient stipulations are determined and examined in order for some formulas of this new operator to achieve subordination. Therefore, these outcomes are an extension for some well known outcomes of starlikeness, convexity, and close to convexity.

2. Preliminaries

Consider the class $\mathcal{H}$ of regular functions in $\mathbb{D} = \{z \in \mathbb{C} : |z| < 1\}$. The function f_1 is named subordinate to f_2 (or f_2 is named superordinate to f_1) and denotes $f_1 \prec f_2$, if there is a regular function ω in $\mathbb{D}$, with $\omega(0) = 0$ and $|\omega(z)| < 1$ and $f_1(z) = f_2(\omega(z))$. If the function f_2 is univalent in $\mathbb{D}$, then

$$f_1 \prec f_2 \Leftrightarrow f_1(0) = f_2(0) \quad \text{and} \quad f_1(\mathbb{D}) \subset f_2(\mathbb{D}).$$

Let Σ represent the class of normalized meromorphic functions $f(z)$ by

$$f(z) = \frac{1}{z} + \sum_{\kappa=1}^{\infty} \eta_{\kappa} z^{\kappa},$$

that are regular in the punctured unit disk

$$\mathbb{D}^* = \{z : z \in \mathbb{C} \text{ and } 0 < |z| < 1\}.$$

Furthermore, it indicates the classes of meromorphic starlike functions of order ζ and meromorphic convex of order ζ by $\Sigma_{\mathcal{S}^*(\zeta)}$ and $\Sigma_{\mathcal{K}(\zeta)}$, $(\zeta \geq 0)$, respectively (see [22,23,27,28]).

The convolution product of two meromorphic functions $f_\ell(z)$ $(\ell = 1, 2)$ in the following formula:

$$f_\ell(z) = \frac{1}{z} + \sum_{\kappa=1}^{\infty} \eta_{\kappa,\ell} z^{\kappa} \quad (\ell = 1, 2),$$

is defined by

$$(f_1 * f_2)(z) = \frac{1}{z} + \sum_{\kappa=1}^{\infty} \eta_{\kappa,1}\, \eta_{\kappa,2}\, z^\kappa. \tag{4}$$

The meromorphic Kummer function $\widetilde{K}(\varrho;\omega,z)$ is formulated as:

$$\widetilde{K}(\varrho;\omega,z) = \frac{1}{z} + \sum_{\kappa=0}^{\infty} \frac{(\varrho)_{\kappa+1}}{(\omega)_{\kappa+1}} \frac{z^\kappa}{(\kappa+1)!}, \tag{5}$$

$(\varrho \in \mathbb{C},\ \omega \in \mathbb{C}\setminus\{0,-1,\ldots\},\ z \in \mathbb{D}^*)$.

Corresponding to (5) and (3), based on a convolution tool, we imposed the following new convolution complex operator for $f(z) \in \Sigma$ as:

$$L_\mu^\varkappa(\varrho,\omega,\wp)f(z) = \widetilde{K}(\varrho;\omega,z) * \mathfrak{A}_{\varkappa,\mu}(z) * f(z)$$

$$= \frac{1}{z} + \sum_{\kappa=1}^{\infty} \frac{(\varrho)_{\kappa+1}\,(\wp)_{\kappa+1}}{(\omega)_{\kappa+1}\,(\kappa+1)!\,(\kappa+1)!} \left(\frac{\mu+1}{\mu+\kappa+1}\right)^\varkappa \eta_\kappa\, z^\kappa, \tag{6}$$

where

$$\mathfrak{A}_{\varkappa,\mu}(z) = (\mu+1)^\varkappa \left[\psi_{\mu,\varkappa}^\wp(z) - \frac{1}{\mu^\varkappa} + \frac{1}{z(\mu+1)^\varkappa} \right]$$

$$= \frac{1}{z} + \sum_{\kappa=1}^{\infty} \frac{(\wp)_{\kappa+1}}{(\kappa+1)!}\left(\frac{\mu+1}{\mu+\kappa+1}\right)^\varkappa z^\kappa \qquad (z \in \mathbb{U}^*) \tag{7}$$

The major goal of this paper is to study the following subordinations:

$$\frac{L_\mu^\varkappa(\varrho+1,\omega,\wp)f(z)}{L_\mu^\varkappa(\varrho,\omega,\wp)f(z)} \prec \frac{h(1-z)}{h-z}, \qquad (h > 1)$$

$$\frac{L_\mu^\varkappa(\varrho,\omega,\wp)f(z)}{z} \prec \frac{1+Ez}{1-z}, \qquad (-1 \leqslant E < 1)$$

and

$$\frac{L_\mu^\varkappa(\varrho,\omega,\wp)f(z)}{z} \prec \frac{h(1-z)}{h-z}, \qquad (h > 1). \tag{8}$$

In particular, we obtain sufficient conditions for which the function $f \in \Sigma$ satisfies such subordination, which extends certain outcomes in this direction concerning starlikeness, convexity, and close to convexity.

The following lemma will be needed to accomplish our proofs. We refer the reader to [29], Theorem 3.4, p. 132, for the proof of this lemma.

Lemma 1. Let $\mathfrak{q}(z)$ be univalent in $\mathbb{D}$ and let Θ and Φ be regular in a domain $\mathfrak{D} \supset \mathfrak{q}(\mathbb{D})$, with $\Phi(\omega) \neq 0$ when $\omega \in \mathfrak{q}(\mathbb{D})$. Set

$$Y(z) = z\mathfrak{q}'(z)\Phi(\mathfrak{q}(z)),\quad \Lambda(z) = \Theta(\mathfrak{q}(z)) + Y(z)$$

Suppose that

(1) $Y(z)$ is starlike in $\mathbb{D}$, and

(2) $\Re \frac{z\Lambda'(z)}{Y(z)} > 0$ for $z \in \mathbb{D}$.

If $\mathfrak{p}(z)$ is regular in $\mathbb{D}$ with $\mathfrak{p}(0) = \mathfrak{q}(0)$, $\mathfrak{p}(\mathbb{D}) \subset \mathfrak{D}$ and

$$\Theta(\mathfrak{p}(z)) + z\mathfrak{p}'(z)\Phi(\mathfrak{p}(z)) \prec \Theta(\mathfrak{q}(z)) + z\mathfrak{q}'(z)\Phi(\mathfrak{q}(z)), \tag{9}$$

then $\mathfrak{p}(z) \prec \mathfrak{q}(z)$ and $\mathfrak{q}(z)$ is the best dominant.

3. Main Outcomes

First, we treat the first subordinate in (8).

Theorem 1. *Let $\varrho > 0$, $h > 1, \zeta \in \mathbb{R}$ and $f \in \Sigma$. Then, if $|\zeta| \leq 1$, $L_\mu^*(\varrho, \omega, \wp) f(z)/z \neq 0$ in $\mathbb{D}^*$ and*

$$\left(\frac{L_\mu^*(\varrho+1, \omega, \wp)f(z)}{L_\mu^*(\varrho, \omega, \wp)f(z)} \right)^\zeta \left((\varrho+1)\frac{L_\mu^*(\varrho+2, \omega, \wp)f(z)}{L_\mu^*(\varrho+1, \omega, \wp)f(z)} - 1 \right) \prec \Lambda(z), \tag{10}$$

where

$$\Lambda(z) = \left(\frac{h(1-z)}{h-z} \right)^{\zeta+1} \left(\varrho - \frac{(h-1)z}{h(1-z)^2} \right),$$

we have

$$\frac{L_\mu^*(\varrho+1, \omega, \wp)f(z)}{L_\mu^*(\varrho, \omega, \wp)f(z)} \prec \frac{h(1-z)}{h-z}.$$

Proof. From (10) and the assumption

$$L_\mu^*(\varrho, \omega, \wp)f(z)/z \neq 0$$

in $\mathbb{D}^*$, we infer that $L_\mu^*(\varrho+1, \omega, \wp)f(z)/z \neq 0$ in $\mathbb{D}^*$. Define

$$\mathfrak{p}(z) = \frac{L_\mu^*(\varrho+1, \omega, \wp)f(z)}{L_\mu^*(\varrho, \omega, \wp)f(z)}.$$

Then, $\mathfrak{p}(z)$ is regular in $\mathbb{D}^*$ and

$$\frac{z\mathfrak{p}'(z)}{\mathfrak{p}(z)} = \frac{z\left(L_\mu^*(\varrho+1, \omega, \wp)f(z) \right)'}{L_\mu^*(\varrho+1, \omega, \wp)f(z)} - \frac{z\left(L_\mu^*(\varrho, \omega, \wp)f(z) \right)'}{L_\mu^*(\varrho, \omega, \wp)f(z)}. \tag{11}$$

By virtue of the identity

$$z\left(L_\mu^*(\varrho, \omega, \wp)f(z) \right)' = \varrho\left(L_\mu^*(\varrho+1, \omega, \wp)f(z) \right) - (\varrho+1)L_\mu^*(\varrho, \omega, \wp)f(z)$$

and (11), we get

$$(\varrho+1)\frac{L_\mu^*(\varrho+2, \omega, \wp)f(z)}{L_\mu^*(\varrho+1, \omega, \wp)f(z)} = 1 + \varrho\mathfrak{p}(z) + \frac{z\mathfrak{p}'(z)}{\mathfrak{p}(z)}. \tag{12}$$

Now, (12) together with (10) imply

$$\varrho(\mathfrak{p}(z))^{\zeta+1} + z\mathfrak{p}'(z)(\mathfrak{p}(z))^{\zeta-1} \prec \Lambda(z). \tag{13}$$

Let
$$\mathfrak{q}(z) := \frac{h(1-z)}{h-z}.$$

Then, $\mathfrak{q}$ is, clearly, convex in $\mathbb{D}^*$ and
$$\Lambda(z) = \varrho(\mathfrak{q}(z))^{\zeta+1} + z\mathfrak{q}'(z)(\mathfrak{q}(z))^{\zeta-1}.$$

Let
$$\Theta(\varpi) = \varrho\varpi^{\zeta+1} \text{ and } \Phi(\varpi) = \varpi^{\zeta-1}.$$

Then, (13) may be written in the form of (9). Denoting $z\mathfrak{q}'(z)\Phi(\mathfrak{q}(z))$ by $Y(z)$ yields
$$Y(z) = \frac{(1-h)zh^{\zeta}(1-z)^{\zeta-1}}{(h-z)^{1+\zeta}},$$

and
$$\Lambda(z) = \Theta(\mathfrak{q}(z)) + Y(z) = \left(\frac{h(1-z)}{h-z}\right)^{1+\zeta}\left(\varrho - \frac{(h-1)z}{h(1-z)^2}\right).$$

But, $h > 1$ and $|\zeta| \leq 1$. Hence,
$$\Re\frac{zY'(z)}{Y(z)} = \Re\left(1 + \frac{z(1-\zeta)}{1-z} + (1-\zeta)\frac{z}{h-z}\right)$$
$$> -1 + \frac{1}{2}(1-\zeta) + \frac{(1+\zeta)h}{1+h}$$
$$= \frac{(1+\zeta)(h-1)}{2(1+h)} > 0.$$

Consequently, $Y(z)$ is starlike. Moreover,
$$\Re\frac{z\Lambda'(z)}{Y(z)} = \varrho(1+\zeta)\Re\frac{h(1-z)}{h-z} + \Re\frac{zY'(z)}{Y(z)} \geq 0.$$

By employing Lemma 1, we gain $\mathfrak{p}(z) \prec \mathfrak{q}(z)$ that is
$$\frac{L_\mu^\varkappa(\varrho+1,\omega,\wp)f(z)}{L_\mu^\varkappa(\varrho,\omega,\wp)f(z)} \prec \frac{h(1-z)}{h-z}.$$

The proof is completed. □

A special case of Theorem 1 is when $\varkappa = \zeta = 0$ and $\varrho = \omega = \wp = 1$, where we get

Corollary 1. *If $h > 1$ and $f \in \Sigma$ attains $f(z)/z \neq 0$ in $\mathbb{D}^*$ and*
$$1 + \frac{zf''(z)}{f'(z)} \prec \frac{h(1-z)}{h-z} - \frac{(h-1)z}{(h-z)(1-z)},$$

then
$$\frac{zf'(z)}{f(z)} \prec \frac{h(1-z)}{h-z}.$$

Remark 1. *When $z \in \mathbb{R}$,*
$$\Lambda(z) = \frac{h(1-z)}{h-z} - \frac{(h-1)z}{(h-z)(1-z)} = \frac{z}{h-z} + \frac{1}{1-z} \in \mathbb{R}.$$

Moreover, $\Lambda(0) = 1$ and $\Lambda(\mathbb{D}) = \Re\Lambda(z) < \frac{(h+1)}{2(h-1)}$ for $1 < h \leq 2$ and $\Re\Lambda(z) < \frac{(5h-1)}{2(h+1)}$ for $2 < h$. Hence, this outcome is a generalization of the outcome obtained in [30].

Note that, when

$$\Lambda(z) = 1 - \frac{(h-1)z}{h(1-z)^2},$$

$$\Lambda(\mathbb{D}) = \mathbb{C} - \left[\frac{5h-1}{4h}, \infty\right].$$

Thus, setting $\varkappa = 0$, $\zeta = -1$ and $\varrho = \omega = \wp = 1$ in the Theorem 1 implies the following outcome:

Corollary 2. *Let $h > 1$ and $f \in \Sigma$ satisfy $f(z)/z \neq 0$ in $\mathbb{D}^*$ and*

$$\Re\left(\frac{1 + \frac{zf''(z)}{f'(z)}}{\frac{zf'(z)}{f(z)}}\right) < \frac{5h-1}{4h}.$$

Then,

$$\frac{zf'(z)}{f(z)} \prec \frac{h(1-z)}{h-z}.$$

Theorem 2. *Let $\varrho > 0$, $-1 \leq \zeta < 0$, $-1 \leq E < 1$ and $f \in \Sigma$. If $L_\mu^\varkappa(\varrho, \omega, \wp)f(z)/z \neq 0$ in $\mathbb{D}^*$ and*

$$\left(\frac{L_\mu^\varkappa(\varrho, \omega, \wp)f(z)}{z}\right)^\zeta \left(\varrho\frac{L_\mu^\varkappa(\varrho+1, \omega, \wp)f(z)}{z}\right) \prec \Lambda(z), \tag{14}$$

for

$$\Lambda(z) = \left(\frac{1+Ez}{1-z}\right)^\zeta \left(\varrho\frac{1+Ez}{1-z} + \frac{(1+E)z}{(1-z)^2}\right),$$

then

$$\frac{L_\mu^\varkappa(\varrho, \omega, \wp)f(z)}{z} \prec \frac{1+Ez}{1-z}.$$

Proof. Let

$$\mathfrak{p}(z) = \frac{L_\mu^\varkappa(\varrho, \omega, \wp)f(z)}{z}. \tag{15}$$

Then, clearly $\mathfrak{p}$ is regular in $\mathbb{D}^*$. Then, it follows by (9), that

$$\varrho\left(L_\mu^\varkappa(\varrho+1, \omega, \wp)f(z)\right)' = z\mathfrak{p}'(z) - (\varrho-1)\mathfrak{p}(z). \tag{16}$$

Thus, (14) becomes

$$\varrho\mathfrak{p}(z)^{1+\zeta} + \mathfrak{p}(z)^\lambda z\mathfrak{p}'(z) \prec \Lambda(z).$$

Now, we define $q(z)$ by

$$q(z) = \frac{1+Ez}{1-z}.$$

Then, $q(z)$ is univalent in $\mathbb{D}$ and $q(\mathbb{D}) = \{z : \Re q(z) > (1-E)/2\}$. Let Θ and Φ be

$$\Theta(\omega) = \varrho\omega^{\zeta+1} \quad \text{and} \quad \Phi(\omega) = \omega^{\zeta}.$$

After that, Φ and Θ are regular in $\mathbb{C}\setminus\{0\}$, and (14) has the form of (9). Moreover, by letting

$$Y(z) = zq'(z)\Phi(q(z)) = \frac{(1+E)z(1+Ez)^{\zeta}}{(1-z)^{2+\zeta}},$$

we have

$$\Lambda(z) = f(q(z)) + Y(z).$$

Next, by the assumptions of the theorem,

$$\Re \frac{z\Lambda'(z)}{Y(z)} = \Re\left[1 + \zeta\frac{Ez}{1+Ez} + (2+\zeta)\frac{z}{1-z}\right]$$

$$> 1 - \frac{\zeta|E|}{1+|E|} - \frac{2+\lambda}{2} = \frac{-\zeta(1-|E|)}{2(1+|E|)} > 0,$$

and

$$\Re \frac{z\Lambda'(z)}{Y(z)} = \Re\left[\frac{f'(q(z))}{f(q(z))} + \frac{zY'(z)}{Y(z)}\right] = \varrho(1+\zeta) + \Re\frac{zY'(z)}{Y(z)} \geq 0.$$

An application of Lemma 1 now yields the result. □

Since the function $\Lambda(z) = \varrho + \frac{1+Ez}{(1-z)(1+Ez)}$ maps real values to real values, $\Lambda(0) = \varrho$, $\Lambda(\mathbb{D})$ is symmetric with respect to the real axis and

$$\Re\Lambda(z) > \varrho + \frac{1}{2} - \frac{1}{1-|E|}, \quad z \in \mathbb{D}^*,$$

we may apply Theorem 2 by letting $\zeta = -1$ to get the following.

Corollary 3. *Let* $-1 < E < 1$, $\varrho > 0$ *and* $f \in \Sigma$. *If* $L_a^t(\nu,\tau)f(z)/z \neq 0$ *in* $\mathbb{D}^*$ *and*

$$\Re\left(\frac{L_\mu^*(\varrho+1,\omega,\wp)f(z)}{L_\mu^*(\varrho,\omega,\wp)f(z)}\right) > 1 + \frac{1}{2\varrho} - \frac{1}{\varrho(1-|E|)},$$

then

$$\frac{L_\mu^*(\varrho,\omega,\wp)f(z)}{z} \prec \frac{1+Ez}{1-z}.$$

Theorem 3. *Let* $\zeta \geq -1$, $h > 1$ *and* $f \in \Sigma$. *If* $L_\mu^*(\varrho,\omega,\wp)f(z)/z \neq 0$ *in* $\mathbb{D}^*$ *and*

$$\left(\frac{L_\mu^*(\varrho,\omega,\wp)f(z)}{z}\right)^{\zeta}\left(\varrho\frac{L_\mu^*(\varrho+1,\omega,\wp)f(z)}{z}\right) \prec \frac{h^{1+\zeta}(1-z)^{\zeta}}{(h-z)^{1+\zeta}}\left(\varrho(1-z) - \frac{h(1-z)}{h-z}\right),$$

then
$$\frac{L_\mu^*(\varrho,\omega,\wp)f(z)}{z} \prec \frac{h(1-z)}{h-z}.$$

Proof. The outcome yields from Lemma 1 by defining the functions Φ and Θ by $\Theta(\varpi) = \varrho\varpi^{-(1+\zeta)}$ and $\Phi(\varpi) = -\varpi^{-(2+\zeta)}$. □

Observe that $\Re\left(1 - \frac{(h-1)z}{(h-z)(1-z)}\right) < \frac{3h-1}{2(h-1)}$ when $z \in \mathbb{D}^*$. Hence, when letting $\varrho = \omega = \wp = 1$ and $\varkappa = 0$ in the above Theorem, we get the following.

Corollary 4. Let $h > 1$ and $f \in \Sigma$. If $f'(z) \neq 0$ in $\mathbb{D}$ and
$$\Re\left(1 + \frac{zf''(z)}{f'(z)}\right) < \frac{3h-1}{2(h-1)},$$
then
$$f'(z) \prec \frac{h(1-z)}{h-z}.$$

4. Conclusions

In this analytic investigation, based on convolution concept, we have defined and applied prosperously a complex linear operator which is associated with the meromorphic Hurwitz–Lerch Zeta type functions and Kummer functions. By utilizing this new linear operator, we have discussed several interesting merits of some new geometric subclasses of meromorphicy univalent functions in the punctured unit disk $\mathbb{D}^*$.

Author Contributions: F.G. writing the original draft; K.A.-S., M.D. and H.F.A.-J. writing review and editing. All authors have read and agreed to the published version of the manuscript.

Funding: This research was funded by Nizwa College of Technology (NCT), Oman, Grant No. BFP/RGP/CBS/18/054-Oman-TRC.

Acknowledgments: The third author is supported by MOHE grant: FRGS/1/2019/STG06/UKM/01/1.

Conflicts of Interest: The authors declare no conflict of interest.

References

1. Duren, P.L. *Univalent Functions*; Springer: New York, NY, USA, 1983.
2. Branges, L.D. A proof of the Bieberbach conjecture. *Acta Math.* **1985**, *154*, 137–152. [CrossRef]
3. Ghanim, F.; Al-Janaby, H.F. An analytical study on Mittag-Leffler-confluent hypergeometric functions with fractional integral operator. *Math. Meth. Appl. Sci.* **2020**, 1–10. [CrossRef]
4. Noor, K.L. Integral operators defined by convolution with hypergeometric functions. *Appl. Math. Comput.* **2006**, *182*, 1872–1881. [CrossRef]
5. El-Ashwah, R.M.; Hassan, A.H. Third-order differential subordination and superordination results by using Fox-Wright generalized hypergeometric function. *Funct. Anal. Theory Meth. Appl.* **2016**, *2*, 34–51.
6. Rassias, M.T.; Yang, B. On an Equivalent Property of a Reverse Hilbert-Type Integral Inequality Related to the Extended Hurwitz-Zeta Function. *J. Math. Inequal.* **2019**, *13*, 315–334. [CrossRef]
7. Rassias, M.T.; Yang, B. On a Hilbert-type integral inequality related to the extended Hurwitz zeta function in the whole plane. *Acta Appl. Math.* **2019**, *160*, 67–80. [CrossRef]
8. Ghanim, F.; Al-Janaby, H.F. A Certain Subclass of Univalent Meromorphic Functions Defined by a Linear Operator Associated with the Hurwitz–Lerch Zeta Function. *Rad Hrvat. Akad. Znan. Umjet. Mat. Znan.* **2019**, *23*, 71–83. [CrossRef]
9. Ghanim, F.; Al-Janaby, H.F. Inclusion and Convolution Features of Schlicht Meromorphic Functions Correlating with Mittag-Leffler Function. *Filomat* **2020**, *34*, 2141–2150.
10. Al-Janaby, H.F.; Ghanim, F.; Darus, M. Some Geometric Properties of Integral Operators Proposed by Hurwitz–Lerch Zeta Function. *IOP Conf. Ser. J. Phys. Conf. Ser.* **2019**, *1212*, 012010.
11. Al-Janaby, H.F.; Ghanim, F.; Agawal, P. Geometric studies on inequalities of harmonic functions in a complex field based on ζ-Generalized Hurwitz–Lerch zeta function. *Iran. J. Math. Sci. Inf.* **2020**, accepted.

12. Cuyt, A.; Petersen, V.B.; Verdonk, B.; Waadeland, H.; Jones, W.B. *Handbook of Continued Fractions for Special Functions*; Springer Science and Business Media: Berlin, Germany, 2008.
13. Laurinčikas, A.; Garunktis, R. *The Lerch Zeta-Function*; Kluwer: Dordrecht, The Netherlands; Boston, MA, USA; London, UK, 2002.
14. Erd'elyi, A.; Magnus, W.; Oberhettinger, F.; Tricomi, F.G. *Tables of Integral Transforms*; McGraw-Hill Book Company: New York, NY, USA; Toronto, ON, Canada; London, UK, 1954; Volume II.
15. Goyal, S.P.; Laddha, R.K. On the generalized Zeta function and the generalized Lambert function. *Ganita Sandesh* **1997**, *11*, 99–108.
16. Srivastava, H.M. Some formulas for the Bernoulli and Euler polynomials at rational arguments. *Math. Proc. Camb. Philos. Soc.* **2000**, *129*, 77–84. [CrossRef]
17. Srivastava, H.M.; Choi, J. *Series Associated with Zeta and Related Functions*; Kluwer Academic Publishers: Dordrecht, The Netherlands; Boston, MA, USA; London, UK, 2001.
18. Srivastava, H.M. Some generalizations and basic (or q-) extensions of the Bernoulli, Euler and Genocchi polynomials. *Appl. Math. Inform. Sci.* **2011**, *5*, 390–444.
19. Srivastava, H.M.; Choi, J. *Zeta and q-Zeta Functions and Associated Series and Integrals*; Elsevier Science Publishers: Amsterdam, The Netherlands, 2012.
20. Ghanim, F. A study of a certain subclass of Hurwitz–Lerch zeta function related to a linear operator. *Abstr. Appl. Anal.* **2013**, *2013*, 763756. [CrossRef]
21. Ghanim, F.; Darus, M. New result of analytic functions related to Hurwitz-Zeta function. *Sci. World J.* **2013**, *2013*, 475643. [CrossRef]
22. Srivastava, H.M.; Gaboury, S.; Ghanim, F. Certain subclasses of meromorphically univalent functions defined by a linear operator associated with the ρ-generalized Hurwitz–Lerch zeta function. *Integral Transform. Spec. Funct.* **2015**, *26*, 258–272. [CrossRef]
23. Srivastava, H.M.; Gaboury, S.; Ghanim, F. Some further properties of a linear operator associated with the ρ-generalized Hurwitz–Lerch zeta function related to the class of meromorphically univalent functions. *Appl. Math. Comput.* **2015**, *259*, 1019–1029.
24. Ghanim, F. Certain Properties of Classes of Meromorphic Functions Defined by A Linear Operator and Associated with Hurwitz–Lerch Zeta Function. *Adv. Stud. Contemp. Math.* **2017**, *27*, 175–180.
25. Xing, S.C.; Jose, L.L. A note on the asymptotic expansion of the Lerch's transcendent. *Integr. Trans. Spec. Funct.* **2019**, *30*, 844–855.
26. Al-Janaby, H.F.; Ghanim, F.; Darus, M. On The Third-Order Complex Differential Inequalities of ξ-Generalized-Hurwitz—Lerch Zeta Functions. *Mathematics* **2020**, *8*, 845. [CrossRef]
27. Yuan, S.-M.; Liu, Z.-M.; Srivastava, H.M. Some inclusion relationships and integral-preserving properties of certain subclasses of meromorphic functions associated with a family of integral operators. *J. Math. Anal. Appl.* **2008**, *337*, 505–515. [CrossRef]
28. Wang, Z.-G.; Srivastava, H.M.; Yuan, S.-M. Some basic properties of certain subclasses of meromorphically starlike functions. *J. Inequal. Appl.* **2014**, *2014*, 29. [CrossRef]
29. Miller, S.S.; Mocanu, P.T. *Differential Subordinations: Theory and Applications*; Series in Pure and Applied Mathematics, 225; Marcel Dekker: New York, NY, USA, 2000.
30. Owa, S.; Nunokawa, M.; Saitoh, H.; Srivastava, H.M. Close-to-convexity, starlikeness, and convexity of certain analytic functions. *Appl. Math. Lett.* **2002**, *15*, 63–69. [CrossRef]

Article

On the Connection Problem for Painlevé Differential Equation in View of Geometric Function Theory

Rabha W. Ibrahim [1,2,†], Rafida M. Elobaid [3,*,†] and Suzan J. Obaiys [4,†]

1. Informetrics Research Group, Ton Duc Thang University, Ho Chi Minh City 758307, Vietnam; rabhaibrahim@tdtu.edu.vn
2. Faculty of Mathematics & Statistics, Ton Duc Thang University, Ho Chi Minh City 758307, Vietnam
3. Department of General Sciences, Prince Sultan University, Riyadh 12435, Saudi Arabia
4. School of Mathematical and Computer Sciences, Heriot-Watt University Malaysia, Putrajaya 62200, Malaysia; s.obaiys@hw.ac.uk
* Correspondence: robaid@psu.edu.sa
† These authors contributed equally to this work.

Received: 4 June 2020; Accepted: 15 July 2020; Published: 21 July 2020

Abstract: Asymptotic analysis is a branch of mathematical analysis that describes the limiting behavior of the function. This behavior appears when we study the solution of differential equations analytically. The recent work deals with a special class of third type of Painlevé differential equation (PV). Our aim is to find asymptotic, symmetric univalent solution of this class in a symmetric domain with respect to the real axis. As a result that the most important problem in the asymptotic expansion is the connections bound (coefficients bound), we introduce a study of this problem.

Keywords: Painlevé differential equation; symmetric solution; asymptotic expansion; univalent function; subordination and superordination; analytic function; open unit disk

1. Introduction

The advantage of the Painlevé differential equation (PV) is widely recognized in mathematics and mathematical physics, subsequently the outcomes indicate a part of the nonlinear explanation of special functions. Successively, various studies for the PVs have been offered from various points of vision, such as traditional outcomes, asymptotic, geometric or algebraic constructions. Asymptotic solution of PV-III is investigated extensively because of its requests in material sciences (see [1]). Shimomura [2] presented an asymptotic expansion formal by iteration, and showed the convergence utilizing a concept of majorant series. Kajiwara and Masuda [3] created the asymptotic expansion solution of PV-III by using an expression for the rational solutions whose entries are the Laguerre polynomials. Later, they extended the PV-III into the q-calculus and created the asymptotic expansion solutions by employing the symmetric affinity Weyl group [4]. Gu et al. studied the meromorphic results of PV-III by employing a technique of complex numbers [5]. Bothner et al. occupied the Bäcklund transformation of PV-III [6]. Fasondini et al. investigated the PV-III in a complex domain [7]. Bonelli et al. presented a generalization of PV-III by utilizing q-deformed calculus [8]. Amster and Rogers examined A Neumann-type boundary value problem for a hybrid PV-III. They established the existence properties of approximate outcomes [9]. Recently, Hong and Tu delivered meromorphic results for several types of q-difference PV-III [10]. Bilman et al. planned the fundamental rogue wave solutions of PV-III [11]. Newly, Zeng and Hu [12] suggested the connection problem of the second nonlinear differential equation involving a type of PV and they considered the asymptotic expansion solution.

In this work, we investigate a special class of generalized PV-III equations in a complex domain. We study the asymptotic expansion solution, univalent solution and approximate solution of this class in view of the geometric function theory. We formulate the PV-III as a boundary value problem in terms of the connection estimates. The consequences here are univalent solution with geometric illustration. The novelty of this work is to study a class of the PV equations analytically. The outcomes are based on the geometric function theory to describe the geometric behavior of these solutions. The upper bound of these solutions is indicated by using Janowski formula. Finally, we construct the symmetric solution by using a convex function in the open unit disk.

2. Methodology

The complex PV-III equation can be formulated by the following structure:

$$\zeta \chi(\zeta) \frac{d^2 \chi(\zeta)}{d\zeta^2} = \zeta \left(\frac{d\chi(\zeta)}{d\zeta} \right)^2 - \chi(\zeta) \frac{d\chi(\zeta)}{d\zeta} + \delta \zeta + \beta \chi(\zeta) + \alpha \chi^3(\zeta) + \gamma \zeta \chi^4(\zeta), \tag{1}$$

where α, β, γ and δ are real constants. Kitaev [13] introduced the following special PV-III equation (see Equation (19), p. 83)

$$\zeta \frac{d^2 \chi(\zeta)}{d\zeta^2} + \frac{d\chi(\zeta)}{d\zeta} = \sin(\chi(\zeta)). \tag{2}$$

Asymptotically, Equation (2) becomes

$$\zeta \frac{d^2 \chi(\zeta)}{d\zeta^2} + \frac{d\chi(\zeta)}{d\zeta} \approx \chi(\zeta), \tag{3}$$

subjected to the boundary condition

$$\left(\chi(\zeta) = \zeta + \chi_2 \zeta^2 + O(\zeta^3), \ |\chi_n| \leq 1, n \geq 2, \zeta \in \cup = \{\zeta \in \mathbb{C} : |\zeta| < 1\} \right), \tag{4}$$

where χ_n are indicated the coefficients of the expansion of $\chi(\zeta)$. We are able to investigate the connection problem (coefficient bounds) of Equation (3) by studying the conforming connection problem of geometric classes in the open unit disk ($\cup$). Our exploration method is selected from the GFT, specific the concept of subordination.

Let $\wedge$ be the family of analytic functions $\chi \in \cup$ and normalized by the conditions $\chi(0) = 0$ and $\chi'(0) = 1$, formulating by

$$\chi(\zeta) = \zeta + \sum_{n=2}^{\infty} \chi_n \zeta^n, \quad \zeta \in \cup. \tag{5}$$

A sub-class of $\wedge$ is the class of univalent functions. Consequently, a function $\chi \in \wedge$ is starlike in $\cup$ if and only if $\Re(\zeta \chi'(\zeta)/\chi(\zeta)) > 0$. In addition, a function $\chi \in \wedge$ is convex in $\cup$ if and only if $1 + \Re(\zeta \chi''(\zeta)/\chi'(\zeta)) > 0$.

It is clear that for functions $\chi \in \wedge$, we have $\sin(\chi) \in \wedge$. For example, the following asymptotic expansions for given functions in $\wedge$ (see Figure 1)

$$\sin \left(\frac{\zeta}{1-\zeta} \right) = \zeta + \zeta^2 + (5\zeta^3)/6 + \zeta^4/2 + \zeta^5/120 - (5\zeta^6)/8 + O(\zeta^7)$$

and

$$\sin \left(\frac{\zeta}{(1-\zeta)^2} \right) = \zeta + 2\zeta^2 + (17\zeta^3)/6 + 3\zeta^4 + (181\zeta^5)/120 - (13\zeta^6)/4 + O(\zeta^7).$$

Definition 1. *For two functions χ and $\mathfrak{X}$ in $\wedge$ are subordinated $\chi \prec \mathfrak{X}$, if a Schwarz function ς with $\varsigma(0) = 0$ and $|\varsigma(\zeta)| < 1$ satisfying $\chi(\zeta) = \mathfrak{X}(\varsigma(\zeta))$, $\zeta \in \cup$ (see [14]). Evidently, $\chi(\zeta) \prec \mathfrak{X}(\zeta)$ equivalents to $\chi(0) = \mathfrak{X}(0)$ and $\chi(\cup) \subset \mathfrak{X}(\cup)$.*

Now, rearrange Equation (3), we have the formal

$$\left(\frac{\zeta \chi'(\zeta)}{\chi(\zeta)}\right)\left(1+\frac{\zeta \chi''(\zeta)}{\chi'(\zeta)}\right)=\rho(\zeta), \quad \zeta \in \cup, \tag{6}$$

subjected to the boundary conditions (4), where $\rho(\zeta) = 1 + \rho_1\zeta + \rho_2\zeta^2 + \dots$.

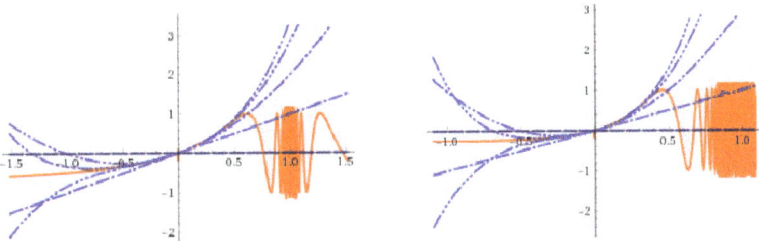

Figure 1. The asymptotic expansions of $\sin\left(\frac{\zeta}{1-\zeta}\right)$ and $\sin\left(\frac{\zeta}{(1-\zeta)^2}\right)$, respectively.

Definition 2. *For a function $\chi \in \wedge$, it is said to be in the class $\mathbf{V}(\rho)$ if and only if*

$$\left(\frac{\zeta \chi'(\zeta)}{\chi(\zeta)}\right)\left(1+\frac{\zeta \chi''(\zeta)}{\chi'(\zeta)}\right) \prec \rho(\zeta), \quad \zeta \in \cup, \tag{7}$$

where $\rho(\zeta) = 1 + \rho_1\zeta + \rho_2\zeta^2 + \rho_3\zeta^3 + \dots$ is convex in $\wedge$ and positive real part with $\rho'(0) > 0, \rho(0) = 1$ (we denote this class by $\mathcal{P}$).

For example, one can suggest the analytic function

$$\rho(\zeta) = \frac{1+\zeta}{1-\zeta} = 1 + 2\zeta^2 + 2\zeta^3 + \dots.$$

Remark 1. *Ma and Minda [15] formulated different sub-classes of starlike and convex functions for which either of the expressions $\frac{\zeta \chi'(\zeta)}{\chi(\zeta)}$ or $1 + \frac{\zeta \chi''(\zeta)}{\chi'(\zeta)}$ are subordinate to an additional common superordinate function. For this class, they presented an analytic function Θ with positive real part in $\cup$, $\Theta(0) = 1, \Theta'(0) > 0$, and Θ maps $\cup$ onto an area starlike with respect to 1 and are symmetric with respect to the real axis. The class of Ma–Minda starlike functions contains function $\chi \in \wedge$ satisfying the subordination $\frac{\zeta \chi'(\zeta)}{\chi(\zeta)} \prec \Theta(\zeta)$. Likewise, the class of Ma–Minda convex functions involves the function $\chi \in \wedge$ fluffing the subordination*

$$1 + \frac{\zeta \chi''(\zeta)}{\chi'(\zeta)} \prec \Theta(\zeta).$$

Moreover, when $\Theta(\zeta) = \frac{1+\zeta}{1-\zeta}$, we obtain the main starlike and convex classes, respectively. Ali et al. [16] combined the two classes in the class

$$\frac{\zeta \chi'(\zeta)}{\chi(\zeta)} + \frac{\zeta \chi''(\zeta)}{\chi'(\zeta)} \prec \Theta(\zeta).$$

3. Connection Bounds

For functions in the class $\mathbf{V}(\rho)$, the following outcome is found.

Theorem 1. *If the function $\chi \in \mathbf{V}(\rho)$ is formulated by (5), then*

$$|\chi_2| \leq \frac{\rho_1}{3}, \quad |\chi_3| \leq \frac{\rho_2 + \rho_1^2/3}{8}, \tag{8}$$

where $\rho(\zeta) = 1 + \rho_1\zeta + \rho_2\zeta^2 + \rho_3\zeta^3 + ...$ is convex in $\wedge$ and positive real coefficients.

Proof. Let $\chi \in \mathbf{V}(\rho)$ having the expansion

$$\chi(\zeta) = \zeta + \chi_2\zeta^2 + \chi_3\zeta^3 + ..., \quad \zeta \in \cup.$$

Then by the definition of the subordination, there subsists a Schwarz function ς with $\varsigma(0) = 0$ and $|\varsigma(\zeta)| < 1$ satisfying

$$\left(\frac{\zeta\chi'(\zeta)}{\chi(\zeta)}\right)\left(1 + \frac{\zeta\chi''(\zeta)}{\chi'(\zeta)}\right) = \rho(\varsigma(\zeta)), \quad \zeta \in \cup.$$

Furthermore, we assume that $|\varsigma(\zeta)| = |\zeta| < 1$, then in view of Schwarz Lemma, there occurs a complex number τ with $|\tau| = 1$ satisfying $\varsigma(\zeta) = \tau\zeta$. Consequently, we obtain

$$\left(\frac{\zeta\chi'(\zeta)}{\chi(\zeta)}\right)\left(1 + \frac{\zeta\chi''(\zeta)}{\chi'(\zeta)}\right) = \left(1 + \chi_2\zeta + (2\chi_3 - \chi_2^2)\zeta^2 + ...\right)$$

$$\times \left(1 + 2\chi_2\zeta + (6\chi_3 - 4\chi_2^2)\zeta^2 + ...\right)$$

$$= 1 + 3\chi_2\zeta + (8\chi_3 - 3\chi_2^2)\zeta^2 + ...$$

$$= 1 + \rho_1\tau\zeta + \rho_2\tau^2\zeta^2 + ...$$

It follows that

$$|\chi_2| \leq \frac{\rho_1|\tau|}{3} = \frac{\rho_1}{3}$$

and

$$|\chi_3| \leq \frac{\rho_2 + \rho_1^2/3}{8}.$$

□

Example 1.

- Let $\rho(\zeta) = \frac{1+\zeta}{1-\zeta} = 1 + 2\zeta^2 + 2\zeta^3 + ...$ then $|\chi_2| \leq \frac{2}{3}$, $|\chi_3| \leq \frac{\rho_2 + \rho_1^2/3}{8} = 0.416$.
- Let $\rho(\zeta) = \left(\frac{1+\zeta}{1-\zeta}\right)^{0.5} = 1 + \zeta + \zeta^2/2 + \zeta^3/2 + (3\zeta^4)/8 + (3\zeta^5)/8 + O(\zeta^6)...$ then

$$|\chi_2| \leq \frac{1}{3}, \quad |\chi_3| \leq \frac{0.5 + 1/3}{8} = 0.104.$$

We have the following consequence.

Corollary 1. *If the function* $\chi \in \mathbf{V}\left(\left(\frac{1+\zeta}{1-\zeta}\right)^\alpha\right)$, $\alpha \in (0, 1]$ *then*

$$|\chi_n| \leq 1, \quad n \geq 2. \tag{9}$$

4. Geometric Behaviors

In this section, we deal with some geometric behaviors of the boundary value problem (6).

Definition 3. *For a function* $\chi \in \wedge$, *it is said to be in the class* $\mathbf{V}(\zeta + \sqrt{\zeta^2 + 1})$ *if and only if*

$$\left(\frac{\zeta\chi'(\zeta)}{\chi(\zeta)}\right)\left(1 + \frac{\zeta\chi''(\zeta)}{\chi'(\zeta)}\right) \prec \zeta + \sqrt{\zeta^2 + 1}, \quad \zeta \in \cup. \tag{10}$$

Note that (see Figure 2)

$$\zeta + \sqrt{\zeta^2 + 1} = 1 + \zeta + \zeta^2/2 - \zeta^4/8 + O(\zeta^6)$$

and that the sub-classes of starlike and convex of the above definition are studied in [17].

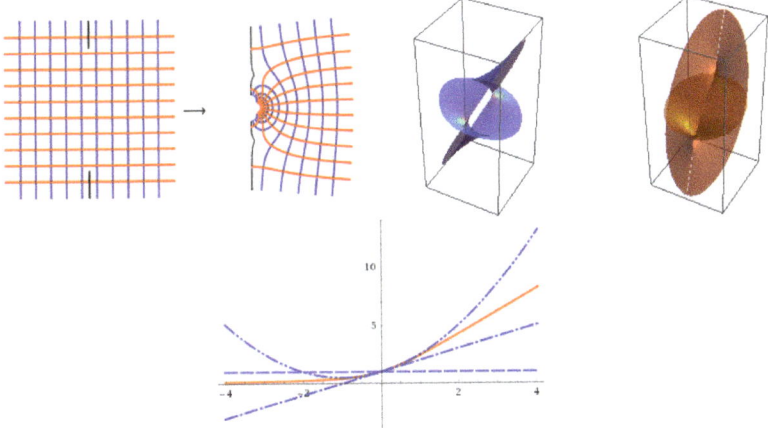

Figure 2. The complex plane, Riemann surface and the asymptotic expansions of $(\zeta + \sqrt{\zeta^2 + 1})$, respectively.

We request the following preliminary, which can be located in [17].

Lemma 1. *If P is analytic in* $\cup$ *and satisfies the subordination*

$$P(\zeta) + \kappa \frac{\zeta P'(\zeta)}{P(\zeta)} \prec (\zeta + \sqrt{\zeta^2 + 1}), \quad \kappa > 0,$$

then $P(\zeta) \prec (\zeta + \sqrt{\zeta^2 + 1})$.

Theorem 2. *If the function* $\chi \in \wedge$ *is formulated by (5) fulfilling the subordination*

$$\left(\frac{\zeta \chi'(\zeta)}{\chi(\zeta)}\right)\left(1 + \frac{\zeta \chi''(\zeta)}{\chi'(\zeta)}\right) + \left(\frac{\zeta\left(1 + \frac{\zeta \chi''(\zeta)}{\chi'(\zeta)}\right)'}{\left(1 + \frac{\zeta \chi''(\zeta)}{\chi'(\zeta)}\right)} + \frac{\zeta\left(\frac{\zeta \chi'(\zeta)}{\chi(\zeta)}\right)'}{\left(\frac{\zeta \chi'(\zeta)}{\chi(\zeta)}\right)}\right) \prec (\zeta + \sqrt{\zeta^2 + 1}) \quad (11)$$

then $\chi \in \mathbf{V}(\zeta + \sqrt{\zeta^2 + 1})$. *Moreover,*

$$|\chi_2| \leq \frac{1}{3}, \quad |\chi_3| \leq \frac{24}{1152}.$$

Proof. Let $\chi \in \wedge$ having the expansion

$$\chi(\zeta) = \zeta + \chi_2 \zeta^2 + \chi_3 \zeta^3 + ..., \quad \zeta \in \cup.$$

Furthermore, we let

$$P(\zeta) := \left(\frac{\zeta \chi'(\zeta)}{\chi(\zeta)}\right)\left(1 + \frac{\zeta \chi''(\zeta)}{\chi'(\zeta)}\right).$$

Thus, in view of Lemma 1 with $\kappa = 1$, we get

$$P(\zeta) + \frac{\zeta P'(\zeta)}{P(\zeta)} = \left(\frac{\zeta \chi'(\zeta)}{\chi(\zeta)}\right)\left(1 + \frac{\zeta \chi''(\zeta)}{\chi'(\zeta)}\right) + \frac{\zeta\left(\left(\frac{\zeta \chi'(\zeta)}{\chi(\zeta)}\right)\left(1 + \frac{\zeta \chi''(\zeta)}{\chi'(\zeta)}\right)\right)'}{\left(\frac{\zeta \chi'(\zeta)}{\chi(\zeta)}\right)\left(1 + \frac{\zeta \chi''(\zeta)}{\chi'(\zeta)}\right)}$$

$$= \left(\frac{\zeta \chi'(\zeta)}{\chi(\zeta)}\right)\left(1 + \frac{\zeta \chi''(\zeta)}{\chi'(\zeta)}\right) + \left(\frac{\zeta\left(1 + \frac{\zeta \chi''(\zeta)}{\chi'(\zeta)}\right)'}{\left(1 + \frac{\zeta \chi''(\zeta)}{\chi'(\zeta)}\right)} + \frac{\zeta\left(\frac{\zeta \chi'(\zeta)}{\chi(\zeta)}\right)'}{\left(\frac{\zeta \chi'(\zeta)}{\chi(\zeta)}\right)}\right)$$

$$\prec (\zeta + \sqrt{\zeta^2 + 1})$$

It follows that $P(\zeta) \prec (\zeta + \sqrt{\zeta^2 + 1})$, which implies that $\chi \in V(\zeta + \sqrt{\zeta^2 + 1})$. Now, a computation implies that

$$\left(\frac{\zeta \chi'(\zeta)}{\chi(\zeta)}\right)\left(1 + \frac{\zeta \chi''(\zeta)}{\chi'(\zeta)}\right) = 1 + 3\chi_2 \zeta + (8\chi_3 - 3\chi_2^2)\zeta^2 + \dots$$

$$= 1 + \zeta + \zeta^2/2 - \zeta^4/8 + O(\zeta^6).$$

A comparison yields that

$$|\chi_2| \leq \frac{1}{3}, \quad |\chi_3| \leq \frac{24}{1152}.$$

□

Next, result can be found in [18].

Lemma 2. *For analytic functions $w, \varpi \in U$, the subordination $w \prec \varpi$ implies that*

$$\int_0^{2\pi} |w(\zeta)|^q d\theta \leq \int_0^{2\pi} |\varpi(\zeta)|^q d\theta,$$

where $\zeta = re^{i\theta}, 0, r < 1$ and q is a positive number.

Theorem 3. *If the function $\chi \in \wedge$ is formulated by (5) achieving the subordination inequality (11). Then*

$$\int_0^{2\pi} \left|\left(\frac{\zeta \chi'(\zeta)}{\chi(\zeta)}\right)\left(1 + \frac{\zeta \chi''(\zeta)}{\chi'(\zeta)}\right)\right|^q d\theta \leq 2\pi \quad (12)$$

for all $q \geq 1$ and $\zeta = re^{i\theta} \in U$ and $0 < r < 1$.

Proof. According to Theorem 2, we have

$$\left(\frac{\zeta \chi'(\zeta)}{\chi(\zeta)}\right)\left(1 + \frac{\zeta \chi''(\zeta)}{\chi'(\zeta)}\right) \prec \zeta + \sqrt{\zeta^2 + 1}.$$

Then in view of Lemma 2, we conclude that

$$\int_0^{2\pi} \left|\left(\frac{\zeta \chi'(\zeta)}{\chi(\zeta)}\right)\left(1 + \frac{\zeta \chi''(\zeta)}{\chi'(\zeta)}\right)\right|^q d\theta \leq \int_0^{2\pi} |\zeta + \sqrt{\zeta^2 + 1}|^q d\theta$$

$$= \int_0^{2\pi} |e^{i\theta} + \sqrt{(e^{i\theta})^2 + 1}|^q d\theta, \quad r \to 1$$

$$= 2\pi.$$

This completes the proof. □

Theorem 3 indicates the periodicity of solutions of the boundary value problem (6). We illustrate the following example (see Figure 3):

Example 2.

- Let $q = 1$, we have

$$\int |e^{i\theta} + \sqrt{(e^{i\theta})^2 + 1}| d\theta = -i(e^{i\theta} + \sqrt{1 + e^{2i\theta}} - \tanh^{-1}(\sqrt{1 + e^{2i\theta}})) + constant$$

- Let $q = 2$ then we get

$$\int |e^{i\theta} + \sqrt{(e^{i\theta})^2 + 1}|^2 d\theta = \theta - ie^{2i\theta} - ie^{i\theta}\sqrt{1 + e^{2i\theta}} - i\sinh^{-1}(e^{i\theta}) + constant.$$

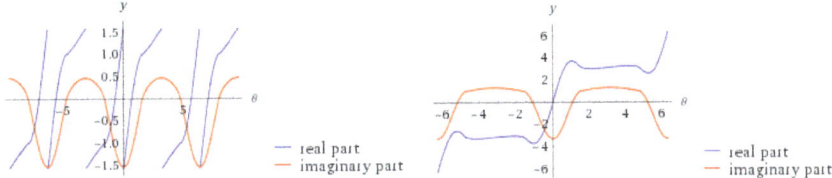

Figure 3. Periodic solution of Equation (6), when $q = 1$ and $q = 2$, respectively.

We proceed to study some geometric behaviors of Equation (6). We need the following concept.

Definition 4. *A majorization of two analytic functions having the asymptotic expansions respectively, $\chi(\zeta) = \sum_{m=0}^{\infty} \chi_m \zeta^m$ and $\mathfrak{X}(\zeta) = \sum_{m=0}^{\infty} \mathfrak{X}_m \zeta^m$ is denoted by $\chi \ll \mathfrak{X}$ and satisfies the connections bounds $|\chi_m| \leq |\mathfrak{X}_m|$, for all m.*

Definition 5. *For a function $\chi \in \wedge$, it is said to be in the class $\mathbf{V}(\frac{1+\varrho_1\zeta}{1+\varrho_2\zeta})$ if and only if*

$$\left(\frac{\zeta\chi'(\zeta)}{\chi(\zeta)}\right)\left(1 + \frac{\zeta\chi''(\zeta)}{\chi'(\zeta)}\right) \prec \frac{1+\varrho_1\zeta}{1+\varrho_2\zeta}, \tag{13}$$

where $\zeta \in \mathbf{U}$ and $\varrho_1, \varrho_2 \in \partial \mathbf{U}$.

Theorem 4. *Let $\chi \in \mathbf{V}(\frac{1+\varrho_1\zeta}{1+\varrho_2\zeta})$. Then there is a probability measure ν on $(\partial \mathbf{U})^2$.*

Proof. Let $\chi \in \mathbf{V}(\frac{1+\varrho_1\zeta}{1+\varrho_2\zeta})$. This yields that

$$\left(\frac{\zeta\chi'(\zeta)}{\chi(\zeta)}\right)\left(1 + \frac{\zeta\chi''(\zeta)}{\chi'(\zeta)}\right) \prec \frac{1+\varrho_1\zeta}{1+\varrho_2\zeta}.$$

A calculation brings that $|\chi_n| \leq 1$ for all $n \geq 1$. Furthermore,

$$\frac{1+\varrho_1\zeta}{1+\varrho_2\zeta} \ll \frac{1+\zeta}{1-\zeta}. \tag{14}$$

According to Theorem 1.11 in [19], we obtain that the function $\frac{1+\varrho_1\zeta}{1+\varrho_2\zeta}$ indicates a probability measure ν in $(\partial U)^2$ achieving

$$\psi(\zeta) = \int_{(\partial U)^2} \left(\frac{1+\varrho_1\zeta}{1+\varrho_2\zeta}\right) d\nu(\varrho_1,\varrho_2), \quad \zeta \in U.$$

Then there is a diffusion constant A satisfying

$$\int_{(\partial U)^2} \left(\frac{1+\varrho_1\zeta}{1+\varrho_2\zeta}\right) d\nu(\varrho_1,\varrho_2) = A \int_{(\partial U)^2} \left(\frac{\zeta\chi'(\zeta)}{\chi(\zeta)}\right)\left(1+\frac{\zeta\chi''(\zeta)}{\chi'(\zeta)}\right) d\nu(\varrho_1,\varrho_2)$$

$$\Big(\zeta \in U, A \in \mathbb{R}, \chi \in \wedge\Big).$$

□

5. Symmetric Solution

In this section, we introduce a study regarding the symmetric solution of (6). For this purpose, we need to define a symmetric class as follows:

Definition 6. *For a function $\chi \in \wedge$, it is said to be in the symmetric class $\mathbf{V}_{symmetric}(\Phi)$, where Φ takes the formula*

$$\Phi(\zeta) = \frac{1}{2}[\rho(\zeta) + \rho(-\zeta)], \quad \zeta \in U, \rho \in \mathcal{P},$$

where ρ is convex in U if and only if

$$\left(\frac{\zeta\chi'(\zeta)}{\chi(\zeta)}\right)\left(1+\frac{\zeta\chi''(\zeta)}{\chi'(\zeta)}\right) \prec \Phi(\zeta), \quad \zeta \in U. \tag{15}$$

In addition, a function $\chi \in \wedge$, is stated to be in the symmetric class $\mathbf{V}_{symmetric}(\Psi)$, where Ψ is formulated by the symmetric construction

$$\Psi(\zeta) = \frac{4\zeta\rho(\zeta)}{\rho(\zeta) - \rho(-\zeta)}, \quad \zeta \in U, \rho \in \mathcal{P},$$

where ρ is convex in U if and only if

$$\left(\frac{\zeta\chi'(\zeta)}{\chi(\zeta)}\right)\left(1+\frac{\zeta\chi''(\zeta)}{\chi'(\zeta)}\right) \prec \Psi(\zeta), \quad \zeta \in U. \tag{16}$$

To establish the existence of symmetric solution of (6), we request the following result (see Theorem 3.2, p. 97 in [14]).

Lemma 3. *Let Φ be convex in U such that $\Phi(0) = 1$. If ρ is the analytic solution of the equation*

$$\rho(\zeta) + \frac{\zeta\rho'(\zeta)}{\rho(\zeta)} = \Phi(\zeta), \quad \rho(0) = 1$$

and if $\Re(\rho) > 0$, then ρ is univalent solution. If $P \in H[1,n]$ (the class of analytic function) achieves the subordination

$$P(\zeta) + \frac{\zeta P'(\zeta)}{P(\zeta)} \prec \Phi(\zeta),$$

then $P \prec \rho$ and ρ is the best dominant.

Theorem 5. *Let $\chi \in \mathbf{V}_{symmetric}(\Phi)$, where $\rho \in \mathcal{P}$ is convex and the functional*

$$\Phi(\zeta) = \frac{1}{2}[\rho(\zeta) + \rho(-\zeta)], \quad \zeta \in U$$

satisfies the inequality

$$\left(\frac{\zeta\chi'(\zeta)}{\chi(\zeta)}\right)\left(1+\frac{\zeta\chi''(\zeta)}{\chi'(\zeta)}\right)+\left(\frac{\zeta\left(1+\frac{\zeta\chi''(\zeta)}{\chi'(\zeta)}\right)'}{\left(1+\frac{\zeta\chi''(\zeta)}{\chi'(\zeta)}\right)}+\frac{\zeta\left(\frac{\zeta\chi'(\zeta)}{\chi(\zeta)}\right)'}{\left(\frac{\zeta\chi'(\zeta)}{\chi(\zeta)}\right)}\right) \prec \Phi(\zeta). \tag{17}$$

Then $\chi \in \mathbf{V}(\rho)$.

Proof. Our aim is to achieve all the conditions of Lemma 3. Since ρ is convex then Φ is convex in $\mathbb{U}$ such that $\Phi(0) = 1$. Moreover, ρ is the univalent solution of the equation

$$\frac{\zeta \rho'(\zeta)}{\rho(\zeta)} = \frac{1}{2}[\rho(-\zeta) - \rho(\zeta)], \quad \rho(0) = 1$$

with $\Re(\rho) > 0$. Suppose that

$$P(\zeta) := \left(\frac{\zeta\chi'(\zeta)}{\chi(\zeta)}\right)\left(1+\frac{\zeta\chi''(\zeta)}{\chi'(\zeta)}\right).$$

Then, we obtain

$$P(\zeta) + \frac{\zeta P'(\zeta)}{P(\zeta)} = \left(\frac{\zeta\chi'(\zeta)}{\chi(\zeta)}\right)\left(1+\frac{\zeta\chi''(\zeta)}{\chi'(\zeta)}\right) + \frac{\zeta\left(\left(\frac{\zeta\chi'(\zeta)}{\chi(\zeta)}\right)\left(1+\frac{\zeta\chi''(\zeta)}{\chi'(\zeta)}\right)\right)'}{\left(\frac{\zeta\chi'(\zeta)}{\chi(\zeta)}\right)\left(1+\frac{\zeta\chi''(\zeta)}{\chi'(\zeta)}\right)}$$

$$= \left(\frac{\zeta\chi'(\zeta)}{\chi(\zeta)}\right)\left(1+\frac{\zeta\chi''(\zeta)}{\chi'(\zeta)}\right) + \left(\frac{\zeta\left(1+\frac{\zeta\chi''(\zeta)}{\chi'(\zeta)}\right)'}{\left(1+\frac{\zeta\chi''(\zeta)}{\chi'(\zeta)}\right)} + \frac{\zeta\left(\frac{\zeta\chi'(\zeta)}{\chi(\zeta)}\right)'}{\left(\frac{\zeta\chi'(\zeta)}{\chi(\zeta)}\right)}\right)$$

$$\prec \Phi(\zeta).$$

By Lemma 3, we have

$$\left(\frac{\zeta\chi'(\zeta)}{\chi(\zeta)}\right)\left(1+\frac{\zeta\chi''(\zeta)}{\chi'(\zeta)}\right) \prec \rho(\zeta).$$

Hence, $\chi \in \mathbf{V}(\rho)$. □

In the similar manner of Theorem 5, we have the following outcome

Theorem 6. Let $\chi \in \mathbf{V}_{symmetric}(\Psi)$, where $\rho \in \mathcal{P}$ is convex and the functional

$$\Psi(\zeta) = \frac{4\zeta \rho(\zeta)}{\rho(\zeta) - \rho(-\zeta)}, \quad \zeta \in \mathbb{U}$$

satisfies the inequality

$$\left(\frac{\zeta\chi'(\zeta)}{\chi(\zeta)}\right)\left(1+\frac{\zeta\chi''(\zeta)}{\chi'(\zeta)}\right)+\left(\frac{\zeta\left(1+\frac{\zeta\chi''(\zeta)}{\chi'(\zeta)}\right)'}{\left(1+\frac{\zeta\chi''(\zeta)}{\chi'(\zeta)}\right)}+\frac{\zeta\left(\frac{\zeta\chi'(\zeta)}{\chi(\zeta)}\right)'}{\left(\frac{\zeta\chi'(\zeta)}{\chi(\zeta)}\right)}\right) \prec \Psi(\zeta). \tag{18}$$

Then $\chi \in \mathbf{V}(\rho)$.

Example 3. Consider the analytic function $\rho(\zeta) = \frac{1+\zeta}{1-\zeta}$, where it maps $\mathbb{U}$ onto the right half-plane convexly. Then $\Phi(\zeta) = \frac{1+\zeta^2}{1-\zeta^2} = 1 + 2\zeta^2 + 2\zeta^4 + O(\zeta^6)$, where $\Phi(0) = 1$ (see Figure 4). By assuming $\chi(\zeta) = \zeta$, we have the subordination $P(\zeta) = \left(\frac{\zeta\chi'(\zeta)}{\chi(\zeta)}\right)\left(1+\frac{\zeta\chi''(\zeta)}{\chi'(\zeta)}\right) \prec \Phi(\zeta)$. Thus, the solution $\chi \in \mathbf{V}_{symmetric}(\frac{1+\zeta^2}{1-\zeta^2})$.

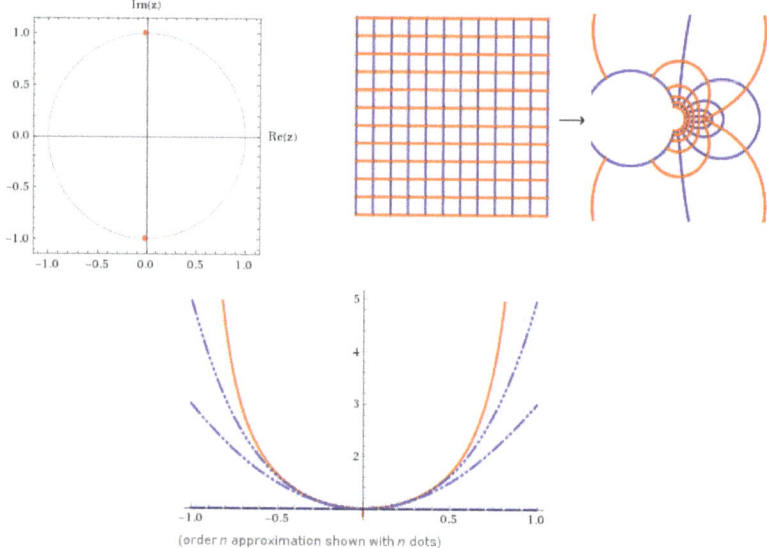

Figure 4. The behavior of $\Phi(\zeta) = \frac{1+\zeta^2}{1-\zeta^2}$ with a symmetric domain for $|\zeta|, 1$.

6. Conclusions

From above, we conclude that the asymptotic behaviors of a special class of Painlevé differential equations (see [13]) can be recognized by using a geometric representation of the equation. From this construction, we introduced the oscillatory, connection bound and other properties of the boundary value problem (6). In addition, Theorem 5 and Theorem 6 indicated that the set $\{\chi : \chi \in \mathbf{V}(\rho)\}$ has symmetric solutions for some symmetric region because $\mathbf{V}_{symmetric}(\Phi) \subset \mathbf{V}(\rho)$ and $\mathbf{V}_{symmetric}(\Psi) \subset \mathbf{V}(\rho)$.

Author Contributions: Conceptualization, R.W.I.; Funding acquisition, R.M.E.; Investigation, R.W.I.; Methodology, R.W.I.; Visualization, R.M.E.; Writing—review & editing, R.M.E. and S.J.O.All authors have read and agreed to the published version of the manuscript.

Funding: This research received no external funding.

Acknowledgments: The authors would like to thank the Editorial Manager for the advising and would like express their sincere appreciation to the reviewers for their very careful review of our paper and rich it in information.

Conflicts of Interest: The authors declare no conflict of interest.

References

1. Dormieux, L.; Ulm, F.J. (Eds.) *Applied Micromechanics of Porous Materials*; Springer Science & Business Media: Berlin/Heidelberg, Germany, 2007; Volume 480.
2. Shimomura, S. A family of solutions of a nonlinear ordinary differential equation and its application to Painlevé equations (III), (V) and (VI). *J. Math. Soc. Jpn.* **1987**, *39*, 649–662. [CrossRef]
3. Kajiwara, K.; Tetsu, M. On the Umemura polynomials for the Painlevé III equation. *Phys. Lett. A* **1999**, *260*, 462–467. [CrossRef]
4. Kajiwara, K.; Kinji, K. On a q-Difference Painlevé III Equation: I. Derivation, Symmetry and Riccati Type Solutions. *J. Nonlinear Math. Phys.* **2003**, *10*, 86–102. [CrossRef]

5. Gu, Y.; Aminakbari, N.; Yuan, W.; Wu, Y.H. Meromorphic solutions of a class of algebraic differential equations related to Painlevé equation III. *Houst. J. Math.* **2017**, *43*, 1045–1055.
6. Bothner, T.; Miller, P.D.; Sheng, Y. Rational solutions of the Painlevé-III equation. *Stud. Appl. Math.* **2018**, *141*, 626–679. [CrossRef]
7. Fasondini, M.; Fornberg, B.; Weideman, J.A.C. A computational exploration of the McCoy-Tracy-Wu solutions of the third Painlevé equation. *Phys. D Nonlinear Phenom.* **2018**, *363*, 18–43. [CrossRef]
8. Bonelli, G.; Grassi, A.; Tanzini, A. Quantum curves and q-deformed Painlevé equations. *Lett. Math. Phys.* **2019**, *109*, 1961–2001. [CrossRef]
9. Amster, P.; Rogers, C. On a Neumann boundary value problem for Ermakov-Painlevé III. *Electron. J. Qual. Theory Differ. Equ.* **2019**, *2019*, 1–10. [CrossRef]
10. Xu, H.Y.; Tu, J. Existence of rational solutions for q-difference Painlevé equations. *Electron. J. Differ. Equ.* **2020**, *2020*, 1–14.
11. Bilman, D.; Ling, L.; Miller, P.D. Extreme superposition: Rogue waves of infinite order and the Painlevé-III hierarchy. *Duke Math. J.* **2020**, *169*, 671–760. [CrossRef]
12. Zeng, Z.Y.; Hu, L. On the connection problem for nonlinear differential equation. *Bound. Value Probl.* **2019**, *2019*, 73. [CrossRef]
13. Kitaev, A.V. Parametric Painlevé equations. *J. Math. Sci.* **2013**, *192*, 81–90. [CrossRef]
14. Miller, S.S.; Mocanu, P.T. *Differential Subordinations: Theory and Applications*; CRC Press: Boca Raton, FL, USA, 2000.
15. Ma, W.; Minda, D. A unified treatment of some special classes of univalent functions. In Proceedings of the Conference on Complex Analysis, Tianjin, China, 19–23 June 1992; pp. 157–169.
16. Ali, R.M.; Lee, S.K.; Ravichandran, V.; Supramaniam, S. Coefficient estimates for bi-univalent Ma-Minda starlike and convex functions. *Appl. Math. Lett.* **2012**, *25*, 344–351. [CrossRef]
17. Raina, R.K.; Sokol, J. Some properties related to a certain class of starlike functions. *Comptes Rendus Math.* **2015**, *353*, 973–978. [CrossRef]
18. Littlewood, J.E. On inequalities in the theory of functions. *Proc. Lond. Math. Soc.* **1925**, *23*, 481–519 [CrossRef]
19. Ruscheweyh, S. *Convolutions in geometric function Theory*; Gaetan Morin Editeur Ltee: Montreal, QC, CA, 1982.

© 2020 by the authors. Licensee MDPI, Basel, Switzerland. This article is an open access article distributed under the terms and conditions of the Creative Commons Attribution (CC BY) license (http://creativecommons.org/licenses/by/4.0/).

Article

Symmetric Conformable Fractional Derivative of Complex Variables

Rabha W. Ibrahim [1,2], Rafida M. Elobaid [3,*] and Suzan J. Obaiys [4]

1. Informetrics Research Group, Ton Duc Thang University, Ho Chi Minh City 758307, Vietnam; rabhaibrahim@tdtu.edu.vn
2. Faculty of Mathematics & Statistics, Ton Duc Thang University, Ho Chi Minh City 758307, Vietnam
3. Department of General Sciences, Prince Sultan University, Riyadh 12435, Saudi Arabia
4. School of Mathematical and Computer Sciences, Heriot-Watt University, Putrajaya 62200, Malaysia; s.obaiys@hw.ac.uk
* Correspondence: relobaid2000@yahoo.com

Received: 19 February 2020; Accepted: 3 March 2020; Published: 6 March 2020

Abstract: It is well known that the conformable and the symmetric differential operators have formulas in terms of the first derivative. In this document, we combine the two definitions to get the symmetric conformable derivative operator (SCDO). The purpose of this effort is to provide a study of SCDO connected with the geometric function theory. These differential operators indicate a generalization of well known differential operator including the Sàlàgean differential operator. Our contribution is to impose two classes of symmetric differential operators in the open unit disk and to describe the further development of these operators by introducing convex linear symmetric operators. In addition, by acting these SCDOs on the class of univalent functions, we display a set of sub-classes of analytic functions having geometric representation, such as starlikeness and convexity properties. Investigations in this direction lead to some applications in the univalent function theory of well known formulas, by defining and studying some sub-classes of analytic functions type Janowski function and convolution structures. Moreover, by using the SCDO, we introduce a generalized class of Briot–Bouquet differential equations to introduce, what is called the symmetric conformable Briot–Bouquet differential equations. We shall show that the upper bound of this class is symmetric in the open unit disk.

Keywords: univalent function; conformable fractional derivative; subordination and superordination; analytic function; open unit disk

MSC: 30C45

---

1. Introduction

The term Symmetry from Greek means arrangement and organization in measurements. In free language, it mentions a concept of harmonious and attractive proportion and equilibrium. In mathematics, it discusses an object that is invariant via certain transformation or rotation or scaling. In geometry, the object has symmetry if there is an operator or transformation that maps the object onto itself [1,2].

Sàlàgean (1983) presented a differential operator for a class of analytic functions (see [3]). Many sub-classes of analytic functions are studied using this operator. Al-Oboudi [4] generalized this operator. These operators are studied widely in the last decade (see [5–10] for recent works). Our investigation is to study classes of analytic functions by using the symmetric differential operator in a complex domain. Recently, Ibrahim and Jahangiri [7] defined a special type of differential operators, which is called a complex conformable differential operator. This operator is an extension of the Anderson–Ulness operator [11].

A conformable calculus (CC) is a branch of the fractional calculus. It develops the term $\chi^{1-\wp}f'(\chi)$. While the complex conformable calculus (CCC) indicates the term $\xi\varphi'(\xi)$, where ξ is a complex variable and φ is a complex valued analytic function. In this work, we present a new SCDO in the open unit disk. We formulate it in some sub-classes of univalent functions. As applications, we generalize a class of Briot–Bouquet differential equations by using SCDO.

2. Methodology

This section deals with the mathematical processing to study the SCDO for some classes of analytic functions in the open unit disk $\cup = \{\xi \in \mathbb{C} : |\xi| < 1\}$. Let $\wedge$ be the following class of analytic functions

$$Y(\xi) = \xi + \sum_{n=2}^{\infty} Y_n \xi^n, \quad \xi \in \cup. \tag{1}$$

A function $Y \in \wedge$ is starlike via the (0,0) (origin in $\cup$) if the linear segment joining the origin to every other point of Y lies entirely in $Y(\xi : |\xi| < 1)$. A univalent function $(Y \in \mathbb{S})$ is convex in $\cup$ if the linear segment joining any two points of $Y(\xi : |\xi| < 1)$ lies entirely in $Y(\xi : |\xi| < 1)$. We denote these classes by $\mathcal{S}^*$ and $\mathcal{C}$ for starlike and convex respectively. In addition, suppose that the class $\mathcal{P}$ involves all functions Y analytic in $\cup$ with a positive real part in $\cup$ achieving $Y(0) = 1$. Mathematically, $Y \in \mathcal{S}^*$ if and only if $\xi Y'(\xi)/Y(\xi) \in \mathcal{P}$ and $Y \in \mathcal{C}$ if and only if $1 + \xi Y''(\xi)/Y'(\xi) \in \mathcal{P}$. Equivalently, $\Re(\xi Y'(\xi)/Y(\xi)) > 0$ for the starlikeness and $1 + \Re(\xi Y''(\xi)/Y'(\xi)) > 0$ for the convexity.

For two functions Y_1 and Y_2 belong to the class $\wedge$, are said to be subordinate, noting by $Y_1 \prec Y_2$, if we can find a Schwarz function τ with $\tau(0) = 0$ and $|\tau(\xi)| < 1$ achieving $Y_1(\xi) = Y_2(\tau(\xi))$, $\xi \in \cup$ (the detail can be located in [12]). Obviously, $Y_1(\xi) \prec Y_2(\xi)$ if $Y_1(0) = Y_2(0)$ and $Y_1(\cup) \subset Y_2(\cup)$.

Lemma 1 ([12]). *Suppose that $a \in \mathbb{C}$, n is a positive integer and $\aleph[a,n] = \{Y : Y(\xi) = a + a_n \xi^n + a_{n+1} \xi^{n+1} + ...\}$ is a set of analytic functions.*

i. If $\ell \in \mathbb{R}$ then $\Re\left(Y(\xi) + \ell \xi Y'(\xi)\right) > 0 \Longrightarrow \Re(Y(\xi)) > 0$. In addition, if $\ell > 0$ and $Y \in \aleph[1,n]$, then there occurs some constants $a > 0$ and $b > 0$ with $b = b(\ell, a, n)$ where

$$Y(\xi) + \ell \xi Y'(\xi) \prec \left(\frac{1+\xi}{1-\xi}\right)^b \Rightarrow Y(\xi) \prec \left(\frac{1+\xi}{1-\xi}\right)^a.$$

ii. If $\eth \in [0,1)$ and $Y \in \aleph[1,n]$ then a constant $k > 0$ exists satisfying $k = k(a,n)$ so that

$$\Re\left(Y^2(\xi) + 2Y(\xi).\xi Y'(\xi)\right) > \eth \Rightarrow \Re(Y(\xi)) > k.$$

iii. If $Y \in \aleph[a,n]$ with $\Re(a) > 0$ then $\Re\left(Y(\xi) + \xi Y'(\xi) + \xi^2 Y''(\xi)\right) > 0$ or for $\imath : \cup \to \mathbb{R}$ with $\Re\left(Y(\xi) + \imath(\xi)\frac{\xi Y'(\xi)}{Y(\xi)}\right) > 0$ then $\Re(Y(\xi)) > 0$.

Lemma 2 ([12]). *Assume that $\hbar$ is a convex function satisfying $\hbar(0) = a$, and let $\Bbbk \in \mathbb{C} \setminus \{0\}$ be a complex number with $\Re(\Bbbk) \geq 0$. If $Y \in \aleph[a,n]$, and*

$$Y(\xi) + (1/\Bbbk)\xi Y'(\xi) \prec \hbar(\xi), \quad \xi \in \cup,$$

then $\Upsilon(\xi) \prec \iota(z) \prec \hbar(z)$, where

$$\iota(z) = \frac{k}{n\xi^{k/n}} \int_0^\xi \hbar(\tau)\tau^{\frac{k}{(n-1)}} d\tau, \quad \xi \in \cup.$$

Lemma 3 ([13]). *Suppose that* $\Upsilon \in \wedge$ *and there occurs a positive constant* $0 < v \leq 1$. *If*

$$\frac{\xi \Upsilon'(\xi) - \xi}{\Upsilon(\xi)} \prec \frac{2v\xi}{1+\xi}$$

then

$$\frac{\Upsilon(\xi)}{\xi} \prec 1 + v\xi, \quad \xi \in \cup.$$

And the result is sharp.

The Operator SCDO

This sections deals with definition of the SCDO as follows:

Definition 1. *Let* $\Upsilon(\xi) \in \wedge$, *and let* $v \in [0,1]$ *be a constant then the SCDO keeps the following operating*

$$S_v^0 \Upsilon(\xi) = \Upsilon(\xi)$$

$$S_v^1 \Upsilon(\xi) = \left(\frac{\kappa_1(v,\xi)}{\kappa_1(v,\xi) + \kappa_0(v,\xi)}\right) \xi \Upsilon'(\xi) - \left(\frac{\kappa_0(v,\xi)}{\kappa_1(v,\xi) + \kappa_0(v,\xi)}\right) \xi \Upsilon'(-\xi)$$

$$= \left(\frac{\kappa_1(v,\xi)}{\kappa_1(v,\xi) + \kappa_0(v,\xi)}\right) \left(\xi + \sum_{n=2}^\infty n \Upsilon_n \xi^n\right) - \left(\frac{\kappa_0(v,\xi)}{\kappa_1(v,\xi) + \kappa_0(v,\xi)}\right) \left(-\xi + \sum_{n=2}^\infty n(-1)^n \Upsilon_n \xi^n\right)$$

$$= \xi + \sum_{n=2}^\infty n \left(\frac{\kappa_1(v,\xi) + (-1)^{n+1}\kappa_0(v,\xi)}{\kappa_1(v,\xi) + \kappa_0(v,\xi)}\right) \Upsilon_n \xi^n$$

$$S_v^2 \Upsilon(\xi) = S_v^1[S_v^1 \Upsilon(\xi)] \tag{2}$$

$$= \xi + \sum_{n=2}^\infty n^2 \left(\frac{\kappa_1(v,\xi) + (-1)^{n+1}\kappa_0(v,\xi)}{\kappa_1(v,\xi) + \kappa_0(v,\xi)}\right)^2 \Upsilon_n \xi^n$$

⋮

$$S_v^k \Upsilon(\xi) = S_v^1[S_v^{k-1} \Upsilon(\xi)]$$

$$= \xi + \sum_{n=2}^\infty n^k \left(\frac{\kappa_1(v,\xi) + (-1)^{n+1}\kappa_0(v,\xi)}{\kappa_1(v,\xi) + \kappa_0(v,\xi)}\right)^k \Upsilon_n \xi^n.$$

so that $\kappa_1(v,\xi) \neq -\kappa_0(v,\xi)$,

$$\lim_{v \to 0} \kappa_1(v,\xi) = 1, \quad \lim_{v \to 1} \kappa_1(v,\xi) = 0, \quad \kappa_1(v,\xi) \neq 0, \forall \xi \in \cup, v \in (0,1),$$

and

$$\lim_{v \to 0} \kappa_0(v,\xi) = 0, \quad \lim_{v \to 1} \kappa_0(v,\xi) = 1, \quad \kappa_0(v,\xi) \neq 0, \forall \xi \in \cup v \in (0,1).$$

The value $v = 0$ indicates the Sàlàgean operator $\mathcal{S}^k \Upsilon(\xi) = \xi + \sum_{n=2}^\infty n^k \Upsilon_n \xi^n$. We proceed to impose a linear differential operator having the SCDO and the Ruscheweyh derivative. For $\Upsilon \in \wedge$, the Ruscheweyh derivative is defined as follows:

$$\mathcal{R}^k \Upsilon(\xi) = \xi + \sum_{n=2}^\infty C_{k+n-1}^k \Upsilon_n \xi^n,$$

where $\mathbb{C}_{k+n-1}^k$ are the combination terms.

Definition 2. *Let $\Upsilon \in \bigwedge, \nu \in [0,1]$ and $0 \leq \alpha \leq 1$. The linear combination operator joining $\mathcal{R}^k \Upsilon(\xi)$ and $\mathcal{S}_\nu^k \Upsilon(\xi)$ is given by the formal*

$$\mathbb{C}_{\nu,\alpha}^k \Upsilon(\xi) = (1-\alpha)\mathcal{R}^k \Upsilon(\xi) + \alpha \mathcal{S}_\nu^k \Upsilon(\xi)$$

$$= \xi + \sum_{n=2}^{\infty} (1-\alpha)\mathbb{C}_{k+n-1}^k + \alpha \left[n \left(\frac{\kappa_1(\nu,\xi) + (-1)^{n+1}\kappa_0(\nu,\xi)}{\kappa_1(\nu,\xi) + \kappa_0(\nu,\xi)} \right) \right]^k \Upsilon_n \xi^n. \quad (3)$$

Remark 1.

- $k=0 \Longrightarrow \mathbb{C}_{\nu,\alpha}^0 \Upsilon(\xi) = \Upsilon(\xi);$
- $\nu=0 \Longrightarrow \mathbb{C}_{1,\alpha}^k \Upsilon(\xi) = \mathcal{L}_\kappa^k \Upsilon(\xi);$ [14] (Lupas operator)
- $\alpha=0 \Longrightarrow \mathbb{C}_{\nu,\alpha}^k \Upsilon(\xi) = \mathcal{R}^k \Upsilon(\xi);$
- $\alpha=1, \kappa=1 \Longrightarrow \mathbb{C}_{0,1}^k \Upsilon(\xi) = \mathcal{S}^k \Upsilon(\xi);$
- $\alpha=1 \Longrightarrow \mathbb{C}_{\nu,1}^k \Upsilon(\xi) = \mathcal{S}_\nu^k \Upsilon(\xi).$

Definition 3. *Let $\epsilon \in [0,1), \nu, \alpha \in [0,1$ and $k \in \mathbb{N}$. A function $\Upsilon \in \bigwedge$ belongs to the set $\mathcal{B}_k(\nu, \alpha, \epsilon)$ if and only if*

$$\Re\left((\mathbb{C}_{\nu,\alpha}^k \Upsilon(\xi))' \right) > \epsilon, \quad \xi \in \mathbb{U}.$$

Definition 4. *The function $\Upsilon \in \bigwedge$ is specified to be in $\mathbb{J}_\nu^\flat(A, B, k)$ if it satisfies the inequality*

$$1 + \frac{1}{\flat}\left(\frac{2\mathcal{S}_\nu^{k+1} \Upsilon(\xi)}{\mathcal{S}_\nu^k \Upsilon(\xi) - \mathcal{S}_\nu^k \Upsilon(-\xi)} \right) \prec \frac{1+A\xi}{1+B\xi},$$

$$\left(\xi \in \mathbb{U}, -1 \leq B < A \leq 1, k = 1,2,\dots, \flat \in \mathbb{C}\setminus\{0\}, \nu \in [0,1] \right).$$

- $\nu = 0 \Longrightarrow [6];$
- $\nu = 0, B = 0 \Longrightarrow [7];$
- $\nu = 0, A = 1, B = -1, \flat = 2 \Longrightarrow [8].$

The class $\mathbb{J}_\nu^\flat(A, B, k)$ is a generalization of the class of the Janowski starlike functions [15]

$$\rho(\xi) \prec \frac{1+A\xi}{1+B\xi}, \quad \xi \in \mathbb{U},$$

where $\rho(0) = 1$, $\rho(\mathbb{U}) \subset \Omega[A, B]$. The domain $\Omega[A, B]$ is a circular domain and it is referring to an open circular disk with center on the real axis and diameter end points $\frac{1-A}{1-B}$, provide that $B \neq -1$. Functions in the class $\mathbb{J}_\nu^\flat(A, B, k)$ have a circular domain with respect to symmetrical points.

Definition 5. *Let $\epsilon \in [0,1), \nu, \alpha \in [0,1$ and $k \in \mathbb{N}_0$. A function $\Upsilon \in \bigwedge$ is in the set $\mathbb{S}_k(\nu, \epsilon)$ if it achieves the real inequality*

$$\Re\left(\frac{\mathcal{S}_\nu^{k+1} \Upsilon(\xi)}{\mathcal{S}_\nu^k \Upsilon(\xi)} \right) > \epsilon, \quad \xi \in \mathbb{U}.$$

Note that $\mathbb{S}_0(\nu, \epsilon) = \mathcal{S}^*$, $\mathbb{S}_1(0, \epsilon) = \mathcal{C}$.

3. The Outcomes

In this section, we study some properties of the SCDO.

Theorem 1. *For $\Upsilon \in \bigwedge$ and $\alpha \in \mathbb{C}\setminus\{0\}$, if one of the sequencing subordination valid*

- The operator $S_\nu^k \Upsilon(\zeta)$ is of bounded turning type;
- Υ satisfies the relation

$$(S_\nu^k \Upsilon(\zeta))' \prec \left(\frac{1+\zeta}{1-\zeta}\right)^b, \quad b > 0, \ \zeta \in U;$$

- Υ fulfilled the inequality

$$\Re\left((S_\nu^k \Upsilon(\zeta))' \frac{S_\nu^k \Upsilon(\zeta)}{\zeta}\right) > \frac{\delta}{2}, \quad \delta \in [0,1), \ \zeta \in U,$$

- Υ admits the inequality

$$\Re\left(\zeta S_\nu^k \Upsilon(\zeta))'' - S_\nu^k \Upsilon(\zeta))' + 2\frac{S_\nu^k \Upsilon(\zeta)}{\zeta}\right) > 0,$$

- Υ confesses the inequality

$$\Re\left(\frac{\zeta S_\nu^k \Upsilon(\zeta))'}{S_\nu^k \Upsilon(\zeta))} + 2\frac{S_\nu^k \Upsilon(\zeta)}{\zeta}\right) > 1,$$

then $\frac{S_\nu^k \Upsilon(\zeta)}{\zeta} \in \mathcal{P}(\epsilon)$, $\epsilon \in [0,1)$.

Proof. Formulate a function σ as pursues:

$$\sigma(\zeta) = \frac{S_\nu^k \Upsilon(\zeta)}{\zeta} \Rightarrow \zeta \sigma'(\zeta) + \sigma(\zeta) = (S_\nu^k \Upsilon(\zeta))'. \tag{4}$$

By the first relation, $S_\nu^k \Upsilon(\zeta)$ is of bounded turning, this indicates that

$$\Re(\zeta \sigma'(\zeta) + \sigma(\zeta)) > 0.$$

Therefore, according to Lemma 1—i, we attain $\Re(\sigma(\zeta)) > 0$ which gets the first term of the theorem. According to second inequality, we indicate the pursuing subordination inequality

$$(S_\nu^k \Upsilon(\zeta))' = \zeta \sigma'(\zeta) + \sigma(\zeta) \prec \left(\frac{1+\zeta}{1-\zeta}\right)^b.$$

Now, by employing Lemma 1—i, there occurs a fixed constant $a > 0$ with $b = b(a)$ with the pursuing property

$$\frac{S_\nu^k \Upsilon(\zeta)}{\zeta} \prec \left(\frac{1+\zeta}{1-\zeta}\right)^a.$$

Consequently, we indicate that $\Re(S_\nu^k \Upsilon(\zeta)/\zeta) > \epsilon$, for values of $\epsilon \in [0,1)$. Lastly, agree with the third relation to get

$$\Re\left(\sigma^2(\zeta) + 2\sigma(\zeta).\zeta \sigma'(\zeta)\right) = 2\Re\left((S_\nu^k \Upsilon(\zeta))' \frac{S_\nu^k \Upsilon(\zeta)}{\zeta}\right) > \delta. \tag{5}$$

According to Lemma 1—ii, there occurs a positive fixed number $\lambda > 0$ achieving the real inequality $\Re(\sigma(\zeta)) > \lambda$, and yielding

$$\sigma(\zeta) = \frac{S_\nu^k \Upsilon(\zeta)}{\zeta} \in \mathcal{P}(\epsilon)$$

for a few value in $\epsilon \in [0,1)$. It indicates from (5) that $\Re\left(\mathcal{S}_\nu^k \Upsilon(\xi))'\right) > 0$; thus, according to Noshiro-Warschawski and Kaplan Lemmas, this leads to $\mathcal{S}_\nu^k \Upsilon(\xi)$ is univalent and of bounded turning in $\mathbb{U}$. Now, via the differentiating (4) and concluding the real case, we indicate that

$$\Re\left(\sigma(\xi) + \xi \sigma'(\xi) + \xi^2 \sigma''(\xi)\right)$$
$$= \Re\left(\xi(\mathcal{S}_\nu^k \Upsilon(\xi))'' - (\mathcal{S}_\nu^k \Upsilon(\xi))' + 2\frac{\mathcal{S}_\nu^k \Upsilon(\xi)}{\xi}\right)$$
$$> 0.$$

Thus, by the conclusion of Lemma 1—ii, we have

$$\Re\left(\frac{\mathcal{S}_\nu^k \Upsilon(\xi)}{\xi}\right) > 0.$$

Taking the logarithmic differentiation (4) and indicating the real, we arrive at the following conclusion:

$$\Re\left(\sigma(\xi) + \frac{\xi \sigma'(\xi)}{\sigma(\xi)} + \xi^2 \sigma''(\xi)\right)$$
$$= \Re\left(\frac{\xi(\mathcal{S}_\nu^k \Upsilon(\xi))'}{\mathcal{S}_\nu^k \Upsilon(\xi)} + 2\frac{\Delta_a^m \Upsilon(\xi)}{\xi} - 1\right)$$
$$> 0.$$

A direct application of Lemma 1—iii, we get the positive real i.e., $\Re(\frac{\mathcal{S}_\nu^k \Upsilon(\xi)}{\xi}) > 0$. This completes the proof. □

Theorem 2. *Suppose that* $\Upsilon \in \mathbb{J}_\alpha^b(A, B, m)$ *then for every function of the form*

$$\mathfrak{X}(\xi) = \frac{1}{2}[\Upsilon(\xi) - \Upsilon(-\xi)], \quad \xi \in \mathbb{U}$$

agrees with the pursuing relation

$$1 + \frac{1}{b}\left(\frac{\mathcal{S}_\nu^{k+1} \mathfrak{X}(\xi)}{\mathcal{S}_\nu^k \mathfrak{X}(\xi)} - 1\right) \prec \frac{1 + A\xi}{1 + B\xi},$$

and

$$\Re\left(\frac{\xi \mathfrak{X}(\xi)'}{\mathfrak{X}(\xi)}\right) \geq \frac{1 - \lambda^2}{1 + \lambda^2}, \quad |\xi| = \lambda < 1,$$

$$\left(\xi \in \mathbb{U}, -1 \leq B < A \leq 1, m = 1, 2, ..., b \in \mathbb{C} \setminus \{0\}, \nu \in [0, 1]\right).$$

Proof. Because the function $\Upsilon \in \mathbb{J}_\alpha^b(A, B, m)$ then there occurs a function $\wp \in \mathbb{J}(A, B)$, where

$$b(\wp(\xi) - 1) = \left(\frac{2\mathcal{S}_\nu^{k+1} \Upsilon(\xi)}{\mathcal{S}_\nu^k \Upsilon(\xi) - \mathcal{S}_\nu^k \Upsilon(-\xi)}\right)$$

and

$$b(\wp(-\xi) - 1) = \left(\frac{-2\mathcal{S}_\nu^{k+1} \Upsilon(-\xi)}{\mathcal{S}_\nu^k \Upsilon(\xi) - \mathcal{S}_\nu^k \Upsilon(-\xi)}\right).$$

This implies that

$$1 + \frac{1}{b}\left(\frac{\mathcal{S}_\nu^{k+1} \mathfrak{X}(\xi)}{\mathcal{S}_\nu^k \mathfrak{X}(\xi)} - 1\right) = \frac{\wp(\xi) + \wp(-\xi)}{2}.$$

Also, since $\wp(\xi) \prec \dfrac{1+A\xi}{1+B\xi}$, where $\dfrac{1+A\xi}{1+B\xi}$ is univalent then by the concept of the subordination, we have

$$1 + \dfrac{1}{b}\left(\dfrac{S_v^{k+1}\mathcal{X}(\xi)}{\Delta_\alpha^m \mathcal{X}(\xi)} - 1\right) \prec \dfrac{1+A\xi}{1+B\xi}.$$

But the function $\mathcal{X}(\xi)$ is starlike in U, which means that

$$\dfrac{\xi \mathcal{X}(\xi)'}{\mathcal{X}(\xi)} \prec \dfrac{1-\xi^2}{1+\xi^2}$$

and there occurs a Schwarz function $\tau \in U, |\tau(\xi)| \leq |\xi| < 1, \tau(0) = 0$ such that

$$\Psi(\xi) := \dfrac{\xi \mathcal{X}(\xi)'}{\mathcal{X}(\xi)} \prec \dfrac{1-\tau(\xi)^2}{1+\tau(\xi)^2}.$$

This implies that there exists $\zeta, |\zeta| = \lambda < 1$ achieving

$$\tau^2(\zeta) = \dfrac{1-\Psi(\zeta)}{1+\Psi(\zeta)}, \quad \zeta \in U.$$

A computation yields

$$\left|\dfrac{1-\Psi(\zeta)}{1+\Psi(\zeta)}\right| = |\tau(\zeta)|^2 \leq |\zeta|^2.$$

Thus, we conclude that

$$\left|\Psi(\zeta) - \dfrac{1+|\zeta|^4}{1-|\zeta|^4}\right|^2 \leq \dfrac{4|\zeta|^4}{(1-|\zeta|^4)^2}$$

or

$$\left|\Psi(\zeta) - \dfrac{1+|\zeta|^4}{1-|\zeta|^4}\right| \leq \dfrac{2|\zeta|^2}{(1-|\zeta|^4)}.$$

Consequently, we obtain

$$\Re(\Psi(\zeta)) \geq \dfrac{1-\lambda^2}{1+\lambda^2}, \quad |\zeta| = \lambda < 1.$$

□

Theorem 3. *Suppose that* $Y \in B_k(v, \alpha, \epsilon)$, *and the convex analytic function g satisfies the integral equation*

$$F(\xi) = \dfrac{2+c}{\xi^{1+c}} \int_0^\xi \tau^c Y(\tau)d\tau, \quad \xi \in U$$

then the subordination

$$\left(C_{v,\alpha}^k Y(\xi)\right)' \prec g(\xi) + \dfrac{(\xi g'(\xi))}{2+c}, \quad c > 0,$$

implies the subordination

$$\left(C_{v,\alpha}^k F(\xi)\right)' \prec g(\xi),$$

and the outcome is sharp.

Proof. Here, we aim to utilize the result of Lemma 2. By the conclusion of $F(\xi)$, we acquire

$$\left(C_{v,\alpha}^k F(\xi)\right)' + \dfrac{\left(C_{v,\alpha}^k F(\xi)\right)''}{2+c} = \left(C_{v,\alpha}^k Y(\xi)\right)'.$$

Following the conditions of the theorem, we get

$$\left(C_{\nu,a}^k F(\xi)\right)' + \frac{\left(C_{\nu,a}^k F(\xi)\right)''}{2+c} \prec g(\xi) + \frac{(\xi g'(\xi))}{2+c}.$$

By assuming

$$\varrho(\xi) := \left(C_{\nu,a}^k F(\xi)\right)',$$

We have

$$\varrho(\xi) + \frac{(\xi \varrho'(\xi))}{2+c} \prec g(\xi) + \frac{(\xi g'(\xi))}{2+c}.$$

According to Lemma 2, we obtain

$$\left(C_{\nu,a}^k F(\xi)\right)' \prec g(\xi),$$

and g is the best dominant. □

Theorem 4. *Let g be convex such that $g(0) = 1$. If*

$$\left(C_{\nu,a}^k \Upsilon(\xi)\right)' \prec g(\xi) + \xi g'(\xi), \quad \xi \in U,$$

then $\dfrac{C_{\nu,a}^k \Upsilon(\xi)}{\xi} \prec g(\xi)$, and this result is sharp.

Proof. Define the following function

$$\varrho(\xi) := \frac{C_{\nu,a}^k \Upsilon(\xi)}{\xi} \in \aleph[1,1]. \tag{6}$$

A direct application of Lemma 1 yields

$$C_{\nu,a}^k \Upsilon(\xi) = \xi \varrho(\xi) \Longrightarrow \left(C_{\nu,a}^k \Upsilon(\xi)\right)' = \varrho(\xi) + \xi \varrho'(\xi).$$

Thus, we introduce the following subordination:

$$\varrho(\xi) + \xi \varrho'(\xi) \prec g(\xi) + \xi g'(\xi).$$

Hence, we conclude that $\dfrac{C_{\nu,a}^k \Upsilon(\xi)}{\xi} \prec g(\xi)$, and g is the best dominant. □

Theorem 5. *If $\Upsilon \in \bigwedge$ fulfills the subordination*

$$(C_{\nu,a}^k \Upsilon(\xi))' \prec \left(\frac{1+\xi}{1-\xi}\right)^b, \quad \xi \in U, \, b > 0,$$

then

$$\Re\left(\frac{C_{\nu,a}^k \Upsilon(\xi)}{\xi}\right) > \epsilon, \quad \epsilon \in [0,1).$$

Proof. Construct ϱ as in (6). Thus, by subordination possessions, we indicate that

$$(C_{\nu,a}^k \Upsilon(\xi))' = \xi \varrho'(\xi) + \varrho(\xi) \prec \left(\frac{1+\xi}{1-\xi}\right)^b.$$

With the help of Lemma 1—i, there occurs a fixed number $a > 0$ with $b = b(a)$ where

$$\frac{C_{\nu,\alpha}^k \Upsilon(\xi)}{\xi} \prec \left(\frac{1+\xi}{1-\xi}\right)^a.$$

This leads to real conclusion $\Re(C_{\nu,\alpha}^k \Upsilon(\xi)/\xi) > \epsilon, \epsilon \in [0,1)$. □

Theorem 6. *If $\Upsilon \in \bigwedge$ fulfills the real inequality*

$$\Re\left((C_{\nu,\alpha}^k \Upsilon(\xi))' \frac{C_{\nu,\alpha}^k \Upsilon(\xi)}{\xi}\right) > \Re\left(\frac{\alpha}{2}\right), \quad \xi \in U, \alpha \in \mathbb{C}$$

then $C_{\nu,\alpha}^k \Upsilon(\xi) \in \mathcal{B}_k(\nu, \alpha, \epsilon)$.

Proof. Formulate ϱ as in (6). A clear evaluation gives

$$\Re\left(\varrho^2(\xi) + 2\varrho(\xi).\xi\varrho'(\xi)\right) = 2\Re\left(C_{\nu,\alpha}^k \Upsilon(\xi))' \frac{C_{\nu,\alpha}^k \Upsilon(\xi)}{\xi}\right) > \Re(\alpha). \quad (7)$$

By the advantage of Lemma 1—ii, there occurs a constant κ concerning on $\Re(\alpha)$ where $\Re(\varrho(\xi)) > \kappa$, this gives $\Re(\varrho(\xi)) > \epsilon, \epsilon \in [0,1)$. By virtue of (7), it implies that $\Re\left(C_{\nu,\alpha}^k \Upsilon(\xi))'\right) > \epsilon$ and hence based on the idea of Noshiro-Warschawski and Kaplan Theorems, $C_{\nu,\alpha}^k \Upsilon(\xi)$ is univalent and of bounded boundary rotation in U. □

Theorem 7. *The set $\mathcal{B}_k(\nu, \alpha, \epsilon)$ is convex.*

Proof. Suppose that $\Upsilon_i \in \mathcal{B}_k(\nu, \alpha, \epsilon)$, $i = 1, 2$ achieve the formulas $\Upsilon_1(\xi) = \xi + \sum_{n=2}^{\infty} a_n \xi^n$ and $\Upsilon_2(\xi) = \xi + \sum_{n=2}^{\infty} b_n \xi^n$ respectively. It is adequate to show that the linear combination function

$$G(\xi) = w_1 \Upsilon_1(\xi) + w_2 \Upsilon_2(\xi), \quad \xi \in U$$

belongs to $\mathcal{B}_k(\nu, \alpha, \epsilon)$, where $w_1 > 0, w_2 > 0$ and $w_1 + w_2 = 1$.

By the definition of $G(\xi)$, a computation yields that

$$G(\xi) = \xi + \sum_{n=2}^{\infty} (w_1 a_n + w_2 b_n) \xi^n$$

then under the formal $C_{\nu,\alpha}^k$, we obtain

$$C_{\nu,\alpha}^k G(\xi) = \xi + \sum_{n=2}^{\infty} (w_1 a_n + w_2 b_n)$$

$$\times \left[(1-\alpha)C_{m+n-1}^m + \alpha \left(\frac{\kappa_1(\nu,\xi) + (-1)^{n+1}\kappa_0(\nu,\xi)}{\kappa_1(\nu,\xi) + \kappa_0(\nu,\xi)}\right)^k\right] \xi^n.$$

By considering the derivative, we have

$$\Re\{(C_{\nu,\alpha}^k G(\xi))'\}$$
$$= 1 + w_1 \Re\left\{\sum_{n=2}^{\infty} n\left[(1-\alpha)C_{m+n-1}^m + \alpha\left(\frac{\kappa_1(\nu,\xi) + (-1)^{n+1}\kappa_0(\nu,\xi)}{\kappa_1(\nu,\xi) + \kappa_0(\nu,\xi)}\right)^k\right]a_n \xi^{n-1}\right\}$$
$$+ w_2 \Re\left\{\sum_{n=2}^{\infty} n\left[(1-\alpha)C_{m+n-1}^m + \alpha\left(\frac{\kappa_1(\nu,\xi) + (-1)^{n+1}\kappa_0(\nu,\xi)}{\kappa_1(\nu,\xi) + \kappa_0(\nu,\xi)}\right)^k\right]b_n \xi^{n-1}\right\}$$
$$> 1 + w_1(\epsilon - 1) + w_2(\epsilon - 1) = \epsilon.$$

□

4. Applications

A set of complex differential equations is an assembly of differential equations with complex variables. The most important study in this direction is to establish the existence and uniqueness results. There are diffident types of techniques including the utility of majors and minors (or subordination and superordination concepts) (see [12]). Investigation of ODEs in the complex domain suggests the detection of novel transcendental special functions, which currently called a Briot–Bouquet differential equation (BBDE)

$$\omega \Upsilon(\xi) + (1-\omega)\frac{\xi(\Upsilon(\xi))'}{\Upsilon(\xi)} = \hbar(\xi),$$

$$\left(\hbar(0) = \Upsilon(0), \omega \in [0,1], \xi \in \mathbb{U}, \Upsilon \in \bigwedge\right).$$

In this place, we shall generalize the BBDE into a symmetric BBDE by using SCDO. Numerous presentations of these comparisons in the geometric function model have recently achieved in [12].

Needham and McAllister [16] presented a two-dimensional complex holomorphic dynamical system, pleasing the 2-D form

$$\xi_t = \Theta(\xi, \omega); \quad \omega_t = \Theta(\xi, w), \quad \xi, \omega \in \mathbb{U}$$

and t is in any real interval. Development application of the BBDE seemed newly, with different approaches (see [17]) to solve the equation of electronic nano-shells (see [18]). Controlled by the situation effort of traditional shell theory, the transposition fields of the nano-shell take the dynamic system

$$\xi_t = \Theta(\xi, \omega) + \Theta_\theta(\xi, \omega); \quad \omega_t = \Theta(\xi, \omega) + \Theta_\theta(\bar{\xi}, \bar{\omega}), \quad \xi, \omega \in \mathbb{U},$$

where θ is the angles between ξ and ω and their conjugates.

Our purpose is to generalize this class of equation by utilizing the SCDO and establish its properties by applying the subordination concept. In view of (2), we have the generalized BBDE

$$\omega \Upsilon(\xi) + (1-\omega)\left(\frac{\xi(S_\nu^k \Upsilon(\xi))'}{S_\nu^k \Upsilon(\xi)}\right) = \hbar(\xi), \quad \hbar(0) = \Upsilon(0), \xi \in \mathbb{U}. \quad (8)$$

The subordination settings and alteration bounds for a session of SCDO specified in the following formula. A trivial resolution of (8) is given when $\omega = 1$. Consequently, our vision is to carry out the situation, $\Upsilon \in \bigwedge$ and $\omega = 0$. We proceed to present the behavior of the solution of (8).

Theorem 8. For $\Upsilon \in \Lambda$, $\alpha \in [0, \infty)$ and $\hbar$ is univalent convex in $\cup$ if

$$\left(\frac{\xi (\mathcal{S}_\nu^k)'}{\mathcal{S}_\nu^k \Upsilon (\xi)}\right) \prec \hbar(\xi), \quad \xi \in \cup \tag{9}$$

then

$$\mathcal{S}_\nu^k \Upsilon (\xi) \prec \xi \exp\left(\int_0^\xi \frac{\hbar(\top(\xi)) - 1}{\ell} d\ell\right),$$

where $\top$ is a Schwarz function in $\cup$. In addition, we have

$$|\xi| \exp\left(\int_0^1 \frac{\hbar(\top(-\sigma)) - 1}{\sigma} d\sigma\right) \leq \left|\Delta_\alpha^m \Upsilon (\xi)\right| \leq |\xi| \exp\left(\int_0^1 \frac{\hbar(\top(\sigma)) - 1}{\sigma} d\sigma\right).$$

Proof. The subordination in (9) implies that there occurs a Schwarz function $\top$ such that

$$\left(\frac{\xi (\mathcal{S}_\nu^k \Upsilon (\xi))'}{\mathcal{S}_\nu^k \Upsilon (\xi)}\right) = \hbar(\top(\xi)), \quad \xi \in \cup.$$

This yields the inequality

$$\left(\frac{\xi (\mathcal{S}_\nu^k \Upsilon (\xi))'}{\mathcal{S}_\nu^k \Upsilon (\xi)}\right) - \frac{1}{\xi} = \frac{\hbar(\top(\xi)) - 1}{\xi}.$$

By making the integrated operating, we have

$$\log\left(\frac{\mathcal{S}_\nu^k \Upsilon (\xi)}{\xi}\right) = \int_0^\xi \frac{\hbar(\top(\ell)) - 1}{\ell} d\ell. \tag{10}$$

Consequently, we have

$$\log \mathcal{S}_\nu^k \Upsilon (\xi) = \left(\int_0^\xi \frac{\hbar(\top(\ell)) - 1}{\ell} d\ell\right) - \log(\xi). \tag{11}$$

A calculation brings the next subordination relation

$$\mathcal{S}_\nu^k \Upsilon (\xi) \prec \xi \exp\left(\int_0^\xi \frac{\hbar(\top(\ell)) - 1}{\ell} d\ell\right).$$

Moreover, the function $\hbar$ translates the disk $0|\xi|\sigma \leq 1$ into a convex symmetric domain toward the x-axis; in other words, we have

$$\hbar(-\sigma|\xi|) \leq \Re(\hbar(\top(\sigma\xi))) \leq \hbar(\sigma|\xi|), \quad \sigma \in (0,1], |\xi| \neq \sigma,$$

which implies the inequalities:

$$\hbar(-\sigma) \leq \hbar(-\sigma|\xi|), \quad \hbar(\sigma|\xi|) \leq \hbar(\sigma)$$

and

$$\int_0^1 \frac{\hbar(\top(-\sigma|\xi|)) - 1}{\sigma} d\sigma \leq \Re\left(\int_0^1 \frac{\hbar(\top(\sigma)) - 1}{\sigma} d\sigma\right) \leq \int_0^1 \frac{\hbar(\top(\sigma|\xi|)) - 1}{\eta} d\sigma.$$

By employing (10) and the last inequality, we arrive at

$$\int_0^1 \frac{\hbar(\top(-\sigma|\xi|)) - 1}{\sigma} d\sigma \leq \log\left|\frac{\mathcal{S}_\nu^k \Upsilon (\xi)}{\xi}\right| \leq \int_0^1 \frac{\hbar(\top(\sigma|\xi|)) - 1}{\sigma} d\sigma.$$

This equivalence to the fact

$$\exp\left(\int_0^1 \frac{\hbar(\top(-\sigma|\xi|))-1}{\sigma}d\sigma\right) \leq \left|\frac{S_v^k \Upsilon(\xi)}{\xi}\right| \leq \exp\left(\int_0^1 \frac{\hbar(\top(\sigma|\xi|))-1}{\sigma}d\sigma\right).$$

□

We note that the condition of Theorem 8, which the BB formula subordinates by a convex univalent function $\hbar$ can be replaced by a general condition as follows:

Theorem 9. Suppose that $\Upsilon \in \bigwedge$, $\alpha \in [0,\infty)$ and $0 < v \leq 1$. If

$$\left(\frac{\xi(S_v^k \Upsilon(\xi))' - \xi}{S_v^k \Upsilon(\xi)}\right) \prec \frac{2v\xi}{1+\xi}, \quad \xi \in U \tag{12}$$

then

$$\left|\frac{S_v^k \Upsilon(\xi)}{\xi} - 1\right| \leq v. \tag{13}$$

Moreover, define the term

$$v := \frac{1}{(1-r)^{\mathfrak{v}}}, \quad 0 < r < 1,$$

for some positive constant $\mathfrak{v}$, then

$$\left|\left(\frac{S_v^k \Upsilon(\xi)}{\xi}\right)'\right| \leq \frac{\mathfrak{v}+1}{(1-r)^{\mathfrak{v}+1}}. \tag{14}$$

Proof. In view of Lemma 3, we have the subordination inequality

$$\frac{S_v^k \Upsilon(\xi)}{\xi} \prec 1 + v\xi.$$

Since the result is sharp, then directly, we obtain the inequality (13). Consequently, by ([19], lemma 5.1.3), we have the inequality (14). □

Author Contributions: Conceptualization, R.M.E.; Formal analysis, R.W.I.; Funding acquisition, R.M.E.; Methodology, R.W.I.; Project administration, R.M.E.; Writing—review and editing, S.J.O. All authors have read and agreed to the published version of the manuscript.

Funding: This research received no external funding.

Acknowledgments: The authors would like to express their thanks to the reviewers to provide us deep comments.

Conflicts of Interest: The authors declare no conflict of interest.

References

1. Duren, P. *Univalent Functions*; Grundlehren der mathematischen Wissenschaften; Springer-Verlag New York Inc.: New York, NY, USA, 1983; Volume 259, ISBN 0-387-90795-5.
2. Goodman, A.W. *Univalent Functions*; Mariner Pub Co.: Bostan, MA, USA, 1983; ISBN 978-0936166100.
3. Sàlxaxgean, G.S. Subclasses of Univalent Functions, Complex Analysis-Fifth Romanian-Finnish Seminar, Part 1 (Bucharest, 1981); Lecture Notes in Math; Springer: Berlin, Germany, 1983; Volume 1013, pp. 362–372.
4. Al-Oboudi, F.M. On univalent functions defined by a generalized Sàlàgean operator. *Int. J. Math. Math. Sci.* **2004**, *27*, 1429–1436. [CrossRef]
5. Ibrahim, R.W. Operator Inequalities Involved Wiener–Hopf Problems in the Open Unit Disk. In *Differential and Integral Inequalities*; Springer: Cham, Switzerland, 2019; Volume 13, pp. 423–433.
6. Ibrahim, R.W.; Darus, M. Subordination inequalities of a new S. Sàlàgean difference operator. *Int. J. Math. Comput. Sci.* **2019**, *14*, 573–582.

7. Ibrahim, R.W.; Jahangiri, J.M. Conformable differential operator generalizes the Briot-Bouquet differential equation in a complex domain. *AIMS Math.* **2019**, *6*, 1582–1595. [CrossRef]
8. Ibrahim, R.W.; Darus, M. New Symmetric Differential and Integral Operators Defined in the Complex Domain. *Symmetry* **2019**, *7*, 906. [CrossRef]
9. Ibrahim, R.W.; Darus, M. Univalent functions formulated by the Salagean-difference operator. *Int. J. Anal. Appl.* **2019**, *4*, 652–658.
10. Ibrahim, R.W. Regular classes involving a generalized shift plus fractional Hornich integral operator. *Bol. Soc. Parana. Mat.* **2020**, *38*, 89–99. [CrossRef]
11. Anderson, D.R.; Ulness, D.J. Newly defined conformable derivatives. *Adv. Dyn. Syst. Appl.* **2015**, *10*, 109–137.
12. Miller, S.S.; Mocanu, P.T. *Differential Subordinations: Theory and Applications*; CRC Press: Boca Raton, FL, USA, 2000.
13. Tuneski, N.; Obradovic, M. Some properties of certain expressions of analytic functions. *Comput. Math. Appl.* **2011**, *62*, 3438–3445. [CrossRef]
14. Lupas, A. Some differential subordinations using Ruscheweyh derivative and S. Sàlàgean operator. *Adv. Differ. Equ.* **2013**, *150*, 1–11.
15. Janowski, W. Some extremal problems for certain families of analytic functions. *Ann. Pol. Math.* **1973**, *28*, 297–326. [CrossRef]
16. Needham, D.J.; McAllister, S. Centre families in two-dimensional complex holomorphic dynamical systems. *Proc. R. Soc. Lond. Ser.* **1998**, *454*, 2267–2278. [CrossRef]
17. Ebrahimi, F.; Mohammadi, K.; Barouti, M.M.; Habibi, M. Wave propagation analysis of a spinning porous graphene nanoplatelet-reinforced nanoshell. *Waves Random Complex Media* **2019**, *27*, 1–27. [CrossRef]
18. Habibi, M.; Mohammadgholiha, M.; Safarpour, H. Wave propagation characteristics of the electrically GNP-reinforced nanocomposite cylindrical shell. *J. Braz. Soc. Mech. Sci. Eng.* **2019**, *41*, 221. [CrossRef]
19. Hormander, L. *Linear Partial Differential Operators*; Springer: Berlin/Heidelberg, Germany, 1963.

© 2020 by the authors. Licensee MDPI, Basel, Switzerland. This article is an open access article distributed under the terms and conditions of the Creative Commons Attribution (CC BY) license (http://creativecommons.org/licenses/by/4.0/).

Article

Taming the Natural Boundary of Centered Polygonal Lacunary Functions—Restriction to the Symmetry Angle Space

Leah K. Mork [1], Keith Sullivan [1] and Darin J. Ulness [2,*]

[1] Department of Mathematics, Concordia College, Moorhead, MN 56562, USA; lmork@cord.edu (L.K.M.); ksulliv1@cord.edu (K.S.)
[2] Department of Chemistry, Concordia College, Moorhead, MN 56562, USA
* Correspondence: ulnessd@cord.edu

Received: 16 March 2020; Accepted: 6 April 2020; Published: 11 April 2020

Abstract: This work investigates centered polygonal lacunary functions restricted from the unit disk onto symmetry angle space which is defined by the symmetry angles of a given centered polygonal lacunary function. This restriction allows for one to consider only the p-sequences of the centered polygonal lacunary functions which are bounded, but not convergent, at the natural boundary. The periodicity of the p-sequences naturally gives rise to a convergent subsequence, which can be used as a grounds for decomposition of the restricted centered polygonal lacunary functions. A mapping of the unit disk to the sphere allows for the study of the line integrals of restricted centered polygonal that includes analytic progress towards closed form representations. Obvious closures of the domain obtained from the spherical map lead to four distinct topological spaces of the "broom topology" type.

Keywords: lacunary function; gap function; centered polygonal numbers; natural boundary; singularities; broom topology

1. Introduction

Analytic functions are of clear importance as an area of mathematics and also in physics, chemistry, engineering, and other applied areas. It is the set of points where analyticity breaks down, in the form of singularities, that often carries the most information about the function and hence about the physical phenomenon it describes. In most applications, the set of singularities is a set of discrete points called isolated singularities. Characteristic of analytic functions is the fact that one can construct a Taylor series representation where the isolated singularities determine the radius of convergence. One is then often able to analytically continue functions outside the radius of convergence by various methods (see References [1,2]).

In certain instances, the singularities are no longer isolated but instead form a curve in the complex plane called a natural boundary. Analytic continuation is not possible through the natural boundary. One set of functions that have a natural boundary are the lacunary functions (see References [1,2]). The Taylor series of Lacunary functions has "gaps" (or "lacunae") in the powers present in the series expansion. One simple example is $f(z) = \sum_{n=1}^{\infty} z^{n^4} = z + z^{16} + z^{81} + z^{256} + \cdots$. In this example, the natural boundary lies on the unit circle and $f(z)$ is analytic in the open unit disk.

Because the natural boundary is difficult to deal with, functions with natural boundaries have not been heavily utilized in physics over the years. Nonetheless, the presence of natural boundaries does result in real physical consequences. Creagh and White showed that in optics, the calculation of evanescent waves extending from elliptical dielectrics can involve functions with natural boundaries (see Reference [3]). In mechanics, particularly integrable/nonintegrable systems, Greene and Percival investigate the role of natural boundaries in the context of Hamiltonian maps (see Reference [4]). Shado and Ikeda have shown that quantum tunneling in some systems can be impacted by natural boundaries which influence instanton orbiting (see Reference [5]). Quite recently, Yamada and Ikeda have studied Anderson-localized states in the Harper model in quantum mechanics and the role of natural boundaries associated with the wavefunctions (see Reference [6]).

Guttmann et al. have proven that any solution of a non-solvable Ising-like model must be expressible in terms of functions having natural boundaries (see References [7,8]). Relatedly, Nickel has shown that natural boundaries appear in the calculation of the magnetic susceptibility in the 2D Ising model (see Reference [9]). In molecular kinetic theory, lacunary functions display characteristics near the natural boundary that are related to Weiner (stochastic) processes. Because of this, lacunary functions have been studied in connection with Brownian motion (see Reference [10]).

More mathematically, Eckstein and Zając investigated heat traces of unbounded operators in Hilbert space (see Reference [11]). Behr et.al. have discussed lacunary generation functions in the context of their rather comprehensive study of Sobolev-Jacobo polynomials (see [12]). And, recently, Kişi, Gümüş, and Savas studied $A^{\mathcal{I}}$-lacunary convergence and Cesàro summability with respect to lacunary sequences (see Reference [13]).

Of the lacunary functions, the family generated by centered polygonal numbers have particularly interesting features. This family is called centered polygonal lacunary functions. Their special properties are mainly due to the unusual symmetry present in this family, compared to an arbitrary lacunary function (see References [14–16]). A class of infinite sequences associated with lacunary functions are called lacunary sequences and recent work has focused on exploring particular bounded sequences of numbers arising at the natural boundary of centered polygonal lacunary sequences (see References [14,15]). These p-sequences, as they are called, have been well characterized and this work has been significantly enhanced by the construction of graphs to represent the p-sequences (see Reference [14]). The graphs that have arisen are interesting in and of themselves, especially in that they reveal self-similarity and scaling that allow for a renormalization approach (see Reference [15]). The self-similarity hints at the fractal character of the centered polygonal lacunary functions. Indeed, explicit investigation of this fractal character in the form of Julia sets has recently been presented (see Reference [16]).

This current contribution builds upon the above-mentioned work and is focused on some of the substructure in the summation terms of the centered polygonal lacunary functions as well as the behavior of these functions on restricted subspaces of the unit disk. The periodic nature of the p-sequences and the fact that there is a well-defined sequence that actually converges to zero at the natural boundary offers an opportunity to make some degree of sense of the centered polygonal lacunary functions at the natural boundary. This is the case, at least, when restricting the domain from the unit disk onto a set of line segments which are determined by the function itself. This restricted space is referred to here as the symmetry angle space and is defined in Section 4. Symmetry angle space, as a topology, is very much like the so-called "broom topology" space (see Reference [17]). Throughout this work, the topology on the unit disk is the normal topology of $\mathbb{C}$ and the topology on the union of line segments is the induced topology, that is, the normal topology of the unit interval. The periodic nature of the p-sequences suggests a natural decomposition of the centered polygonal lacunary functions on symmetry angle space.

Further, there is a convenient surjective mapping of the unit disk to the sphere such that the natural boundary maps to a single point. Symmetry angle space then consists of the union of longitudinal lines on the surface of the manifold of the 2-sphere, S^2 [18]. Obvious closures of the mapped symmetry angle space allow inclusion of the natural boundary as a single point. Line integrals are investigated which include loops "through" the natural boundary.

The ultimate goal of the current work is to provide some useful insight into the nature of the natural boundary of centered polygonal lacunary functions. All visualizations of functions in this work were calculated and produced using MATHEMATICA (see Reference [19]).

2. Centered Polygonal Lacunary Functions

Definitions, notation, and some theorems from References [14,15] are briefly collected here for the convenience of the reader.

The Nth member of a lacunary sequence of functions is defined here as

$$f_N(z) = \sum_{n=1}^{N} z^{g(n)}, \qquad (1)$$

where $g(n)$ is a function of n, a positive integer, that follows the criteria of Hadamard's gap theorem (see Reference [2]). (Note that the sum starts at $n = 1$ for convenience but not necessity.) Following References [14,15], we use the notation

$$\mathfrak{L}(g;z) \equiv \left\{ \sum_{n=1}^{N} z^{g(n)} \right\}, \qquad (2)$$

to represent the particular lacunary sequence described by $g(n)$, in complex variable z. The lacunary function associated with the sequence $\mathfrak{L}(g(n);z)$ is $f(z) = \lim_{N\to\infty} f_N(z)$. One particularly important representation of this example function is shown in the bottom left panel of Figure 1 for the example case of $g(n) = T(n)$, the well-known triangular numbers. Figure 1 shows the modulus of $f(z)$, $|f(z)|$, where the graph is limited to $0 \leq |f(z)| \leq 1$. That is, the graph is truncated at the unity level set. This is done to better expose the symmetry features of the functions otherwise the divergence at the natural boundary obscures the view of these features.

A $g(n)$ family of note that yields particularly interesting lacunary functions are the centered polygonal numbers. The centered polygonal numbers are a sequence of numbers arising from considering points on an polygonal lattice (see References [20–23]). The centered k-gonal numbers are defined by the formula (for positive integer $k < 0$)

$$C^{(k)}(n) = k \frac{n(n-1)}{2} + 1, \quad n \geq 1. \qquad (3)$$

When $g(n) = C^{(k)}(n)$ is the nth centered k-gonal number, then $f(z) = \sum_{n=1}^{\infty} z^{C^{(k)}(n)}$ is the centered polygonal lacunary function. Also, $\mathfrak{L}(C^{(k)};z)$ is the centered polygonal lacunary sequence associated with f.

It turns out that nearly all of the structural features of centered polygonal lacunary functions are independent of the choice of k (see References [14,15]). This is because the centered polygonal numbers are related to the triangular numbers (see Reference [24]) in a simple way. The set of triangular numbers is

$$T = \left\{ \frac{n(n+1)}{2} \right\}. \qquad (4)$$

For convenience, lemmas, theorems, and corollaries are proven in Reference [14] and are stated here without proof. A couple of definitions from Reference [14] are included as well.

Lemma 1.

$$\frac{C^{(k)}(n+1) - 1}{k} = T(n), \qquad (5)$$

where $C^{(k)}(n)$ and $T(n)$ mean the nth member of the respective sequences.

Lemma 2. *The sequence of triangular numbers mod p is a $2p$-cycle. The sequence is symmetric about the midpoint of the $2p$-cycle. The $2p$th member of the $2p$-cycle is zero.*

Definition 1. *Primary symmetry. The rotational symmetry of the $N = 2$ member of $|\mathfrak{L}(g;z)|$, $|f_2(z)|$, is called the primary symmetry.*

Theorem 1. *The primary symmetry of $|\mathfrak{L}(g(n);z)|$ is $k = g(2) - g(1)$.*

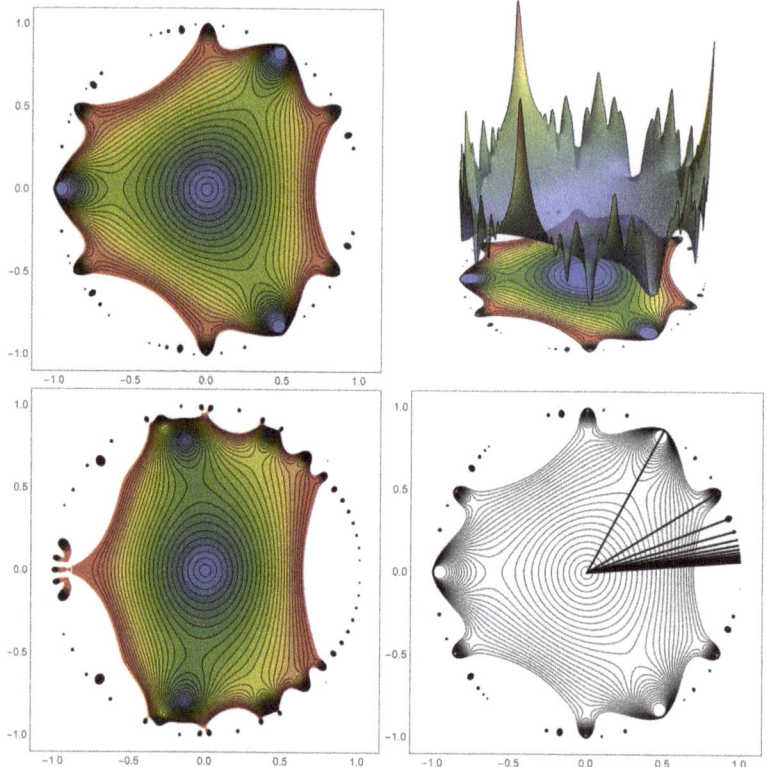

Figure 1. A particularly illustrative way to present graphs of $\mathfrak{L}(C^{(k)};z)$. The representation shown here is especially useful for this work. The contour plot is truncated at the unity level set (blue shading represents low values and red shading represents high values). The top left panel shows the example of $\mathfrak{L}(C^{(3)};z)$ where a plot of $|f_{16}(z)|$. The top right panel shows a superposition of the contour plot and a three-dimensional rendering. The truncated contour plot more clearly exposes the true rotational symmetry of the centered polygonal lacunary functions. The bottom left graph shows the case of $\mathfrak{L}(T_n;z)$, where $T(n)$ are the well-known triangular numbers, again for $|f_{16}(z)|$. Despite the intimate relationship between the centered polygonal numbers and the triangular numbers, the plots are strikingly different. The bottom right panel shows an unshaded contour plot of the same function shown in the left panel of Figure 1. The superimposed black lines indicate the symmetry angles. The first 15 symmetry angles are shown (see text for details).

3. The p–Sequences

The centered polygonal lacunary functions have very interesting organizational structure at the natural boundary (see References [14,15]). Of particular interest are the p-sequences (see Reference [14]). These arise when considering the value of the centered polygonal lacunary function on the line segment that runs from the origin to the natural boundary at an angle of $\phi = \frac{\pi}{kp}$, $p \in \mathbb{Z}^+$. Interestingly, in the limit of $\rho \to 1_-$, the sequence $\mathfrak{L}\left(C^{(k)}; \rho e^{\frac{i\pi}{kp}}\right)$ becomes a bounded $4p$ cycle of complex numbers (see Reference [14]).

Definition 2. *Symmetry angle. Let the primary symmetry be k-fold. The first symmetry angle is $\alpha_1 = \frac{\pi}{k}$, $k \in \mathbb{Z}$. The pth symmetry angle is $\alpha_p = \frac{\pi}{pk}$, $p, k \in \mathbb{Z}$. The primary symmetry angle is α_1.*

At the natural boundary, the p-sequences have intricate structure (see Reference [14]) that is a manifestation of Lemma 1. Because of Lemma 2, the values of $f_N(e^{i\alpha_p})$ oscillate. Further, they take on the value of zero at values of $N = 4mp$, where m is a positive integer. This allows for a convergent sub-sequence which is discussed in Section 5.

This section concludes with an additional theorem specific to centered polygonal numbers proven here.

Theorem 2. *The following rearrangement holds for $f_N(\rho e^{i\alpha_p})$.*

$$f_N(\rho e^{i\alpha_p}) = \sum_{n=1}^{N}(\rho e^{\frac{i\pi}{kp}})^{C^k(n)} = \sum_{n=1}^{N}(-1)^{\left\lfloor \frac{C^{(k)}(n)}{kp} \right\rfloor}(-1)^{\frac{C^{(k)}(n) \mod (kp)}{(kp)}} \rho^{C^{(k)}(n)}, \tag{6}$$

where $\lfloor x \rfloor$ indicates the floor function.

Proof. The identity of Equation (6) follows directly from the identity $e^{i\pi} = -1$ and the well-known quotient-remainder formula, $a = b\lfloor \frac{a}{b} \rfloor + a \mod b$, where $a, b, b \neq 0$ are any integers. Each term in the summation is then,

$$\rho^{C^{(k)}(n)}(e^{\frac{i\pi}{kp}})^{C^k(n)} = (-1)^{\left\lfloor \frac{C^{(k)}(n)}{kp} \right\rfloor}(-1)^{\frac{C^{(k)}(n) \mod (kp)}{(kp)}} \rho^{C^{(k)}(n)}. \tag{7}$$

Thus Equation (6) holds and Theorem 2 is proven. □

This theorem has real practical use in that it radically speeds up certain calculations and simplifies certain expressions on MATHEMATICA.

4. Symmetry Angle Spaces

The focus of this work is to restrict the centered polygonal functions, which are analytic on the open complex disk, to a topological space consisting of the union of the line segments lying along the symmetry angles which run from the origin to the natural boundary (located on the unit circle).

Let $\mathcal{D}$ be the open unit disk in the complex plane and let let $\tilde{\mathcal{D}}$ be the closed unit disk. Further, one can define $\mathcal{I}_p \equiv \rho e^{i\alpha_p}$ for $0 \leq \rho < 1$ (that, is the line segment along the pth symmetry angle, α_p. One likewise define the closure of $\mathcal{I}_p$ as $\tilde{\mathcal{I}}_p$, where now $0 \leq \rho \leq 1$.

The symmetry angle space is then defined as

$$\mathcal{P} \equiv \bigcup_{p=1}^{\infty} \mathcal{I}_p, \tag{8}$$

and its closure,

$$\tilde{\mathcal{P}} \equiv \bigcup_{p=1}^{\infty} \tilde{\mathcal{I}}_p. \tag{9}$$

Note that as p approaches ∞ the symmetry line approaches the real axis. Thus one needs to consider a second type of closure. If the real line is included, one denotes the subspaces as $\check{\mathcal{P}}$ and $\check{\tilde{\mathcal{P}}}$.

Thus, there are four related subspaces upon which the centered lacunary functions are restricted: $\mathcal{P}$, $\tilde{\mathcal{P}}$, $\check{\mathcal{P}}$, and $\check{\tilde{\mathcal{P}}}$. These subspaces are related to the so-called broom topological spaces (see

Reference [17]). They naturally take on a subspace topology, that is the normal topology for a line segment. All four of these subspaces are arc-connected and, in fact, star-connected through the origin. In Section 7, subspaces $\mathcal{P}, \check{\mathcal{P}}, \acute{\mathcal{P}}$, and $\mathring{\mathcal{P}}$ are homeomorphically mapped to longitudinal lines of the sphere. This allows for closed form expressions for integrals of $f(z)$ along paths in these mapped spaces.

5. Cyclic Decomposition

Along the symmetry angle, the resultant p-sequence has a $4p$ cycle, and, in fact, the $4p$ cycle further breaks into a $2p$-cycle at the modulus level as discussed in Section 3. Finally, by Lemma 2 the $2p^{\text{th}}$ member of the the $2p$ cycle is zero. Because of this, it is natural to consider a subset of $\mathcal{L}\left(C^{(k)}, \rho e^{\frac{i\pi}{kp}}\right)$ for which $N = 2pm$, $m \in \mathbb{Z}^+$; call this subsequence $\hat{\mathcal{L}}\left(C^{(k)}, \rho e^{\frac{i\pi}{kp}}\right)$. For every member of this subsequence $\lim_{\rho \to 1^-}$ is zero.

One can express the jth cycle as

$$f_j^{(k)}(\rho e^{\frac{i\pi}{kp}}) = \sum_{n=pj+1}^{2p(j+1)} \rho e^{\frac{i\pi}{kp}}, \tag{10}$$

where j is any non-negative integer. Thus, the full function can be decomposed into the cyclic summations,

$$f^{(k)} = \sum_{j=1}^{\infty} f_j^{(k)}. \tag{11}$$

Figure 2 shows the cyclic summation decomposition of $f^{(3)}(\rho e^{\frac{i\pi}{3p}})$ for the examples of $p = 1$ and $p = 3$. The fundamental component, $f_0^{(3)}$, captures much of the full function, but deviates significantly as $\rho \simeq 0.9$. The actual peak occurs at $\rho = \rho_{\max}$. An inspection of Figure 2 shows that $\rho_{\max}$ increases with increasing k as the curves are skewed towards the natural boundary.

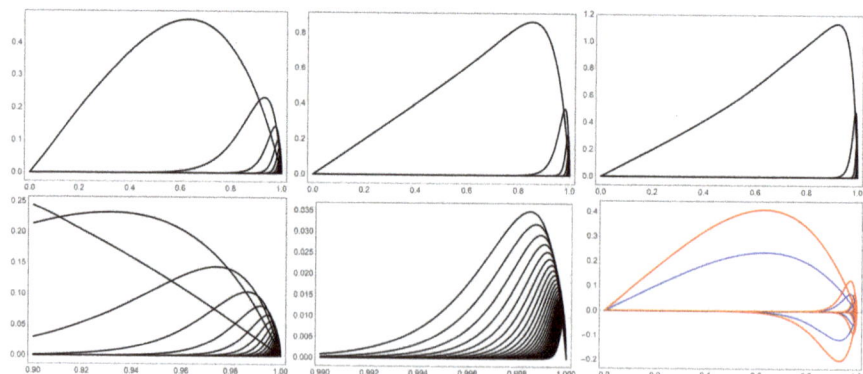

Figure 2. Cyclic decompositions for the centered polygonal lacunary functions along three of the line segments shown in the bottom right panel of Figure 1, that is, $k = 3$. The first 40 f_j are shown. The top row shows $|f_{40p}(\rho e^{\frac{i\pi}{3p}})|$: left panel $p = 1$, middle panel $p = 2$, right panel $p = 3$. The bottom row focuses on the $p = 1$ case in more detail. The left and middle panels show a sequential blow up near the natural boundary of the top left graph (note the displayed domain on the ρ axis). For better clarity, the first 10 f_j are not shown in the left panel and the first 20 f_j are not shown in the middle panel. Finally, the bottom right panel shows the real (blue) and imaginary (red) parts of $f_{40}(\rho e^{\frac{i\pi}{3}})$, that is, $k = 3, p = 1$.

The higher components $j \geq 1$ contribute very little for low values of ρ. Each of the subsequent higher components begin to make significant contributions to the full function closer and closer to the natural boundary. One notices in Figure 2 that both the real (blue curve in figure) and imaginary (red curve) parts of the component cyclic summations alternate signs. Figure 3 shows cyclic decompositions for the centered polygonal lacunary functions along the line segments at α_1 for several different values of k.

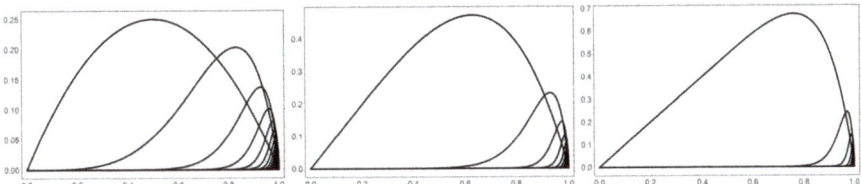

Figure 3. Cyclic decompositions for the centered polygonal lacunary functions along the line segments at α_1 for (left-to-right) $k = 1$, $k = 3$ (sames as in Figure 2), and $k = 8$. The other parameters are the same as in Figure 2. Increasing k skews the graph towards the natural boundary.

6. Parametric Curves

The centered polygonal lacunary functions on $\mathcal{P}$ can be represented in a visually instructive way via the parametric curves:

$$\mathsf{P}^{(k)}(\rho;p) = \left(\mathrm{Re}\left[f^{(k)}(\rho e^{\frac{i\pi}{kp}}) \right], \mathrm{Im}\left[f^{(k)}(\rho e^{\frac{i\pi}{kp}}) \right] \right). \tag{12}$$

The parametric curves for $k = 1, 2, 4, 8$ are shown in Figure 4. Here values of $p \in \{1, ..., 10\}$ are shown for each k. (Note, $\mathsf{P}^{(k)}$ is plotted in an auxiliary $\mathbb{R}^2$ plane, not in the original complex plane containing $\mathcal{P}$.)

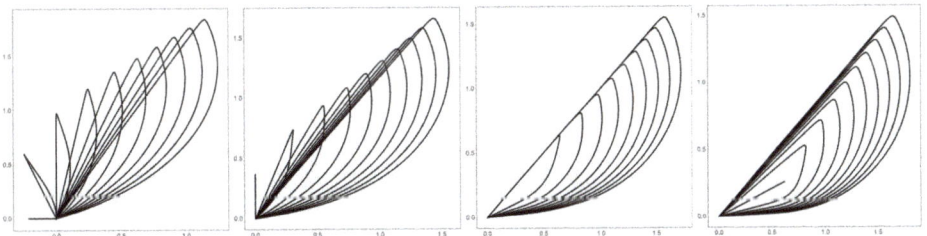

Figure 4. Parametric curves, $\mathsf{P}^{(3)}(\rho;p)$ from Equation (12), of $f_N(z)$ for four different values of k read left-to-right, top-to-bottom: $k = 1$, $k = 2$, $k = 4$, $k = 8$. Shown are the first 10 values of p. The case of $p = 1$ has no interior points and is directed at an angle equal to α_1 in $\mathbb{R}^2$. Increasing values of p lead to closed curves which are bigger and have greater interior area. As ρ goes from 0 to 1 the curve is traversed in a counterclockwise direction.

The most obvious feature is that these produce a closed curve in the plane starting at the origin for $\rho = 0$ and returning to the origin for $\rho = 1$. Note that the curves $\mathsf{P}^{(k)}(\rho;1)$ are all degenerate meaning that the encircled area is zero. Higher values of p give rise to larger and larger enclosed areas (Figure 4). Hand-in-hand with increasing area is increasing arclength which is also shown in Figures 4 and 5.

A more subtle view of the closed curves reveals an "acceleration" with ρ and this acceleration increases with increasing p. The "velocity" is represented as red tangent vectors in Figure 5. One notices a slow acceleration along the lower arc of the curve (for $\rho < \rho_{\max}$). Acceleration then rapidly increases at the apex of the curve and along the return path ($\rho \geq \rho_{\max}$). The change in acceleration at the apex corresponds to an abrupt change in arclength with ρ (see the bottom left panel of Figure 5).

An incidental observation regarding arclength is that it closely fits an empirical curve of the form $h(p) = A\sqrt{p} + c$ regardless of k (see bottom right panel of Figure 5).

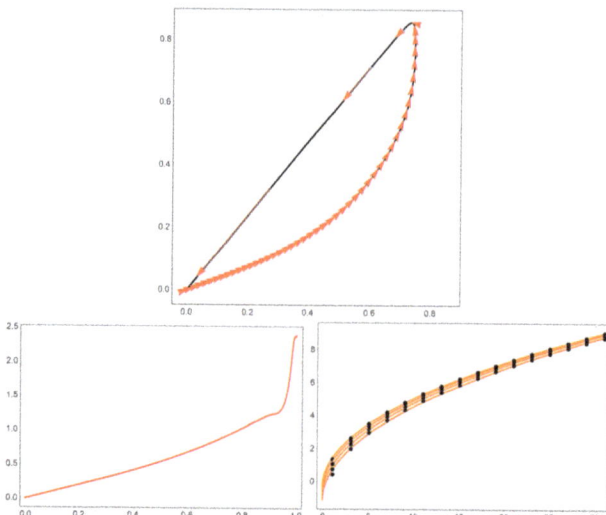

Figure 5. Top: Parametric plot $\mathsf{P}^{(3)}(\rho;p)$ (black curve) superimposed with red vectors indicating the "velocity" along the curve. One notices a modest "acceleration" until the curve turns back towards the origin whereupon the acceleration is markedly increased. Bottom left: the arclength (ordinate) versus ρ (abscissa) for $\mathsf{P}^{(3)}(\rho;p)$. Bottom right: Arclength of $\mathsf{P}^{(k)}(\rho;p)$ for $p=1,2,\ldots,30$ (black dots) associated with the parametric plots shown in Figure 4 fitted to $A\sqrt{p}+c$ (orange curve). The top curve is for $k=1$ and the bottom curve is for $k=8$. Fit parameters (A,c) for $k=1, k=2, k=4, k=8$ respectively: $(1.7880, -1.1038)$, $(1.7571, -0.7926)$, $(1.7175, -0.5052)$, $(1.6794, -0.2701)$.

Perhaps more interesting, however, is the geometrical behavior of the curves. The initial angle of the curve at $\rho = 0$ is $\alpha_p = \frac{\pi}{kp}$. This is intuitive and quick to prove.

Theorem 3. *The initial angle of* $\mathsf{P}^{(k)}(\rho;p)$ *is* α_p.

Proof. One can make use of the fact that for small ρ, $f_N(z)$ is dominated by the first term in the sum. This goes as

$$\lim_{\rho \to 0} f_N^{(k)}(\rho e^{\frac{i\pi}{kp}}) = \lim_{\rho \to 0} \sum_{n=1}^{N} (\rho e^{\frac{i\pi}{kp}})^{C^{(k)}(n)}$$

$$= \lim_{\rho \to 0} \left(\rho e^{\frac{i\pi}{kp}} + \rho^{k+1} e^{\frac{i\pi(k+1)}{kp}} + \rho^{\frac{3k+2}{2}} e^{\frac{i\pi(3k+2)}{2kp}} + \cdots \rho^{\frac{(N^2-N)k+2}{2}} e^{\frac{i\pi((N^2-N)k+2)}{2kp}} \right). \tag{13}$$

The asymptotic form as $\rho \to 0$ is

$$\lim_{\rho \to 0} f_N^{(k)}(\rho e^{\frac{i\pi}{kp}}) \sim \rho e^{\frac{i\pi}{kp}}. \tag{14}$$

The phase is $\frac{i\pi}{kp} = \alpha_k$, which completes the proof. □

Less intuitive is the behavior of the return angle as $\rho \to 1_-$. First after $\rho_{\max}$ the curve is nearly a straight line. Further, the angle of that line is $\frac{\pi}{k}$ for $p = 1$, but, interestingly, it asymptotically goes to $\frac{\pi}{4}$ as $p \to \infty$. The return angle becomes independent of k. The proof of this statement is probabilistic in nature and is wanting of a more rigorous proof.

Theorem 4. *The return angle of* $\mathsf{P}^{(k)}(\rho;p)$ *for* $p=1$ *is* $\frac{\pi}{k}$.

Proof. From Theorem 2 and $p=1$, one has

$$f_N(\rho e^{i\alpha_1}) = \sum_{n=1}^{N}(-1)^{\left\lfloor \frac{C^{(k)}(n)}{k} \right\rfloor}(-1)^{\frac{C^{(k)}(n) \mod k}{k}} \rho^{C^{(k)}(n)}. \tag{15}$$

Now, from Equation (3)

$$\left\lfloor \frac{C^{(k)}(n)}{k} \right\rfloor = \left\lfloor \frac{n^2-n}{2} + \frac{1}{k} \right\rfloor$$

$$= \left\lfloor \frac{n^2-n}{2} \right\rfloor = m(n) \in \mathbb{N} \tag{16}$$

and

$$\frac{C^{(k)}(n) \mod k}{k} = \frac{1}{k}\left(\frac{k(n^2-n)}{2}+1 \mod k\right)$$

$$= \frac{1}{k}. \tag{17}$$

So this reduces $f_N(\rho e^{i\alpha_1})$ to

$$f_N(\rho e^{i\alpha_1}) = \sum_{n=1}^{N}(-1)^{m(n)}(-1)^{\frac{1}{k}}\rho^{C^{(k)}(n)}$$

$$= (-1)^{\frac{1}{k}} \sum_{n=1}^{N}(-1)^{m(n)}\rho^{C^{(k)}(n)}. \tag{18}$$

The sum is now pure real and setting $(-1)^{\frac{1}{k}} = e^{\frac{i\pi}{k}}$. Hence, the return angle is $\frac{\pi}{k}$. □

Conjecture 1. *For p a positive integer,*

$$\lim_{p \to \infty} \mathsf{P}^{(k)}(\rho;p) = \frac{\pi}{4}. \tag{19}$$

Remark 1. *The proof is subtle and an analytic one remains elusive. Nonetheless, the conjecture is understandable on probabilistic grounds. Unfortunately, the limit of $\rho = 1$ is not helpful since the function is identically zero and information about the approach angle is lost. As opposed to the case of $\lim_{\rho \to 0}$, the case of $\lim_{\rho \to 1^-}$ now activates many terms in the summation of $f_N^{(k)}(\rho e^{\frac{i\pi}{kp}})$. In between ρ_{max} and 1 there is not equal weighting of the terms in the cyclic summation, but the weights of the higher terms are no longer negligible. Thus, the limit is a (non-zero) weighted average of many terms. For large p values, the weighed average of many $C^{(k)}(n)$ ultimately gives rise to $\mathrm{Re}\left[f_N^{(k)}\right] = \mathrm{Im}\left[f_N^{(k)}\right]$ and, hence, the return angle is $\frac{\pi}{4}$.*

Because these parametric curves produce enclosed regions, the area within the curves can be calculated. This area is found through a numerical integration of the curve, however the area of every value of k for $p = 1$ will be zero, as the parametric graph of $p = 1$ is a straight line. Figure 6 is a graph of the area of the associated parametric curves for $1 \leq k \leq 5$ and $1 \leq p \leq 10$. Each set of points shows the area for a distinct k value, with the bottom set being the area of $k = 1$, and the top being the area of $k = 5$. As the p value increases a linear trend appears, however the equation for what p approaches to does not seem to have a general trend.

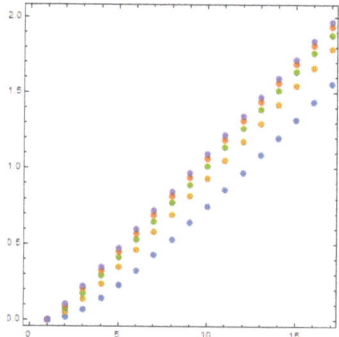

Figure 6. Plot of area for $1 \leq p \leq 17$ and $1 \leq k \leq 5$. Where the lowest set is the area of $k = 1$ and each successively higher line corresponds to the next greatest k value. The area for each p value approaches to a distinct line for each k value.

7. Whole Sphere Mapping

Due to the natural boundary of the centered polygonal lacunary functions sitting on the unit circle in the complex plane, there is no reason to consider the domain outside of the closed disk. There is an interesting and convenient mapping that maps the disk to S^2, which is embedded in $\mathbb{R}^3$ as the unit sphere centered at the origin, such that the entire unit circle is mapped to the south pole $((0,0,-1))$. As will be seen, this, in some sense, compresses the natural boundary to an isolated singularity. Further, the symmetry angle spaces map to longitudinal arcs and, given the nature of the p-sequences, this singularity is, again in some sense, removed.

Specifically, the above map is a parametric mapping that will take point, $z = \rho e^{i\phi}$ from $\bar{\mathcal{D}}$ into $\mathbb{R}^3$, such that the set of all points in $\bar{\mathcal{D}}$ cover the unit sphere centered at the origin. It is convenient to use spherical polar coordinates to describe the parametric surface. This is done with the identification,

$$\begin{aligned} \phi &= \phi & 0 \leq \phi < 2\pi \\ \theta &= 2\arcsin\rho & 0 \leq \theta < \pi. \end{aligned} \quad (20)$$

With this identification, ρ is expressed as a function of the zenith angle (θ) from spherical polar coordinates, $\rho = \sin(\frac{\theta}{2})$. Likewise, ϕ in the complex plane corresponds directly to azimuthal angle (ϕ) of spherical polar coordinates. The parametric mapping on S^2, which is embedded in $\mathbb{R}^3$ parametrically as $(\cos(\phi)\sin(\theta), \sin(\phi)\sin(\theta), \cos(\theta))$, can be written as

$$\hat{S} : \bar{\mathcal{D}} \to S^2$$
$$z = \rho e^{i\phi} \mapsto (\cos(\phi)\sin(2\arcsin\rho), \sin(\phi)\sin(2\arcsin\rho), \cos(2\arcsin\rho)). \quad (21)$$

Under this mapping, $f_N(z)$ becomes $f_N(\phi, \theta)$. The example of $|f_{16}(\phi, \theta)|$ is shown in Figure 7. The mapping is that of the centered polygonal lacunary function shown in the upper left panel of Figure 1. So the centered polygonal functions is the wrapped over the sphere such that the natural boundary gets pinched into the south pole.

Then the restriction to, for example, $\bar{\mathcal{P}}$ induces the map

$$\hat{S}_{\bar{\mathcal{P}}} : \bar{\mathcal{D}} \to S^2_{\bar{\mathcal{P}}}$$
$$z \mapsto (\cos(\alpha_p)\sin(2\arcsin\rho), \sin(\alpha_p)\sin(2\arcsin\rho), \cos(2\arcsin\rho)). \quad (22)$$

Here $S^2_{\bar{\mathcal{P}}}$ is the restricted domain of longitudinal arcs; an example is shown in Figure 8. $S^2_{\bar{\mathcal{P}}}$ (as well as $S^2_{\bar{\mathcal{P}}'}$) are star-connected through both the north pole (origin) and south pole (contracted unit circle). Because of this, one can define loops on $S^2_{\bar{\mathcal{P}}}$ and $S^2_{\bar{\mathcal{P}}'}$, with the north pole as the base-point, that traverse one longitudinal arc $S^2_{\hat{\mathcal{I}}_i}$ and return along another $S^2_{\hat{\mathcal{I}}_j}$. The fundamental group (in the homotopy sense) is $\pi_1 = \prod^\infty *\mathbb{Z} = \mathbb{Z}*\mathbb{Z}*\cdots$, where $*$ is the loop product, that is, the concatenation of loops [18].

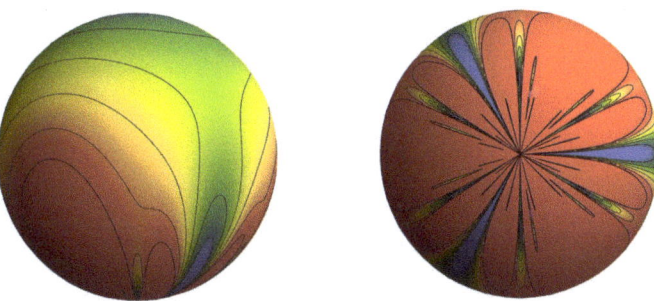

Figure 7. The whole sphere mapping of $\tilde{\mathcal{D}}$ onto S^2 (see text for the equations of the map). The mapping is that of the centered polygonal lacunary function shown in Figure 1 under $\hat{S}$. Two different viewpoints of the same function ($|f^{(3)}_{16}(\phi,\theta)|$) are shown. The left panel shows a "front" view such that the north pole $(0,0,1)$ is located directly on top and the south pole $(0,0,-1)$ directly on the bottom. The right panel shows the "bottom" view such that the south pole is directly in the center of the image. The unit circle maps to the single point at the south pole.

Figure 8. The superposition of the line segments shown in the bottom right panel of Figure 1 onto the sphere shown in Figure 7 under the mapping $\hat{S}$.

The spaces $S^2_{\hat{\mathcal{P}}}$ and $S^2_{\hat{\mathcal{P}}}$ offer an interesting opportunity to explore closed-loop path integrals of $f_N(z)$. Call the path along the p symmetry angle running from the north pole to south pole in $S^2_{\hat{\mathcal{P}}}$, Γ_p. Then a closed-loop can be obtained by considering $\Gamma_{ij} \equiv \Gamma_{p_i} - \Gamma_{p_j}$. The integral along Γ_p is expressed as

$$I_p(k,N) = \int_0^\pi f_N\left(e^{i\alpha_p}\sin\left(\frac{\theta}{2}\right)\right)\frac{e^{i\alpha_p}\cos\left(\frac{\theta}{2}\right)}{2}d\theta. \tag{23}$$

The second factor accounts for the appropriate integration metric along angle α_p. This integral can be evaluated and one has the following theorem.

Theorem 5. Let k, m, p be any positive integers and let $N = 4mp$. Then,

$$I_p(k, 4mp) = e^{\frac{2\pi i}{kp}} \sum_{n=1}^{4mp} \frac{(-1)^{\frac{n^2-n}{2p}}}{C^{(k)}(n)+1}. \tag{24}$$

Proof. Now, Equation (23) is

$$I_p(k, 4mp) = \frac{1}{2} \int_0^{\pi} f_N\left(e^{i\alpha_p} \sin\left(\frac{\theta}{2}\right)\right) e^{i\alpha_p} \cos\left(\frac{\theta}{2}\right) d\theta. \tag{25}$$

Expressing $f_{N(=4mp)}$ in summation form and interchanging the summation and the integration gives

$$I_p(k, 4mp) = \frac{1}{2} \sum_{n=1}^{4mp} e^{i\alpha_p(C^{(k)}(n)+1)} \int_0^{\pi} \left(\sin\left(\frac{\theta}{2}\right)\right)^{C^{(k)}(n)} \cos\left(\frac{\theta}{2}\right) d\theta. \tag{26}$$

Using,

$$\int_0^{\pi} \left(\sin\frac{\theta}{2}\right)^n \cos\frac{\theta}{2} d\theta = \frac{2}{n+1}, \tag{27}$$

the integral is quickly evaluated.

$$I_p(k, 4mp) = \frac{1}{2} \sum_{n=1}^{4mp} e^{i\alpha_p(C^{(k)}(n)+1)} \frac{2}{C^{(k)}(n)+1}$$

$$= \sum_{n=1}^{4mp} e^{\frac{i\pi}{k}(k\frac{n^2+n}{2}+2)} \frac{1}{C^{(k)}(n)+1}$$

$$= e^{\frac{2\pi i}{kp}} \sum_{n=1}^{4mp} \frac{(-1)^{\frac{n^2-n}{2p}}}{C^{(k)}(n)+1}. \tag{28}$$

where α_p and $C^{(k)}(n)$ were expressed in their functional form as well as expressing $e^{i\pi} = -1$. □

Conjecture 2. Let k and m be positive integers and $p = 1$,

$$\lim_{m \to \infty} I_1(k, m) \equiv I_1(k) = \frac{e^{\frac{2\pi i}{k}}}{2k\Delta_k} \left[-\psi\left(\frac{1-\Delta_k}{8}\right) + \psi\left(\frac{3-\Delta_k}{8}\right) + \psi\left(\frac{5-\Delta_k}{8}\right) \right.$$
$$\left. -\psi\left(\frac{7-\Delta_k}{8}\right) + \psi\left(\frac{1+\Delta_k}{8}\right) - \psi\left(\frac{3+\Delta_k}{8}\right) \right.$$
$$\left. -\psi\left(\frac{5+\Delta_k}{8}\right) + \psi\left(\frac{7+\Delta_k}{8}\right) \right]. \tag{29}$$

This can also be written as

$$I_1(k) = \frac{e^{\frac{2\pi i}{k}}}{k\Delta_k} \left[\pi \sec\left(\frac{\pi}{4}(1+\Delta_k)\right) - \pi \csc\left(\frac{\pi}{4}(1+\Delta_k)\right) \right], \tag{30}$$

where ψ is the digamma function (see Reference [25]) and $\Delta_k \equiv \sqrt{\frac{k-16}{k}}$.

Remark 2. When $p = 1$, Equation (28) becomes

$$\lim_{m \to \infty} I_1(k) = \sum_{n=1}^{\infty} \frac{(-1)^{\frac{n^2-n}{2}}}{C^{(k)}(n)+1} = \sum_{n=1}^{\infty} \frac{2(-1)^{\frac{n^2-n}{2}}}{kn^2 - kn + 4}. \tag{31}$$

This summation yields to a closed form which is Equation (29) (see Reference [19]). Using the relations for the digamma function built in to MATHEMATICA, this simplifies to Equation (30) (see Reference [19]).

The digamma function has many applications in physics and even in the life sciences (see the review by Hăşmăşanu et al. [26])

I_1 versus k is shown in Figure 9. I_1 approaches a k-dependent limit value.

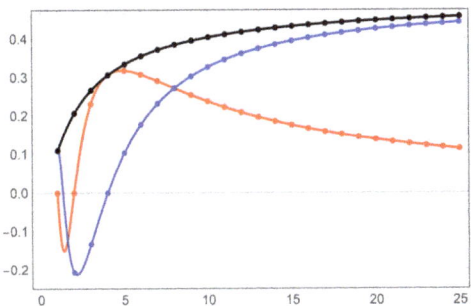

Figure 9. I_1 versus k (dots: $|I_1|$ - black, Re$[I_1]$ - blue, Im$[I_1]$ - red). The curves arise from Equation (30) in Corollary 2.

Corollary 1. Let k, m, p be positive integers and $k \neq 16$. Then, on $S^2_{\tilde{p}}$,

$$\lim_{p \to \infty} \lim_{m \to \infty} I_p(k, 4mp) \equiv I_\infty(k) = \frac{2\pi \tan\left(\frac{1}{2}\pi\Delta_k\right)}{k\Delta_k}, \quad (32)$$

where $\Delta_k \equiv \sqrt{\frac{k-16}{k}}$.

Proof. Beginning with the summation formula of the tangent function,

$$\pi \tan(\pi x) = 8x \sum_{n=1}^{\infty} \frac{1}{(2n-1)^2 - 4x^2} \quad (33)$$

(see Reference [25,27]), and $x = \frac{\Delta_k}{2}$. Starting with the right hand side of Equation (32),

$$\frac{2\pi \tan\left(\frac{1}{2}\pi\Delta_k\right)}{k\Delta_k} = \frac{16}{k\Delta_k} \frac{\Delta_k}{2} \sum_{n=1}^{\infty} \frac{1}{(2n-1)^2 - 4\frac{\Delta_k^2}{4}}. \quad (34)$$

Simplify and manipulating gives,

$$\frac{2\pi \tan\left(\frac{1}{2}\pi\Delta_k\right)}{k\Delta_k} = \frac{8}{k}\sum_{n=1}^{\infty} \frac{1}{(2n-1)^2 - \Delta_k^2}$$
$$= \frac{8}{k}\sum_{n=1}^{\infty} \frac{1}{(2n-1)^2 - \frac{k-16}{k}}$$
$$= 8\sum_{n=1}^{\infty} \frac{1}{k(2n-1)^2 - k - 16}$$
$$= 8\sum_{n=1}^{\infty} \frac{1}{4(kn^2 - kn + 4)}$$
$$= \sum_{n=1}^{\infty} \frac{1}{k\frac{n^2-n}{2} + 2} = \sum_{n=1}^{\infty} \frac{1}{C^{(k)}(n) + 1}. \tag{35}$$

This completes the proof. □

Corollary 2. *Let k be a positive integer, then $\lim_{k\to\infty} I_1(k) = \frac{1}{2}$ and $\lim_{k\to\infty} I_\infty(k) = \frac{1}{2}$.*

Proof. The proof follows from Equations (30) and (32) by first making a change of variable $k = \frac{1}{x}$. Upon doing this and performing a bit of algebraic simplification, Equation (30) becomes

$$I_1(x) = \frac{\pi e^{2i\pi x}\left(\csc\left(\frac{1}{4}\pi\left(\sqrt{1-16x}+1\right)\right) - \sec\left(\frac{1}{4}\pi\left(\sqrt{1-16x}+1\right)\right)\right)}{\sqrt{1-16x}}. \tag{36}$$

One can then expand this expression in a Taylor series about $x = 0$ to get

$$I_1(x) = \frac{1}{2} + (2 - (1-i)\pi)x + \left(24 - (8-4i)\pi - \left(\frac{2}{3} + 2i\right)\pi^2\right)x^2 + O\left(x^3\right). \tag{37}$$

Which in the limit of $x \to 0$ becomes $\frac{1}{2}$.

By a similar procedure, Equation (32) becomes

$$I_\infty(x) = \frac{2\pi x \tan\left(\frac{1}{2}\pi\sqrt{1-16x}\right)}{\sqrt{1-16x}}. \tag{38}$$

Series expansion gives

$$I_\infty(x) = \frac{1}{2} + 2x + \left(24 - \frac{8\pi^2}{3}\right)x^2 + O\left(x^3\right), \tag{39}$$

which, again is $\frac{1}{2}$ in the limit of $x \to 0$. □

Based on Corollary 2 the following unproven conjecture is proposed.

Conjecture 3. *Let k be a positive integer, then $\lim_{k\to\infty} I_p(k) = \frac{1}{2}$.*

Remark 3. $k = 16$ is special and Equations (30) and (32) must be evaluated using limits of $k \to 16$ and L'Hospital's rule. When $k = 16$, Equation (30) becomes $I_1 = \frac{\sqrt[8]{-1}\pi^2}{16\sqrt{2}}$ and Equation (32) becomes $I_\infty = \frac{\pi^2}{16}$.

The closed-loop integral is then on $S_{\tilde{p}}^2$ and $S_{\tilde{p}'}^2$,

$$L_{ij}(k) \equiv I_i(k) - I_j(k). \tag{40}$$

Of special interest is L_{p1}, where the return path is along $-\Gamma_1$. The left panel of Figure 10 shows the behavior of L_{p1} for $p = 1$ through $p = 20$ and $k = 1$. A finite limiting values is reached for $L_{\infty 1} \equiv \lim_{p \to \infty} L_{p1}$. It is natural to consider a normalized version of $L_{\infty 1}$ to compare different values of k. This is done by multiplying by $\frac{k}{\pi}$ and a graph is shown in the right panel of Figure 10. The dashed line in the figure represent the limiting value of $\frac{k}{\pi} L_{\infty 1}$ as $k \to \infty$ as given by the following theorem.

Theorem 6. Let k be any positive integer. On $S^2_{\frac{p}{p'}}$,

$$\lim_{k \to \infty} \frac{k}{\pi} L_{\infty 1}(k) = 1 - i \qquad (41)$$

Proof. One considers

$$\lim_{k \to \infty} \frac{k}{\pi}(I_\infty(k) - I_1(k)), \qquad (42)$$

(c.f., Equations (30) and (32)) and uses the same strategy as in the proof of those expressions. Change of variable $k = \frac{1}{x}$ and simplification gives,

$$\frac{1}{x\pi}(I_\infty(x) - I_1(x)) = \frac{\frac{2\pi x \tan\left(\frac{1}{2}\pi\sqrt{1-16x}\right)}{\sqrt{1-16x}} + \frac{\pi e^{2i\pi x} x \left(\csc\left(\frac{1}{4}\pi\left(\sqrt{1-16x}+1\right)\right) - \sec\left(\frac{1}{4}\pi\left(\sqrt{1-16x}+1\right)\right)\right)}{\sqrt{1-16x}}}{\pi x}. \qquad (43)$$

And series expansion yields

$$\frac{1}{x\pi}(I_\infty(x) - I_1(x)) = (1 - i) + ((8 - 4i) - (2 - 2i)\pi)x + (-32 + 16i)(\pi - 3)x^2 + O\left(x^3\right). \qquad (44)$$

Thus the limit as $x \to 0$ is $1 + i$. □

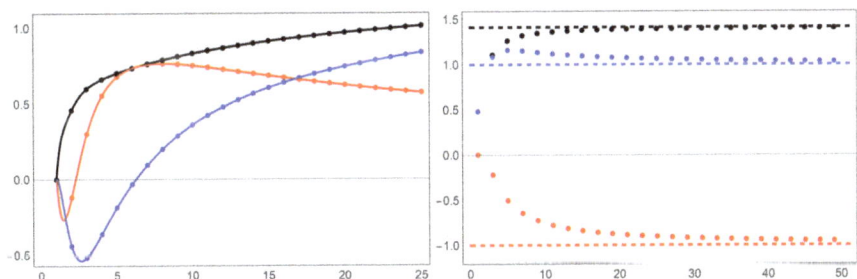

Figure 10. Left Panel: L_{p1} versus p (dots: $|L_{p1}|$ - black, Re$[L_{p1}]$ - blue, Im$[L_{p1}]$ - red) for the case of $k = 1$. The curves arise from Equation (30) in Corollary 2. Right Panel: $\frac{k}{\pi} L_{\infty 1}$ versus k ($|\frac{k}{\pi} L_{\infty 1}|$ - black, Re$[\frac{k}{\pi} L_{\infty 1}]$ - blue, Im$[\frac{k}{\pi} L_{\infty 1}]$ - red). The dashed lines represent the $\lim_{k \to \infty} \frac{k}{\pi} L_{\infty 1} = 1 - i$.

8. Conclusions

This work focused on the centered polygonal lacunary functions restricted to symmetry angle space. The periodicity of the p-sequences and the existence of a convergent subsequence provided a framework for decomposition of the centered polygonal lacunary functions. This decomposition could be potentially useful in renormalization procedures as one approaches the natural boundary.

The surjective spherical mapping of the unit disk such that the natural boundary is mapped to the south pole was useful in investigating line integrals of the centered polygonal lacunary functions. Closed form functional representations were achieved in some cases.

It is hoped that this work provides useful insight into the nature of the natural boundary of centered polygonal lacunary functions, both on the full unit disk and also restricted to symmetry angle space. Statistical mechanics is the most promising link of this work to physics. This is for two reasons. First is simply a counting application, for example, the canonical partition function. Second is the self-similarity of these functions and the possible use in renormalization schemes applied to phase transitions. Optics may be the closest experimental link either via simple signal processing scheme or, more interestingly, in application to, for example the Talbot effect or other such phenomena.

Author Contributions: L.K.M., K.S. and D.J.U. conceived of and designed the investigation; K.S. and D.J.U. provided background for the investigation; D.J.U. wrote the MATHEMATICA code to perform the investigation; L.K.M., K.S., and D.J.U. analyzed the data; D.J.U. wrote the original draft of manuscript; L.K.M., K.S. and D.J.U. edited the manuscript. All authors have read and agreed to the published version of the manuscript.

Funding: This research was funded by the Concordia College Chemistry Endowment Fund.

Acknowledgments: Douglas R. Anderson, Trenton Vogt and Drew Rutherford are acknowledged for valuable discussion.

Conflicts of Interest: The authors declare no conflict of interest. The funders had no role in the design of the study; in the collection, analyses, or interpretation of data; in the writing of the manuscript, or in the decision to publish the results.

References

1. Hille, E. *Analytic Function Theory, Vol. I*; Ginn and Company: Boston, MA, USA, 1959.
2. Hille, E. *Analytic Function Theory, Vol. II*; Ginn and Company: Boston, MA, USA, 1962.
3. Creagh, S.C.; White, M.M. Evanescent escape from the dielectric ellipse. *J. Phys. A* **2010**, *43*, 465102. [CrossRef]
4. Greene, J.M.; Percival, I.C. Hamiltonian maps in the complex plane. *Physica* **1981**, *3D*, 530–548.
5. Shudo, A.; Ikeda, K.S. Tunneling effect and the natural boundary of invariant tori. *Phys. Rev. Lett.* **2012**, *109*. [CrossRef] [PubMed]
6. Yamada, H.S.; Ikeda, K.S. Analyticity of quantum states in one-dimen-sional tight-binding model. *Eur. Phys. J. B* **2014**, *87*, 208. [CrossRef]
7. Guttmann, A.J.; Enting, I.G. Solvability of some statistical mechanical systems. *Phys. Rev. Lett.* **1996**, *76*, 344–347. [CrossRef] [PubMed]
8. Orrick, W.P.; Nickel, B.G.; Guttmann, A.J.; Perk, J.H.H. Critical behavior of the two-dimensional Ising susceptibility. *Phys. Rev. Lett.* **2001**, *86*, 4120–4123. [CrossRef] [PubMed]
9. Nickel, B. On the singularity structure of the 2D Ising model susceptibility. *J. Phys. A Math. Gen.* **1999**, *32*, 3889–3906. [CrossRef]
10. Jensen, G.; Pommerenke, C.; Ramirez, J.M. On the path properties of a lacunary power series. *Proc. Am. Math. Soc.* **2014**, *142*, 1591–1606. [CrossRef]
11. Eckstein, M.; Zając A. Asymptotic and exact expansion of heat traces. *Math. Phys. Anal. Geom.* **2015**, *18*, 28. [CrossRef]
12. Behr, N.; Dattoli, G.; Duchamp, G.H.E.; Licciardi, S.; Penson K.A. Operational methods in the study of Sobolev-Jacobi polynomials. *Mathematics* **2019**, *7*, 124. [CrossRef]
13. Kişi, Ö.; Gümüş, H.; Savas, E. New definitions about $A^\mathcal{I}$-statistical convergence with respect to a sequence of modulus functions and lacunary sequences. *Axioms* **2018**, *7*, 24. [CrossRef]
14. Sullivan, K.; Rutherford, D.; Ulness, D.J. Centered polygonal lacunary sequences. *Mathematics* **2019**, *7*, 943. [CrossRef]
15. Sullivan, K.; Rutherford, D.; Ulness, D.J. Centered polygonal lacunary Graphs: A graph theoretic approach to p-sequences of centered polygonal lacunary functions. *Mathematics* **2019**, *7*, 1021. [CrossRef]
16. Mork, L.K.; Vogt, T.; Sullivan, K.; Rutherford, D.; Ulness, D.J. Exploration of filled-in Julia sets arising from centered polygonal lacunary functions. *Fract. Fract.* **2019**, *3*, 42. [CrossRef]
17. Steen, L.A.; Seebach, J.A., Jr. *Counterexamples in Topology*; Dover Publications: New York, NY, USA, 1995.
18. Munkres, J.R. *Topology*; Pearson: London, UK, 2017.
19. *Mathematica 11*; Wolfram Research: Champaign, IL, USA, 2018.
20. Schlicker, S.J. Numbers simultaneously polygonal and centered polygonal. *Math. Mag.* **2011**, *84*, 339–350. [CrossRef]
21. Teo, B.K.; Sloane, J.A. Magic numbers in polygonal clusters. *Inorg. Chem.* **1985**, *24*, 4545–4558. [CrossRef]
22. Deza, E.; Deza, M.-M. *Figurate Numbers*; World Scientific: Hackensack, NJ, USA, 2012.
23. Edgar, T. Visual decompositions of polygonal number. *College Math. J.* **2020**, *51*, 9–12. [CrossRef]
24. Hoggatt, V.E., Jr.; Bicknell, M. Triangular Numbers. *Fibonacci Q.* **1974**, *12*, 221–230.

25. Abromowitz, M.; Stegun, I.A. *Handbook of Mathematical Functions*; Dover Publications: New York, NY, USA, 1972.
26. Hăşmăşanu, M.; Bolboacă, S.D.; Jäntschi, L. Bose-Einstein and Fermi-Dirac distributions and their use in biological sciences. *Bull. UASVM Vet. Med.* **2014**, *71*, 114–123.
27. Frietag, E. *Complex Analysis*, 2nd ed.; Springer: New York, NY, USA, 2009.

© 2020 by the authors. Licensee MDPI, Basel, Switzerland. This article is an open access article distributed under the terms and conditions of the Creative Commons Attribution (CC BY) license (http://creativecommons.org/licenses/by/4.0/).

Article

Best Subordinant for Differential Superordinations of Harmonic Complex-Valued Functions

Georgia Irina Oros

Department of Mathematics and Computer Sciences, Faculty of Informatics and Sciences, University of Oradea, 410087 Oradea, Romania; georgia_oros_ro@yahoo.co.uk or goros@uoradea.ro

Received: 17 September 2020; Accepted: 11 November 2020; Published: 16 November 2020

Abstract: The theory of differential subordinations has been extended from the analytic functions to the harmonic complex-valued functions in 2015. In a recent paper published in 2019, the authors have considered the dual problem of the differential subordination for the harmonic complex-valued functions and have defined the differential superordination for harmonic complex-valued functions. Finding the best subordinant of a differential superordination is among the main purposes in this research subject. In this article, conditions for a harmonic complex-valued function p to be the best subordinant of a differential superordination for harmonic complex-valued functions are given. Examples are also provided to show how the theoretical findings can be used and also to prove the connection with the results obtained in 2015.

Keywords: differential subordination; differential superordination; harmonic function; analytic function; subordinant; best subordinant

MSC: 30C80; 30C45

---

1. Introduction and Preliminaries

Since Miller and Mocanu [1] (see also [2]) introduced the theory of differential subordination, this theory has inspired many researchers to produce a number of analogous notions, which are extended even to non-analytic functions, such as strong differential subordination and superordination, differential subordination for non-analytic functions, fuzzy differential subordination and fuzzy differential superordination.

The notion of differential subordination was adapted to fit the harmonic complex-valued functions in the paper published by S. Kanas in 2015 [3]. In that paper, considering Ω and Δ any sets in the complex plane $\mathbb{C}$ and taking the functions $\varphi : \mathbb{C}^3 \times U \to \mathbb{C}$ and p, a harmonic complex-valued function in the unit disc U of the form $p(z) = p_1(z) + \overline{p_2(z)}$, where p_1 and p_2 are analytic in U properties of the function p were determined such that p satisfies the differential subordination

$$\psi(p(z), Dp(z), D^2p(z); z) \subset \Omega \Rightarrow p(U) \subset \Delta.$$

Inspired by the idea provided by Miller and Mocanu [1], and following the research in [3,4], the notion of differential superordination for harmonic complex—valued functions was introduced in [5]. In that paper, properties of the harmonic complex-valued function p of the form $p(z) = p_1(z) + \overline{p_2(z)}$, with p_1 and p_2 analytic in U, such that p satisfies the differential superordination

$$\Omega \subset \psi(p(z), Dp(z), D^2p(z); z) \Rightarrow \Delta \subset p(U).$$

Continuing the study on differential superordinations for harmonic complex-valued functions started in paper [5], the problem of finding the best subordinant of a differential superordination for harmonic complex-valued functions is studied in the present paper and a method for finding the best subordinant is provided in a theorem and few corollaries in the Main Results section. Examples are also given using those original and new theoretical findings.

The well-known definitions and notations familiar to the field of complex analysis are used. The unit disc of the complex plane is denoted by U. $\mathcal{H}(U)$ stands for the class of analytic functions in the unit disc and the classical definition for class A_n is applied, and it is known that it contains all functions from class $\mathcal{H}(U)$, which have the specific form

$$f(z) = z + a_{n+1}z^{n+1} + \ldots,$$

with $z \in U$ and A_1 written simply A. All the functions in class A which are univalent in U form the class denoted by S. In particular, the functions in class A who have the property that

$$\operatorname{Re} \frac{zf''(z)}{zf'(z)} + 1 > 0$$

represent the class of convex functions K.

A harmonic complex-valued mapping of the simply connected region Ω is a complex-valued function of the form

$$f(z) = h(z) + \overline{g(z)}, \tag{1}$$

where h and g are analytic in Ω, with $g(z_0) = 0$, for some prescribed point $z_0 \in \Omega$.

We call h and g analytic and co-analytic parts of f, respectively. If f is (locally) injective, then f is called (locally) univalent. The Jacobian and the second complex dilatation of f are given by

$$J_f(z) = |h'(z)|^2 - |g'(z)|^2 \quad \text{and} \quad w(z) = \frac{g'(z)}{h'(z)}, \quad z \in U,$$

respectively. If $J_f(z) > 0, z \in U$, then f is a local sense-preserving diffeomorphism.

A function $f \in C^2(\Omega)$, $f(z) = u(z) + iv(z)$, which satisfies

$$\Delta f = \frac{\partial^2 f}{\partial x^2} + \frac{\partial^2 f}{\partial y^2} = 0$$

is called harmonic function.

By $Har(U)$ we denote the class of complex-valued, sense-preserving harmonic mappings in U. For $f \in Har(U)$, let the differential operator D be defined as follows

$$Df = z \cdot \frac{\partial f}{\partial z} - \overline{z} \frac{\partial f}{\partial \overline{z}} = zh'(z) - \overline{zg'(z)}, \tag{2}$$

where $\frac{\partial f}{\partial z}$ and $\frac{\partial f}{\partial \overline{z}}$ are the formal derivatives of function f

$$\frac{\partial f}{\partial z} = \frac{1}{2}\left(\frac{\partial f}{\partial x} - i\frac{\partial f}{\partial y}\right) \quad \text{and} \quad \frac{\partial f}{\partial \overline{z}} = \frac{1}{2}\left(\frac{\partial f}{\partial x} + i\frac{\partial f}{\partial y}\right). \tag{3}$$

The conditions (3) are satisfied for any function $f \in C'(\Omega)$ not necessarily harmonic, nor analytic. Moreover, we define the n-th order differential operator by recurrence relation

$$D^2 f = D(Df) = Df + z^2 h'' - \overline{z^2 q''}, \quad D^n f = D(D^{n-1}f). \tag{4}$$

Remark 1. *If $f \in \mathcal{H}(U)$ (i.e., $g(z) = 0$) then $Df(z) = zf'(z)$.*

In order to prove the main results of this paper, we use the following definitions and lemmas:

Definition 1 ([3] Definition 2.2). By Q, we denote the set of functions

$$q(z) = q_1(z) + \overline{q_2(z)},$$

harmonic complex-valued and univalent on $\overline{U} - E(q)$, where

$$E(q) = \left\{ \zeta \in \partial U : \lim_{z \to \zeta} q(z) = \infty \right\}.$$

Moreover, we assume that $D(q(\zeta)) \neq 0$, for $\zeta \in \partial U \setminus E(q)$.
The set $E(q)$ is called an exception set. We note that the functions

$$q(z) = \overline{z}, \quad q(z) = \frac{1+\overline{z}}{1-\overline{z}}$$

are in Q, therefore Q is a nonempty set.

Definition 2 ([5] Definition 2.2). Let $\varphi : \mathbb{C}^3 \times U \to \mathbb{C}$ and let h be harmonic univalent in U. If p and $\varphi(p(z), Dp(z), D^2(p(z)))$ are harmonic univalent in U, and satisfy the second-order differential superordination for harmonic complex-valued functions

$$h(z) \prec \varphi(p(z), Dp(z), D^2 p(z); z) \tag{5}$$

then p is called a solution of the differential superordination.

A harmonic univalent function q is called a subordinant of the solutions of the differential superordination for harmonic complex-valued functions, or more simply a subordinant if $q \prec p$, for all p satisfying (5).

An univalent harmonic subordinant $\tilde{q}$ that satisfies $q \prec \tilde{q}$ for all subordinants q of (5) is said to be the best subordinant. The best subordinant is unique up to a rotation of U.

Lemma 1 ([5] Theorem 3.2). Let h, q be harmonic and univalent functions in U, $\varphi : \mathbb{C}^2 \times \overline{U} \to \mathbb{C}$, and suppose that

$$\varphi(q(z), tDq(z); \zeta) \in h(U),$$

for $z \in U$, $\zeta \in \partial U$ and $0 < t \leq \dfrac{1}{m} \leq 1$, $m \geq 1$.

If $p \in Q$ and $\varphi(p(z), D(p(z)); z \in U)$ is univalent in U, then

$$h(z) \prec \varphi(p(z), Dp(z); z \in U)$$

implies

$$q(z) \prec p(z), \ z \in U.$$

Furthermore, if $\varphi(q(z), Dq(z); z \in U) = h(z)$, has an univalent solution $q \in Q$, then q is the best subordinant.

Let $f : U \to \mathbb{C}$. We consider the special set

$$E(U) = \{f : f \in C'(U), \ Df \in C'(U)\} \supset C^2(U).$$

Lemma 2 ([6] Theorem 7.2.2, p. 131). If the function $f \in E(U)$ satisfies

(i) $f(0) = 0$, $f(z) \cdot Df(z) \neq 0$, $z \in \dot{U}$;

(ii) $Jf(z) = \left| \dfrac{\partial f}{\partial z} \right|^2 - \left| \dfrac{\partial f}{\partial \overline{z}} \right|^2 > 0$, $z \in U$;

(iii) $\operatorname{Re} \dfrac{D^2 f(z)}{Df(z)} > 0$, $z \in \dot{U}$

then the function f is convex in U. Furthermore $f(U_r)$ is a convex domain for any $r \in (0,1)$.

2. Main Results

In Definitions 1 and 2, just like in the hypothesis of Lemma 1, the function q must have a "nice" behavior on the border of the unit disc. If this condition is not satisfied or if the behavior of function q on the border of the domain is unknown, then the superordination $q(z) \prec p(z)$ can be proven by using a limiting procedure.

The next theorem and the corollaries give the sufficient conditions for obtaining the best subordinant for the differential superordination.

Theorem 1. *Let h be a convex harmonic complex-valued function in U, with $h(0) = a$, and let $\theta : D \subset \mathbb{C} \to \mathbb{C}$, $\phi : D \subset \mathbb{C} \to \mathbb{C}$ be a harmonic complex-valued function in a domain D. Suppose that the differential equation*

$$\theta[q(z)] + Dq(z) \cdot \phi[q(z)] = h(z), \ z \in U, \tag{6}$$

has an univalent harmonic solution q that satisfies $q(0) = a$, $q(U) \subset D$ and

$$\theta[q(z)] \prec h(z), \ z \in U. \tag{7}$$

Let p be a harmonic complex-valued univalent function with $p(0) = h(0) = \theta[p(0)]$, $p \in Q$ and $p(U) \subset D$. Then

$$h(z) \prec \theta[p(z)] + Dp(z) \cdot \phi[p(z)], \tag{8}$$

implies

$$q(z) \prec p(z), \ z \in U.$$

The function q is the best subordinant.

Proof. We can assume that h, p and q satisfy the conditions of the theorem on the closed disc $\overline{U}$, and $Dq(\zeta) \neq 0$, for $|\zeta| = 1$. If not, we can replace h, p and q by $h(\rho z)$, $p(\rho z)$ and $q(\rho z)$, where $0 < \rho < 1$.

These new functions have the desired properties on $\overline{U}$, and we can use them in the proof of the theorem. Theorem 1 would then follow by letting $\rho \to 1$. We will use Lemma A to prove this result.

Let $\varphi : \mathbb{C}^2 \times U \to \mathbb{C}$, where

$$\varphi(r, s) = \theta(r) + s \cdot \phi(r). \tag{9}$$

For $r = p(z)$, $s = Dp(z)$, relation (9) becomes

$$\varphi(p(z), Dp(z)) = \theta[p(z)] + Dp(z) \cdot \phi[p(z)], \tag{10}$$

and the superordination (8) becomes

$$h(z) \prec \varphi[p(z)] + Dp(z) \cdot \phi[p(z)]. \tag{11}$$

For $r = q(z)$ and $s = Dq(z)$, relation (9) becomes

$$\varphi(q(z), Dq(z)) = \theta[q(z)] + Dq(z) \cdot \phi[q(z)], \ z \in U \tag{12}$$

and (6) is equivalent to

$$\varphi(q(z), Dq(z)) = h(z), \ z \in U.$$

For $r = q(z)$ and $s = tDq(z)$, $0 \leq t \leq 1$, relation (9) becomes

$$\varphi(q(z), tDq(z)) = \theta[q(z)] + tDq(z) \cdot \phi[q(z)], \ 0 \leq t \leq 1. \tag{13}$$

From (6), we have
$$Dq(z) \cdot \phi[q(z)] = h(z) - \theta[q(z)]. \tag{14}$$

Using (14) in (13), we have
$$\varphi(q(z), tDq(z)) = (1-t)\theta[q(z)] + th(z), \ 0 \le t \le 1. \tag{15}$$

Since h is a convex function, $f(U)$ is a convex domain and using (7), we have
$$\varphi(q(z), tDq(z)) \in h(U), \text{ for } 0 \le t \le 1.$$

Since the conditions from Lemma A are satisfied, we have
$$q(z) \prec p(z), \ z \in U.$$

Since q is the solution of Equation (6) we get that q is the best subordinant. □

In the special case when $\theta(w) = w$, and
$$\phi(w) = \frac{1}{\beta w + \gamma}, \ w = q(z), \ z \in U,$$

we obtain the following result for the Briot–Bouquet differential superordination.

Corollary 1. *Let $\beta, \gamma \in \mathbb{C}, \beta \ne 0$, and let h be a convex harmonic complex-valued function in U, with $h(0) = a$. Suppose that the differential equation*
$$q(z) + \frac{Dq(z)}{\beta q(z) + \gamma} = h(z), \ z \in U$$

has an univalent harmonic complex-valued solution q that satisfies $q(0) = a$ and $q(z) \prec h(z)$. If $p \in Q$ and $p(z) + \frac{Dp(z)}{\beta p(z) + \gamma}$ is harmonic complex-valued univalent in U, then
$$h(z) \prec p(z) + \frac{Dp(z)}{\beta p(z) + \gamma}$$

implies $q(z) \prec p(z), z \in U$. The function q is the best subordinant.

If $\theta(w) = w$ and $\phi(w) = \beta w + \gamma, \beta \ne 0, w = q(z), \gamma \in \mathbb{C}$, we obtain the following result.

Corollary 2. *Let $\beta, \gamma \in \mathbb{C}, \beta \ne 0$, and let h be a convex harmonic-valued function in U, with $h(0) = a$. Suppose that the differential equation*
$$q(z) + Dq(z)[\beta q(z) + \gamma] = h(z), \ z \in U,$$

has an univalent harmonic complex-valued solution q that satisfies $q(0) = a$ and $q(z) \prec h(z)$. If $p \in Q$ and $p(z) + Dp(z)[\beta p(z) + \gamma]$ is univalent harmonic complex valued in U, then
$$h(z) \prec p(z) + Dp(z)[\beta p(z) + \gamma]$$

implies $q(z) \prec p(z)$. The function q is the best subordinant.

If $\theta(w) = w, \phi(w) = \frac{1}{\gamma}, \gamma \ne 0, w = q(z)$, we obtain the following result.

Corollary 3. Let h be a convex harmonic complex-valued function in U, with $h(0) = a$. Let $\gamma \neq 0$, with $\operatorname{Re} \gamma \geq 0$. Suppose that the differential equation

$$q(z) + \frac{1}{\gamma} Dq(z) = h(z), \; z \in U$$

has an univalent harmonic complex-valued solution q that satisfies $q(0) = a$ and $q(z) \prec p(z)$.
If $p \in Q$ and $p(z) + \frac{1}{\gamma} Dp(z)$ is univalent harmonic complex-valued in U, then

$$h(z) \prec p(z) + \frac{1}{\gamma} \cdot Dp(z), \; z \in U$$

implies

$$q(z) \prec p(z), \; z \in U.$$

The function q is the best subordinant.

Example 1. For $\gamma = 2$, the univalent harmonic complex-valued function $q(z) = 6z - 4\bar{z}$, is the solution of the equation

$$q(z) + \frac{1}{2} Dq(z) = h(z) = 9z - 2\bar{z}.$$

We next prove that h is a harmonic non-analytic function.

$$h(z) = 9(x + iy) - 2(x - iy) = 7x + 11iy.$$

We have

$$\frac{\partial h}{\partial x} = 7, \; \frac{\partial^2 h}{\partial x^2} = 0, \; \frac{\partial h}{\partial y} = 11, \; \frac{\partial^2 h}{\partial y^2} = 0, \; \frac{\partial^2 h}{\partial x^2} + \frac{\partial^2 h}{\partial y^2} = 0.$$

We obtain that h is univalent harmonic complex-valued function and since $\frac{\partial h}{\partial x} \neq \frac{\partial h}{\partial y}$, we conclude that it is not analytic.

We next prove that the harmonic function h is also convex.
In order to do that, we show that it satisfies the conditions in the hypothesis of Lemma 2.
We calculate:

$$Dh(z) = z \frac{\partial h}{\partial z} - \bar{z} \frac{\partial h}{\partial \bar{z}} = 9z + 2\bar{z},$$

$$D^2 h(z) = D(Dh(z)) = 9z - 2\bar{z},$$

(i) $h(0) = 0$, $h(z) \cdot Dh(z) = (9z - 2\bar{z})(9z + 2\bar{z}) \neq 0$, $z \in \dot{U}$;

(ii) $Jh(z) = \left|\frac{\partial h}{\partial z}\right|^2 - \left|\frac{\partial h}{\partial \bar{z}}\right|^2 = 77 > 0$;

(iii) $\operatorname{Re} \frac{D^2 h(z)}{Dh(z)} = \operatorname{Re} \frac{9z - 2\bar{z}}{9z + 2\bar{z}} = \frac{77x^2 + 77y^2}{121x^2 + 49y^2} > 0, z \in \dot{U}.$

As can be seen, all the conditions in Lemma B are satisfied, hence h is a harmonic convex function.
Using Corollary 3, we have:
If $p \in Q$, $p(0) = q(0) = 0$ and $p(z) + \frac{Dp(z)}{z}$ is univalent harmonic complex-valued in U, then

$$9z - 2\bar{z} \prec p(z) + \frac{Dp(z)}{z}$$

implies

$$6z - 4\bar{z} \prec p(z), \; z \in U.$$

The function $q(z) = 6z - 4\bar{z}$ is the best subordinant.

Example 2. *For $\gamma = 1$, the univalent harmonic complex-valued function $q(z) = 1 + 2z - 4\bar{z}$ is the solution of the equation:*

$$q(z) + Dq(z) = 1 + 4z = h(z).$$

We next prove that h is a harmonic complex-valued function.

$$h(z) = 1 + 4(x + iy) = 1 + 4x + i \cdot 4y.$$

We have

$$\frac{\partial h}{\partial x} = 4, \quad \frac{\partial^2 h}{\partial x^2} = 0, \quad \frac{\partial h}{\partial y} = 4, \quad \frac{\partial^2 h}{\partial y^2} = 0.$$

From $\frac{\partial^2 h}{\partial x^2} + \frac{\partial^2 h}{\partial y^2} = 0$, we have $h(z) = 1 + 4z$, is a harmonic complex-valued function.

We next prove that the harmonic function is also convex.
In order to that, we show that it satisfies the conditions in the hypothesis of Lemma B.
We calculate:

$$Dh(z) = z \cdot \frac{\partial h}{\partial z} - \bar{z} \cdot \frac{\partial h}{\partial \bar{z}} = 4z,$$

$$D^2 h(z) = D(Dh(z)) = z \cdot \frac{Dh(z)}{z} - \bar{z} \cdot \frac{Dh(z)}{\partial \bar{z}} = 4z.$$

(i) $h(0) = 1$, $h(z) \cdot Dh(z) = 4z + 16z^2 \neq 0$, $z \in \dot{U}$;

(ii) $Jh(z) = \left|\frac{\partial h}{\partial z}\right|^2 - \left|\frac{\partial h}{\partial \bar{z}}\right|^2 = 16 > 0$;

(iii) $\operatorname{Re} \frac{D^2 h(z)}{Dh(z)} = \operatorname{Re} \frac{4z}{4z} = \operatorname{Re} 1 = 1 > 0$, $z \in \dot{U}$.

As can be seen, all the conditions in Lemma B are satisfied, hence h is a harmonic convex function.
Using Corollary 3 we have:
If $p \in Q$, $p(0) = q(0) = 1$ and $p(z) + Dp(z)$ is univalent harmonic complex-valued in U, then

$$1 + 4z \prec p(z) + Dp(z)$$

implies

$$1 + 2z - 4\bar{z} \prec p(z), \quad z \in U.$$

The function $q(z) = 1 + 2z - 4\bar{z}$ is the best subordinant.

Remark 2. *Using Example 2 and Example 2.4 in [3], we can write the following sandwich type result:*
If $p \in Q$, $p(0) = q(0) = 1$ and $p(z) + Dp(q)$ is univalent harmonic complex-valued in U, then

$$1 + 4z \prec p(z) + Dp(q) \prec \frac{1+z}{1-z} + \frac{\bar{z}}{1-\bar{z}}$$

implies

$$1 + 2z - 4\bar{z} \prec p(z) \prec \frac{1+z}{1-z} + \frac{\bar{z}}{1-\bar{z}}, \quad z \in U.$$

3. Conclusions

The notion of differential superordination for harmonic complex-valued functions is a new topic emerged in the theory of differential superordinations. It contributes to further developing the theory of differential superordinations. The study done related to the research of this topic is just starting, so the present paper provides essential means for continuing this idea. The original and new results contained in the Main Results section of the present paper are important, since the problem of finding the best subordinant of the differential superordination for harmonic complex-valued functions is essential for the study related to the topic as it is well-known from the classical theory of differential superordinations. No further findings can be done without having a method for finding the best subordinant. A method is given in Theorem 1 and in the corollaries that follow. Using those results, researchers interested in the topic should be able to obtain further original results. Two examples are also enclosed, giving a better view on the idea.

The second example contains a sandwich-type result which makes the connection of the original results in this paper with the results previously obtained by S. Kanas [3]. The examples are useful by inspiring researchers in using the theoretical results contained in the theorem and corollaries for further studies on the subject.

Funding: This research received no external funding.

Conflicts of Interest: The author declares no conflict of interest.

References

1. Miller, S.S.; Mocanu, P.T. Subordinants of differential superordinations. *Complex Var.* **2003**, *48*, 815–826. [CrossRef]
2. Miller, S.S.; Mocanu, P.T. Briot-Bouquet differential superordinations and sandwich theorems. *J. Math. Anal. Appl.* **2007**, *329*, 327–335. [CrossRef]
3. Kanas, S. Differential subordinations for harmonic complex-valued functions. *arXiv* **2015**, arXiv:1509.037511V1.
4. Bulboacă, T. *Differential Subordinations and Superordinations. Recent Results*; Casa Cărții de Știință: Cluj-Napoca, Romania, 2005.
5. Oros, G.I.; Oros, G. Differential superordination for harmonic complex-valued functions. *Stud. Univ. Babeș-Bolyai Math.* **2019**, *64*, 487–496. [CrossRef]
6. Mocanu, P.T.; Bulboacă, T.; Sălăgean, S.G. *Geometric Function Theory*; Casa Cărții de Știință: Cluj-Napoca, Romania, 1999.

Publisher's Note: MDPI stays neutral with regard to jurisdictional claims in published maps and institutional affiliations.

© 2020 by the author. Licensee MDPI, Basel, Switzerland. This article is an open access article distributed under the terms and conditions of the Creative Commons Attribution (CC BY) license (http://creativecommons.org/licenses/by/4.0/).

Article

Coefficient Related Studies for New Classes of Bi-Univalent Functions

Ágnes Orsolya Páll-Szabó [1,†] and Georgia Irina Oros [2,*,†]

1. Faculty of Mathematics and Computer Science, Babes-Bolyai University, 400084 Cluj Napoca, Romania; pallszaboagnes@math.ubbcluj.ro
2. Department of Mathematics and Computer Sciences, Faculty of Informatics and Sciences, University of Oradea, 410087 Oradea, Romania
* Correspondence: georgia_oros_ro@yahoo.co.uk
† These authors contributed equally to this work.

Received: 4 June 2020; Accepted: 6 July 2020; Published: 6 July 2020

Abstract: Using the recently introduced Sălăgean integro-differential operator, three new classes of bi-univalent functions are introduced in this paper. In the study of bi-univalent functions, estimates on the first two Taylor–Maclaurin coefficients are usually given. We go further in the present paper and bounds of the first three coefficients $|a_2|$, $|a_3|$ and $|a_4|$ of the functions in the newly defined classes are given. Obtaining Fekete–Szegő inequalities for different classes of functions is a topic of interest at this time as it will be shown later by citing recent papers. So, continuing the study on the coefficients of those classes, the well-known Fekete–Szegő functional is obtained for each of the three classes.

Keywords: bi-univalent functions; Sălăgean integral and differential operator; coefficient bounds; Fekete–Szegő problem

MSC: 30C45; 30C50

1. Introduction

Let $\mathcal{A}$ denote the class of functions of the form:

$$f(z) = z + \sum_{k=2}^{\infty} a_k z^k, \qquad (1)$$

which are analytic in the open unit disk $U = \{z \in \mathbb{C} : |z| < 1\}$ and normalized by the conditions $f(0) = 0, f'(0) = 1$. Let $\mathcal{S} \subset \mathcal{A}$ denote the class of all functions in $\mathcal{A}$ which are univalent in U.

The Koebe One-Quarter Theorem [1] ensures that the image of the unit disk under every $f \in \mathcal{S}$ function contains a disk of radius $\frac{1}{4}$. Thus every univalent function f has an inverse f^{-1}, which is defined by

$$f^{-1}(f(z)) = z, \quad (z \in U),$$

and

$$f(f^{-1}(w)) = w, \quad \left(|w| < r_0(f);\ r_0(f) \geq \frac{1}{4}\right),$$

where

$$g(w) = f^{-1}(w) = w - a_2 w^2 + (2a_2^2 - a_3) w^3 - (5a_2^3 - 5a_2 a_3 + a_4) w^4 + \cdots. \qquad (2)$$

A function $f \in \mathcal{A}$ is said to be bi-univalent in U if $U \subset f(U)$ and if both f and f^{-1} are univalent in U. Let Σ denote the class of bi-univalent functions in U given by (1).

Studying the class of bi-univalent functions begun some time ago, around the year 1970 as it can be seen from papers [2–4]. The topic resurfaced as interesting in the last decade, many papers being published since 2011, for example, [5,6]. Interesting results related to coefficient estimates for certain special classes of univalent functions appeared like the ones published in [7–13].

The operators have been used ever since the beginning of the study of complex functions. Many known results have been proved easier by using them and new results could be obtained especially related to starlikeness and convexity of certain functions. Introducing new classes of analytic functions is the most common outcome of the study that involves operators.

The study of bi-univalent functions using operators is also an approach that is in trend nowadays as it can be seen in the very recent results from papers [14,15] and a particular interest is shown to obtaining the Fekete–Szegő functional for the special classes that are being introduced as it can be seen in the very recent paper [16].

The study on coefficients of the functions in certain special classes is a topic that has its origin at the very beginning of the study of univalent functions. A main result in the theory of univalent functions is Gronwall's Area Theorem stated in 1914 and used for obtaining bounds on the coefficients of the class of meromorphic functions. An analogous problem for the class S was solved by Bieberbach and its famous conjecture stated in 1916, only proven in 1984, has stimulated the development of different methods in the geometric theory of functions of a complex variable. Just as in the case of the classes studied by Gronwall and Bieberbach, in the study of bi-univalent functions, estimates on the first two Taylor–Maclaurin coefficients are usually given. We extend the study and manage to give estimates on the fourth coefficient too, concerning the functions in the classes introduced in the present paper.

Another aspect of the novelty of the results contained in the present paper is given by the operator used in defining the three new classes for which coefficient estimates are obtained. The operator was previously defined in the paper [17] as a new type of operator introduced by mixing the two forms of the well-known Sălăgean operator, its differential and integral forms.

Definition 1. *[18] For $f \in \mathcal{A}$, $n \in \mathbb{N}_0 = \mathbb{N} \cup \{0\} = \{0, 1, 2, \ldots\}$, the Sălăgean differential operator $\mathcal{D}^n$ is defined by*

$$\mathcal{D}^n : \mathcal{A} \to \mathcal{A},$$

$$\mathcal{D}^0 f(z) = f(z),$$

$$\mathcal{D}^{n+1} f(z) = z \left(\mathcal{D}^n f(z) \right)', z \in \mathcal{U}.$$

Remark 1. *If $f \in \mathcal{A}$ and $f(z) = z + \sum_{k=2}^{\infty} a_k z^k$, then*

$$\mathcal{D}^n f(z) = z + \sum_{k=2}^{\infty} k^n a_k z^k, z \in \mathcal{U}.$$

Definition 2. *[18] For $f \in \mathcal{A}, n \in \mathbb{N}_0$, the Sălăgean integral operator I^n is defined by*

$$I^0 f(z) = f(z),$$

$$I^1 f(z) = I f(z) = \int_0^z f(t) \, t^{-1} dt, \ldots,$$

$$I^{n+1} f(z) = I \left(I^n f(z) \right), z \in \mathcal{U}.$$

The I^1 is the Alexander operator used for the first time in [19], the I^n operator is called the generalized Alexander operator.

Remark 2. If $f \in \mathcal{A}$ and $f(z) = z + \sum_{k=2}^{\infty} a_k z^k$, then

$$I^n f(z) = z + \sum_{k=2}^{\infty} \frac{a_k}{k^n} z^k, \tag{3}$$

$z \in U$, $(n \in \mathbb{N}_0)$ and $z \left(I^{n+1} f(z)\right)' = I^n f(z)$.

Remark 3. We have $\mathcal{D}^n I^n f(z) = I^n \mathcal{D}^n f(z) = f(z)$, $f \in \mathcal{A}$, $z \in U$.

Definition 3. [17] Let $\tilde{\zeta} \geq 0, n \in \mathbb{N}_0$. Denote by $\mathcal{D}I^n$ the operator given by

$$\mathcal{D}I^n : \mathcal{A} \to \mathcal{A},$$

$$\mathcal{D}I^n f(z) = \left(1 - \tilde{\zeta}\right) \mathcal{D}^n f(z) + \tilde{\zeta} I^n f(z), z \in U.$$

Remark 4. [17] If $f \in \mathcal{A}$ and $f(z) = z + \sum_{k=2}^{\infty} a_k z^k$, then

$$\mathcal{D}I^n f(z) = z + \sum_{k=2}^{\infty} \left[k^n \left(1 - \tilde{\zeta}\right) + \tilde{\zeta} \frac{1}{k^n}\right] a_k z^k = z + \sum_{k=2}^{\infty} \Gamma_k a_k z^k, z \in U, \tag{4}$$

where $\Gamma_k = k^n \left(1 - \tilde{\zeta}\right) + \tilde{\zeta} \frac{1}{k^n}$, $k \geq 2$.

This generalized operator is the linear combination of the Sălăgean differential and Sălăgean integral operator.

In 1933, Fekete and Szegő [20] proved that

$$\left|a_3 - \mu a_2^2\right| \leq \begin{cases} 4\mu - 3, & \mu \geq 1, \\ 1 + 2\exp\left[\frac{-2\mu}{1-\mu}\right], & 0 \leq \mu < 1, \\ 3 - 4\mu, & \mu < 0, \mu \in \mathbb{R}. \end{cases}$$

holds for any normalized univalent function and the result is sharp. The problem of maximizing the absolute value of the functional $|a_3 - \mu a_2^2|$ is called the Fekete–Szegő problem. Many authors obtained Fekete–Szegő inequalities for different classes of functions: [21–23].

In order to prove the original results from the main results part of the paper, the following lemmas are used:

We denote by $\mathcal{P}$ the class of Carathéodory functions analytic in the open unit disk U, for example,

$$\mathcal{P} = \{f \in \mathcal{A} | \ f(0) = 1, \ \Re f(z) > 0, \ z \in U\}.$$

Lemma 1. [24] If $h \in \mathcal{P}$ then $|c_k| \leq 2$, $\forall k$, where $h(z) = 1 + c_1 z + c_2 z^2 + \cdots$ for $z \in U$.

Lemma 2. [1] Let $p \in \mathcal{P}$ be of the form $p(z) = 1 + c_1 z + c_2 z^2 + \ldots$ then

$$\left|c_2 - \frac{c_1^2}{2}\right| \leq 2 - \frac{|c_1|^2}{2} \text{ and } |c_k| \leq 2, \ \forall k \in \mathbb{N}.$$

Lemma 3. [25] If $p(z) = 1 + c_1 z + c_2 z^2 + \ldots, z \in U$ is a function with positive real part in U and μ is a complex number, then

$$\left|c_2 - \mu c_1^2\right| \leq 2 \max\{1; |2\mu - 1|\}.$$

The result is sharp for the function given by

$$p(z) = \frac{1+z^2}{1-z^2} \text{ and } p(z) = \frac{1+z}{1-z}, \; z \in U.$$

2. Main Results

Using the operator shown in Definition 3, we introduce three new classes as follows:

Definition 4. *For $0 < \alpha \leq 1, 0 \leq \lambda \leq 1$ a function $f(z)$ given by (1) is said to be in the class $\mathcal{P}_\Sigma^\alpha(\lambda)$ if the following conditions are satisfied:*

$$\left| \arg \left(\frac{z \left(\mathscr{D}I^n f(z) \right)' + \lambda z^2 \left(\mathscr{D}I^n f(z) \right)''}{(1-\lambda) \mathscr{D}I^n f(z) + \lambda z \left(\mathscr{D}I^n f(z) \right)'} \right) \right| < \frac{\alpha \pi}{2} \tag{5}$$

and

$$\left| \arg \left(\frac{w \left(\mathscr{D}I^n g(w) \right)' + \lambda w^2 \left(\mathscr{D}I^n g(w) \right)''}{(1-\lambda) \mathscr{D}I^n g(w) + \lambda w \left(\mathscr{D}I^n g(w) \right)'} \right) \right| < \frac{\alpha \pi}{2}, \tag{6}$$

where $z, w \in U$ and the function g is given by (2).

Example 1. *If $\lambda = n = 0$ we have the well-known class of strongly bi-starlike functions of order α:*

$$\left| \arg \frac{z \left(f(z) \right)'}{f(z)} \right| < \frac{\alpha \pi}{2}, \quad \left| \arg \frac{w \left(g(w) \right)'}{g(w)} \right| < \frac{\alpha \pi}{2}, \quad 0 < \alpha \leq 1.$$

Example 2. *If $\lambda = 1$ and $n = 0$ we have the class of strongly bi-convex functions of order α:*

$$\left| \arg \left(1 + \frac{z \left(f(z) \right)''}{\left(f(z) \right)'} \right) \right| < \frac{\alpha \pi}{2}, \quad \left| \arg \left(1 + \frac{w \left(g(w) \right)''}{\left(g(w) \right)'} \right) \right| < \frac{\alpha \pi}{2}, \quad 0 < \alpha \leq 1.$$

Definition 5. *For $0 \leq \beta < 1, 0 \leq \lambda \leq 1$ a function $f(z)$ given by (1) is said to be in the class $\mathcal{Q}_\Sigma^\beta(\lambda)$ if the following conditions are satisfied:*

$$\Re \left(\frac{z \left(\mathscr{D}I^n f(z) \right)' + \lambda z^2 \left(\mathscr{D}I^n f(z) \right)''}{(1-\lambda) \mathscr{D}I^n f(z) + \lambda z \left(\mathscr{D}I^n f(z) \right)'} \right) > \beta \tag{7}$$

and

$$\Re \left(\frac{w \left(\mathscr{D}I^n g(w) \right)' + \lambda w^2 \left(\mathscr{D}I^n g(w) \right)''}{(1-\lambda) \mathscr{D}I^n g(w) + \lambda w \left(\mathscr{D}I^n g(w) \right)'} \right) > \beta, \tag{8}$$

where $z, w \in U$ and the function g is given by (2).

Example 3. *If $\lambda = n = 0$ we have the well-known class of bi-starlike functions of order β:*

$$\Re \left(\frac{z \left(f(z) \right)'}{f(z)} \right) > \beta, \quad \Re \left(\frac{w \left(g(w) \right)'}{g(w)} \right) > \beta, \quad 0 \leq \beta < 1.$$

Example 4. *If $\lambda = 1$ and $n = 0$ we have the class of bi-convex functions of order β:*

$$\Re \left(1 + \frac{z \left(f(z) \right)''}{\left(f(z) \right)'} \right) > \beta, \quad \Re \left(1 + \frac{w \left(g(w) \right)''}{\left(g(w) \right)'} \right) > \beta, \quad 0 \leq \beta < 1.$$

Definition 6. Let $h, l : U \to \mathbb{C}$ be analytic functions and

$$\min\{\Re(h(z)), \Re(l(z))\} > 0, \quad (z \in U) \quad h(0) = l(0) = 1.$$

A function $f(z)$ given by (1) is said to be in the class $\mathcal{P}_\Sigma^{h,l}$ if the following conditions are satisfied:

$$\frac{z(\mathcal{D}I^n f(z))' + \lambda z^2 (\mathcal{D}I^n f(z))''}{(1-\lambda)\mathcal{D}I^n f(z) + \lambda z (\mathcal{D}I^n f(z))'} \in h(U) \qquad (9)$$

and

$$\frac{w(\mathcal{D}I^n g(w))' + \lambda w^2 (\mathcal{D}I^n g(w))''}{(1-\lambda)\mathcal{D}I^n g(w) + \lambda w (\mathcal{D}I^n g(w))'} \in l(U), \qquad (10)$$

where $z, w \in U$ and the function g is given by (2).

Remark 5. If we let $h(z) = \left(\frac{1+z}{1-z}\right)^\alpha$ and $l(z) = \left(\frac{1-z}{1+z}\right)^\alpha$, $0 < \alpha \leq 1$ then the class $\mathcal{P}_\Sigma^{h,l}$ reduces to the class denoted by $\mathcal{P}_\Sigma^\alpha(\lambda)$.

Remark 6. If we let $h(z) = \frac{1+(1-2\beta)z}{1-z}$ and $l(z) = \frac{1-(1-2\beta)z}{1+z}$, $0 \leq \beta < 1$ then the class $\mathcal{P}_\Sigma^{h,l}$ reduces to the class denoted by $\mathcal{Q}_\Sigma^\beta(\lambda)$.

Remark 7. The classes introduced in this paper are defined in the classical way. All subclasses of bi-univalent functions are defined, the connection with the classes of bi-starlike and bi-convex functions being illustrated in the examples above. Being defined using relations related to arguments and real part of the functions contained, a geometric interpretation could be given for the classes. For the class in Definition 4, the geometrical image is in the first trigonometric dial, the section between two lines that converge at the origin having its maximum image the entire dial. The class in Definition 5 has its image in the half right plane. The first two classes defined are connected through the relation obtained for $\alpha = 1$ and $\beta = 0$, $\mathcal{P}_\Sigma^1(\lambda) = \mathcal{Q}_\Sigma^0(\lambda)$. The results for the class of functions $\mathcal{P}_\Sigma^{h,l}$ would generalize and improve the results for the classes of functions from Definitions 4 and 5. For special uses of parameters, new conditions for bi-starlikeness and bi-convexity could be established. Future interpretations are left to the imagination of interested researchers.

3. Coefficient Estimates

First, we give the coefficient estimates for the class $\mathcal{P}_\Sigma^\alpha(\lambda)$ given in Definition 4.

Theorem 1. Let $0 < \alpha \leq 1$, $0 \leq \lambda \leq 1$ and let $f(z)$ given by (1) be in the class $\mathcal{P}_\Sigma^\alpha(\lambda)$. Then

$$|a_2| \leq \frac{2\alpha}{\sqrt{\left|4\alpha\Gamma_3(1+2\lambda) + \Gamma_2^2(1+\lambda)^2(1-3\alpha)\right|}}, \qquad (11)$$

$$|a_3| \leq \frac{\alpha}{\Gamma_3(1+2\lambda)} + \frac{4\alpha^2}{\Gamma_2^2(1+\lambda)^2} \qquad (12)$$

and

$$|a_4| \leq \left| \frac{2\alpha(2\alpha^2+1)}{9\Gamma_4(1+3\lambda)} - \frac{10\alpha(2\alpha-1)}{3[2\Gamma_2\Gamma_3(1+\lambda)(1+2\lambda) - 5\Gamma_4(1+3\lambda)]} \right. \\ \left. + \frac{8\alpha^3\Gamma_2(1+\lambda)[3(1+2\lambda)\Gamma_3 - (1+\lambda)^2\Gamma_2^2]}{3\Gamma_4(1+3\lambda)[4\alpha\Gamma_3(1+2\lambda) + \Gamma_2^2(1+\lambda)^2(1-3\alpha)]\sqrt{\left|4\alpha\Gamma_3(1+2\lambda) + \Gamma_2^2(1+\lambda)^2(1-3\alpha)\right|}} \right|, \qquad (13)$$

where Γ_k, $k \geq 2$ are defined in (4).

Proof. It follows from (5) and (6) that

$$\frac{z\left(\mathscr{D}I^n f(z)\right)' + \lambda z^2 \left(\mathscr{D}I^n f(z)\right)''}{(1-\lambda)\mathscr{D}I^n f(z) + \lambda z \left(\mathscr{D}I^n f(z)\right)'} = [p(z)]^\alpha \tag{14}$$

and

$$\frac{w\left(\mathscr{D}I^n g(w)\right)' + \lambda w^2 \left(\mathscr{D}I^n g(w)\right)''}{(1-\lambda)\mathscr{D}I^n g(w) + \lambda w \left(\mathscr{D}I^n g(w)\right)'} = [q(z)]^\alpha, \tag{15}$$

where $p(z)$ and $q(w)$ are in $\mathcal{P}$ and have the forms

$$p(z) = 1 + p_1 z + p_2 z^2 + \cdots \tag{16}$$

and

$$q(w) = 1 + q_1 w + q_2 w^2 + \cdots. \tag{17}$$

Equating the coefficients in (14) and (15), we get

$$(1+\lambda)\Gamma_2 a_2 = \alpha p_1 \tag{18}$$

$$2(1+2\lambda)\Gamma_3 a_3 - \Gamma_2^2 a_2^2 (1+\lambda)^2 = \frac{1}{2}\left[\alpha(\alpha-1)p_1^2 + 2\alpha p_2\right] \tag{19}$$

$$3\Gamma_4 a_4 (1+3\lambda) - 3\Gamma_2 \Gamma_3 a_2 a_3 (1+\lambda)(1+2\lambda) + \Gamma_2^3 a_2^3 (1+\lambda)^3 =$$

$$= \frac{1}{6}p_1^3 (\alpha-2)(\alpha-1)\alpha + p_1 p_2 (\alpha-1)\alpha + p_3 \alpha \tag{20}$$

$$-(1+\lambda)\Gamma_2 a_2 = \alpha q_1 \tag{21}$$

$$2(1+2\lambda)\Gamma_3\left(2a_2^2 - a_3\right) - \Gamma_2^2 a_2^2 (1+\lambda)^2 = \frac{1}{2}\left[\alpha(\alpha-1)q_1^2 + 2\alpha q_2\right] \tag{22}$$

$$-3\Gamma_4\left(5a_2^3 - 5a_2 a_3 + a_4\right)(1+3\lambda) + 3\Gamma_2\Gamma_3 a_2\left(2a_2^2 - a_3\right)(1+\lambda)(1+2\lambda) -$$

$$-\Gamma_2^3 a_2^3 (1+\lambda)^3 = \frac{1}{6}q_1^3 (\alpha-2)(\alpha-1)\alpha + q_1 q_2 (\alpha-1)\alpha + q_3 \alpha. \tag{23}$$

From (18) and (21), we get

$$p_1 = -q_1 \tag{24}$$

and

$$2\Gamma_2^2 a_2^2 (1+\lambda)^2 = \alpha^2 \left(p_1^2 + q_1^2\right). \tag{25}$$

From (19), (22) and (25), we obtain

$$a_2^2 = \frac{\alpha^2 (p_2 + q_2)}{4\alpha\Gamma_3 (1+2\lambda) + \Gamma_2^2 (1+\lambda)^2 (1-3\alpha)}.$$

Applying Lemma 1 for the coefficients p_2 and q_2, we get (11).
To find the bound on $|a_3|$, first we substract (22) from (19):

$$4a_3\Gamma_3 (1+2\lambda) - 4\Gamma_3 (1+2\lambda) a_2^2 = \alpha(p_2 - q_2) + \frac{\alpha(\alpha-1)}{2}\left(p_1^2 - q_1^2\right). \tag{26}$$

From (24), (25) and (26) follows that

$$a_3 = \frac{\alpha(p_2 - q_2)}{4\Gamma_3 (1+2\lambda)} + \frac{\alpha^2 \left(p_1^2 + q_1^2\right)}{2\Gamma_2^2 (1+\lambda)^2}, \tag{27}$$

and applying Lemma 1 we get (12).
To find the bound on $|a_4|$, first we substract (23) from (20) and using (24) we get

$$6(1+3\lambda)\Gamma_4 a_4 + 15\Gamma_4(1+3\lambda)a_2\left(a_2^2 - a_3\right) - 6\Gamma_2\Gamma_3 a_2^3(1+\lambda)(1+2\lambda) +$$

$$+ 2\Gamma_2^3 a_2^3(1+\lambda)^3 = \frac{1}{3}p_1^3(\alpha-2)(\alpha-1)\alpha + p_1\alpha(\alpha-1)(p_2+q_2) + \alpha(p_3-q_3). \qquad (28)$$

Now we add (20) and (23) and using (24) we get

$$-15\Gamma_4(1+3\lambda)a_2\left(a_2^2 - a_3\right) + 6\Gamma_2\Gamma_3(1+\lambda)(1+2\lambda)a_2\left(a_2^2 - a_3\right) =$$

$$= p_1\alpha(\alpha-1)(p_2 - q_2) + \alpha(p_3 + q_3),$$

or equivalently

$$a_2\left(a_2^2 - a_3\right) = \frac{p_1\alpha(\alpha-1)(p_2-q_2) + \alpha(p_3+q_3)}{3\left[2\Gamma_2\Gamma_3(1+\lambda)(1+2\lambda) - 5\Gamma_4(1+3\lambda)\right]}. \qquad (29)$$

Substituting (29) in (28) and applying Lemma 1 we get (13). □

Now we calculate the Fekete–Szegő functional for the the class $\mathcal{P}_{\Sigma}^{\alpha}(\lambda)$.

Theorem 2. *Let f of the form (1) be in the class $\mathcal{P}_{\Sigma}^{\alpha}(\lambda)$. Then*

$$\left|a_3 - \tilde{\xi}a_2^2\right| \le \begin{cases} \frac{\alpha}{\Gamma_3(1+2\lambda)}; & \left|\frac{\alpha(1-\tilde{\xi})}{4\alpha\Gamma_3(1+2\lambda)+\Gamma_2^2(1+\lambda)^2(1-3\alpha)}\right| \le \frac{1}{4\Gamma_3(1+2\lambda)}, \\ \left|\frac{4\alpha^2(1-\tilde{\xi})}{4\alpha\Gamma_3(1+2\lambda)+\Gamma_2^2(1+\lambda)^2(1-3\alpha)}\right|; & \left|\frac{\alpha(1-\tilde{\xi})}{4\alpha\Gamma_3(1+2\lambda)+\Gamma_2^2(1+\lambda)^2(1-3\alpha)}\right| \ge \frac{1}{4\Gamma_3(1+2\lambda)}. \end{cases}$$

Proof. From Theorem 1 we use the value of a_2^2 and a_3 to calculate $a_3 - \tilde{\xi}a_2^2$.

$$a_3 - \tilde{\xi}a_2^2 = \alpha\left[p_2\left(h\left(\tilde{\xi}\right) + \frac{1}{4\Gamma_3(1+2\lambda)}\right) + q_2\left(h\left(\tilde{\xi}\right) - \frac{1}{4\Gamma_3(1+2\lambda)}\right)\right],$$

where $h\left(\tilde{\xi}\right) = \left(1-\tilde{\xi}\right)\dfrac{\alpha}{4\alpha\Gamma_3(1+2\lambda) + \Gamma_2^2(1+\lambda)^2(1-3\alpha)}.$

Then

$$\left|a_3 - \tilde{\xi}a_2^2\right| \le \begin{cases} \frac{\alpha}{\Gamma_3(1+2\lambda)}; & \left|h\left(\tilde{\xi}\right)\right| \le \frac{1}{4\Gamma_3(1+2\lambda)}, \\ 4\alpha\left|h\left(\tilde{\xi}\right)\right|; & \left|h\left(\tilde{\xi}\right)\right| \ge \frac{1}{4\Gamma_3(1+2\lambda)}. \end{cases}$$

□

Theorem 3. *Let $0 \le \beta < 1, 0 \le \lambda \le 1$ and let $f(z)$ given by (1) be in the class $\mathcal{Q}_{\Sigma}^{\beta}(\lambda)$. Then*

$$|a_2| \le \sqrt{\frac{2(1-\beta)}{\left|2(1+2\lambda)\Gamma_3 - \Gamma_2^2(1+\lambda)^2\right|}}, \qquad (30)$$

$$|a_3| \le \frac{1-\beta}{\Gamma_3(1+2\lambda)} + \frac{4(1-\beta)^2}{\Gamma_2^2(1+\lambda)^2}. \qquad (31)$$

and

$$|a_4| \leq \frac{2(1-\beta)}{3\Gamma_4(1+3\lambda)} - \frac{10(1-\beta)}{3[2\Gamma_2\Gamma_3(1+\lambda)(1+2\lambda) - 5\Gamma_4(1+3\lambda)]} +$$
$$+ \frac{4\Gamma_2(1-\beta)(1+\lambda)[3\Gamma_3(1+2\lambda) - \Gamma_2^2(1+\lambda)^2]}{3\Gamma_4(1+3\lambda)[2\Gamma_3(1+2\lambda) - \Gamma_2^2(1+\lambda)^2]} \sqrt{\frac{1-\beta}{|4(1+2\lambda)\Gamma_3 - 2\Gamma_2^2(1+\lambda)^2|}}, \quad (32)$$

where Γ_k, $k \geq 2$ are defined in (4).

Proof. It follows from (5) and (6) that

$$\frac{z\left(\mathscr{D}I^n f(z)\right)' + \lambda z^2 \left(\mathscr{D}I^n f(z)\right)''}{(1-\lambda)\mathscr{D}I^n f(z) + \lambda z \left(\mathscr{D}I^n f(z)\right)'} = \beta + (1-\beta)p(z) \quad (33)$$

and

$$\frac{w\left(\mathscr{D}I^n g(w)\right)' + \lambda w^2 \left(\mathscr{D}I^n g(w)\right)''}{(1-\lambda)\mathscr{D}I^n g(w) + \lambda w \left(\mathscr{D}I^n g(w)\right)'} = \beta + (1-\beta)q(w), \quad (34)$$

where $p(z)$ and $q(w)$ have the forms (16) and (17).

Equating the coefficients in (33) and (34), we get

$$(1+\lambda)\Gamma_2 a_2 = (1-\beta)p_1 \quad (35)$$

$$2(1+2\lambda)\Gamma_3 a_3 - \Gamma_2^2 a_2^2 (1+\lambda)^2 = (1-\beta)p_2 \quad (36)$$

$$3\Gamma_4 a_4 (1+3\lambda) - 3\Gamma_2\Gamma_3 a_2 a_3 (1+\lambda)(1+2\lambda) + \Gamma_2^3 a_2^3 (1+\lambda)^3 = (1-\beta)p_3 \quad (37)$$

$$-(1+\lambda)\Gamma_2 a_2 = (1-\beta)q_1 \quad (38)$$

$$2(1+2\lambda)\Gamma_3\left(2a_2^2 - a_3\right) - \Gamma_2^2 a_2^2 (1+\lambda)^2 = (1-\beta)q_2 \quad (39)$$

$$-3\Gamma_4\left(5a_2^3 - 5a_2 a_3 + a_4\right)(1+3\lambda) + 3\Gamma_2\Gamma_3 a_2 \left(2a_2^2 - a_3\right)(1+\lambda)(1+2\lambda) -$$
$$- \Gamma_2^3 a_2^3 (1+\lambda)^3 = (1-\beta)q_3. \quad (40)$$

From (35) and (38), we get

$$p_1 = -q_1 \quad (41)$$

and

$$2\Gamma_2^2 a_2^2 (1+\lambda)^2 = (1-\beta)^2\left(p_1^2 + q_1^2\right). \quad (42)$$

From (36) and (39), we obtain

$$a_2^2 = \frac{(1-\beta)(p_2 + q_2)}{4\Gamma_3(1+2\lambda) - 2\Gamma_2^2(1+\lambda)^2}.$$

Applying Lemma 1 for the coefficients p_2 and q_2, we get (30).
To find the bound on $|a_3|$, first we subtract (39) from (36):

$$4a_3\Gamma_3(1+2\lambda) - 4\Gamma_3(1+2\lambda)a_2^2 = (1-\beta)(p_2 - q_2). \quad (43)$$

From (42) and (43) follows that

$$a_3 = \frac{(1-\beta)(p_2 - q_2)}{4\Gamma_3(1+2\lambda)} + \frac{(1-\beta)^2\left(p_1^2 + q_1^2\right)}{2\Gamma_2^2(1+\lambda)^2}, \quad (44)$$

and applying Lemma 1 we get (31).

To find the bound on $|a_4|$, first we subtract (40) from (37) and using (41) we get

$$6(1+3\lambda)\Gamma_4 a_4 + 15\Gamma_4(1+3\lambda)a_2\left(a_2^2 - a_3\right) - 6\Gamma_2\Gamma_3 a_2^3(1+\lambda)(1+2\lambda) +$$
$$+ 2\Gamma_2^3 a_2^3(1+\lambda)^3 = (1-\beta)(p_3+q_3). \tag{45}$$

Now we add (37) and (40) and using (41) we get

$$-15\Gamma_4(1+3\lambda)a_2\left(a_2^2 - a_3\right) + 6\Gamma_2\Gamma_3(1+\lambda)(1+2\lambda)a_2\left(a_2^2 - a_3\right) =$$
$$= (1-\beta)(p_3 - q_3),$$

or equivalently

$$a_2\left(a_2^2 - a_3\right) = \frac{(1-\beta)(p_3+q_3)}{3\left[2\Gamma_2\Gamma_3(1+\lambda)(1+2\lambda) - 5\Gamma_4(1+3\lambda)\right]}. \tag{46}$$

Substituting (46) in (45) and applying Lemma 1 we get (32). □

Theorem 4. *Let f of the form (1) be in the class $\mathcal{Q}_\Sigma^\beta(\lambda)$. Then*

$$\left|a_3 - \tilde{\xi}a_2^2\right| \le \begin{cases} \dfrac{1-\beta}{\Gamma_3(1+2\lambda)}; & \left|\dfrac{1-\tilde{\xi}}{4\alpha\Gamma_3(1+2\lambda)-2\Gamma_2^2(1+\lambda)^2}\right| \le \dfrac{1}{4\Gamma_3(1+2\lambda)}, \\[2mm] \left|\dfrac{4(1-\beta)(1-\tilde{\xi})}{4\alpha\Gamma_3(1+2\lambda)-2\Gamma_2^2(1+\lambda)^2}\right|; & \left|\dfrac{1-\tilde{\xi}}{4\alpha\Gamma_3(1+2\lambda)-2\Gamma_2^2(1+\lambda)^2}\right| \ge \dfrac{1}{4\Gamma_3(1+2\lambda)}. \end{cases}$$

Proof. From Theorem 3 we use the value of a_2^2 and a_3 to calculate $a_3 - \tilde{\xi}a_2^2$.

$$a_3 - \tilde{\xi}a_2^2 = (1-\beta)\left[p_2\left(h\left(\tilde{\xi}\right) + \frac{1}{4\Gamma_3(1+2\lambda)}\right) + q_2\left(h\left(\tilde{\xi}\right) - \frac{1}{4\Gamma_3(1+2\lambda)}\right)\right],$$

where $h\left(\tilde{\xi}\right) = \left(1-\tilde{\xi}\right)\dfrac{1}{4\Gamma_3(1+2\lambda) - 2\Gamma_2^2(1+\lambda)^2}$.

Then

$$\left|a_3 - \tilde{\xi}a_2^2\right| \le \begin{cases} \dfrac{1-\beta}{\Gamma_3(1+2\lambda)}; & \left|h\left(\tilde{\xi}\right)\right| \le \dfrac{1}{4\Gamma_3(1+2\lambda)} \\ 4(1-\beta)\left|h\left(\tilde{\xi}\right)\right|; & \left|h\left(\tilde{\xi}\right)\right| \ge \dfrac{1}{4\Gamma_3(1+2\lambda)}. \end{cases}$$

□

Theorem 5. *Let $0 \le \lambda \le 1$ and let $f(z)$ given by (1) be in the class $\mathcal{P}_\Sigma^{h,l}$. Then*

$$|a_2| \le \min\left\{\sqrt{\frac{|h'(0)|^2 + |l'(0)|^2}{2\Gamma_2^2(1+\lambda)^2}}, \sqrt{\frac{|h''(0)| + |l''(0)|}{4\left|2\Gamma_3(1+2\lambda) - \Gamma_2^2(1+\lambda)^2\right|}}\right\} \tag{47}$$

$$|a_3| \le \min\left\{\frac{|h'(0)|^2+|l'(0)|^2}{2\Gamma_2^2(1+\lambda)^2} + \frac{|h''(0)|+|l''(0)|}{8\Gamma_3(1+2\lambda)},\right.$$
$$\left.\frac{|h''(0)|\left|4\Gamma_3(1+2\lambda)-\Gamma_2^2(1+\lambda)^2\right|+|l''(0)|\Gamma_2^2(1+\lambda)^2}{8\Gamma_3(1+2\lambda)\left|2\Gamma_3(1+2\lambda)-\Gamma_2^2(1+\lambda)^2\right|}\right\} \tag{48}$$

and

$$|a_4| \leq \min\left\{\frac{|h'''(0)|+|l'''(0)|}{36}\left|\frac{1}{\Gamma_4(1+3\lambda)} - \frac{5}{2\Gamma_2\Gamma_3(1+\lambda)(1+2\lambda)-5\Gamma_4(1+3\lambda)}\right| + \right.$$
$$+ \frac{|h'(0)|^2+|l'(0)|^2}{\Gamma_2^2(1+\lambda)^2}\sqrt{\frac{|h'(0)|^2+|l'(0)|^2}{2}} \frac{|3\Gamma_3(1+2\lambda)-\Gamma_2^2(1+\lambda)^2|}{6\Gamma_4(1+3\lambda)},$$
$$\frac{|h'''(0)|+|l'''(0)|}{36}\left|\frac{1}{\Gamma_4(1+3\lambda)} - \frac{5}{2\Gamma_2\Gamma_3(1+\lambda)(1+2\lambda)-5\Gamma_4(1+3\lambda)}\right| +$$
$$\left. \frac{|h''(0)|+|l''(0)|}{|2\Gamma_3(1+2\lambda)-\Gamma_2^2(1+\lambda)^2|}\sqrt{\frac{|h''(0)|+|l''(0)|}{|2\Gamma_3(1+2\lambda)-\Gamma_2^2(1+\lambda)^2|}} \frac{\Gamma_2(1+\lambda)|3\Gamma_3(1+2\lambda)-\Gamma_2^2(1+\lambda)^2|}{24\Gamma_4(1+3\lambda)}\right\}, \quad (49)$$

where Γ_k, $k \geq 2$ are defined in (4).

Proof. For a start, we write the equivalent forms of the argument inequalities in (9) and (10).

$$\frac{z(\mathscr{D}I^n f(z))' + \lambda z^2 (\mathscr{D}I^n f(z))''}{(1-\lambda)\mathscr{D}I^n f(z) + \lambda z (\mathscr{D}I^n f(z))'} = h(z) \quad (50)$$

and

$$\frac{w(\mathscr{D}I^n g(w))' + \lambda w^2 (\mathscr{D}I^n g(w))''}{(1-\lambda)\mathscr{D}I^n g(w) + \lambda w (\mathscr{D}I^n g(w))'} = l(w), \quad (51)$$

where $h(z)$ and $l(w)$ satisfy the conditions of Definition 6 and have the following Taylor–Maclaurin series expansions:

$$h(z) = 1 + h_1 z + h_2 z^2 + \cdots, \quad (52)$$

$$l(w) = 1 + l_1 w + l_2 w^2 + \cdots. \quad (53)$$

Substituting from (52) and (53) into (50) and (51), respectively, and equating the coefficients, we get

$$(1+\lambda)\Gamma_2 a_2 = h_1 \quad (54)$$

$$2(1+2\lambda)\Gamma_3 a_3 - \Gamma_2^2 a_2^2 (1+\lambda)^2 = h_2 \quad (55)$$

$$3\Gamma_4 a_4 (1+3\lambda) - 3\Gamma_2\Gamma_3 a_2 a_3 (1+\lambda)(1+2\lambda) + \Gamma_2^3 a_2^3 (1+\lambda)^3 = h_3 \quad (56)$$

$$-(1+\lambda)\Gamma_2 a_2 = l_1 \quad (57)$$

$$2(1+2\lambda)\Gamma_3 \left(2a_2^2 - a_3\right) - \Gamma_2^2 a_2^2 (1+\lambda)^2 = l_2 \quad (58)$$

$$-3\Gamma_4 \left(5a_2^3 - 5a_2 a_3 + a_4\right)(1+3\lambda) + 3\Gamma_2\Gamma_3 a_2 \left(2a_2^2 - a_3\right)(1+\lambda)(1+2\lambda) -$$
$$- \Gamma_2^3 a_2^3 (1+\lambda)^3 = l_3. \quad (59)$$

From (54) and (57), we get

$$h_1 = -l_1 \quad (60)$$

and

$$2\Gamma_2^2 a_2^2 (1+\lambda)^2 = h_1^2 + l_1^2. \quad (61)$$

Adding (55) and (58), we obtain

$$4\Gamma_3 a_2^2 (1+2\lambda) - 2\Gamma_2^2 a_2^2 (1+\lambda)^2 = h_2 + l_2. \quad (62)$$

Therefore, from (61) and (62), we get

$$a_2^2 = \frac{h_1^2 + l_1^2}{2\Gamma_2^2(1+\lambda)^2} \tag{63}$$

and

$$a_2^2 = \frac{h_2 + l_2}{2\left[2\Gamma_3(1+2\lambda) - \Gamma_2^2(1+\lambda)^2\right]}. \tag{64}$$

We find from (63) and (64) that

$$|a_2|^2 \leq \frac{|h'(0)|^2 + |l'(0)|^2}{2\Gamma_2^2(1+\lambda)^2}$$

and

$$|a_2|^2 \leq \frac{|h''(0)| + |l''(0)|}{4\left[2\Gamma_3(1+2\lambda) - \Gamma_2^2(1+\lambda)^2\right]}.$$

So we get the desired estimate on the coefficient $|a_2|$ as asserted in (47).
Next, in order to find the bound on the coefficient $|a_3|$, by substracting (58) from (55), we get

$$4\Gamma_3(1+2\lambda)a_3 - 4\Gamma_3(1+2\lambda)a_2^2 = h_2 - l_2. \tag{65}$$

Substituting the value of a_2^2 from (63) into (65), it follows that

$$a_3 = \frac{h_2 - l_2}{4\Gamma_3(1+2\lambda)} + \frac{h_1^2 + l_1^2}{2\Gamma_2^2(1+\lambda)^2}.$$

So

$$|a_3| \leq \frac{|h'(0)|^2 + |l'(0)|^2}{2\Gamma_2^2(1+\lambda)^2} + \frac{|h''(0)| + |l''(0)|}{8\Gamma_3(1+2\lambda)}.$$

On the other hand, upon substituting the value of a_2^2 from (64) into (65), it follows that

$$a_3 = \frac{h_2\left[4\Gamma_3(1+2\lambda) - \Gamma_2^2(1+\lambda)^2\right] + l_2\Gamma_2^2(1+\lambda)^2}{4\Gamma_3(1+2\lambda)\left[2\Gamma_3(1+2\lambda) - \Gamma_2^2(1+\lambda)^2\right]}.$$

Consequently, we have

$$|a_3| \leq \frac{|h''(0)|\left|4\Gamma_3(1+2\lambda) - \Gamma_2^2(1+\lambda)^2\right| + |l''(0)|\Gamma_2^2(1+\lambda)^2}{8\Gamma_3(1+2\lambda)\left|2\Gamma_3(1+2\lambda) - \Gamma_2^2(1+\lambda)^2\right|}.$$

To find the bound on $|a_4|$, first we add (56) and (59) and using (60) we get

$$a_2\left(a_2^2 - a_3\right) = \frac{h_3 + l_3}{3\left[2\Gamma_2\Gamma_3(1+\lambda)(1+2\lambda) - 5\Gamma_4(1+3\lambda)\right]}. \tag{66}$$

Now we substract (59) from (56) and using (60) the result is

$$6(1+3\lambda)\Gamma_4 a_4 + 15\Gamma_4(1+3\lambda)a_2\left(a_2^2 - a_3\right) - 6\Gamma_2\Gamma_3 a_2^2(1+\lambda)(1+2\lambda) + 2\Gamma_2^3 a_2^3(1+\lambda)^3 = h_3 - l_3,$$

if we substitute (66) we have

$$a_4 = \frac{h_3 - l_3}{6\Gamma_4(1+3\lambda)} - \frac{5(h_3 + l_3)}{6[2\Gamma_2\Gamma_3(1+\lambda)(1+2\lambda) - 5\Gamma_4(1+3\lambda)]} +$$

$$+ 2\Gamma_2(1+\lambda)\left[3\Gamma_3(1+2\lambda) - \Gamma_2^2(1+\lambda)^2\right]\frac{1}{6\Gamma_4(1+3\lambda)} \cdot a_2^3. \qquad (67)$$

Finally, if we use (63) then (64) in (67) the result is (49). □

4. Conclusions

The original results of this paper are about coefficient estimates given the three original classes defined here. The classes are defined in the paper using an interesting new type of integro-differential operator, Sălăgean integro-differential operator. Since the only study done on them was related to coefficient estimates, they could be of particular interest for further studies related to different other aspects.

As it can be seen in Examples 1–4, for certain use of the parameters of the class given in Definition 4, strongly bi-starlikeness and strongly bi-convexity is proven. Similar studies related to starlikeness, convexity, and close-to-convexity of all the classes defined in the paper using values for the parameters can be conducted. With these studies, more could be found out about an intuitive or high level interpretation of the three function classes defined.

With the introduction of Definitions 1 and 2, it is worth investigating the possibility of applying the Lie algebra method in the work [26] to the complex plane. In the present paper, estimates for coefficient $|a_4|$ are given going further than estimates for coefficients $|a_2|$ and $|a_3|$ which are usually obtained in the study of bi-univalent functions.

It remains an open problem to obtain estimates on bound of $|a_n|, (n \in \mathbb{R} - \{1,2,3,4\})$ for the classes that have been introduced here. Particular uses of coefficient estimates could lead to potentially interesting new results. The results from this paper could also inspire further research related to integro-differential operators used for introducing new classes of bi-univalent functions.

Author Contributions: These authors contributed equally to this work. All authors have read and agreed to the published version of the manuscript.

Funding: This research received no external funding.

Conflicts of Interest: The authors declare no conflict of interest.

References

1. Duren, P.L. *Univalent Functions, Grundlehren der Mathematischen Wissenschaften*; Springer: New York, NY, USA; Berlin/Hiedelberg, Germany; Tokyo, Japan, 1983.
2. Brannan, D.A.; Clunie, J.; Kirwan, W.E. Coefficient estimates for a class of starlike functions. *Can. J. Math.* **1970**, *22*, 476–485. [CrossRef]
3. Lewin, M. On a coefficient problem for bi-univalent functions. *Proc. Am. Math. Soc.* **1967**, *18*, 63–68. [CrossRef]
4. Tan, D.L. Coefficient estimates for bi-univalent functions. *Chin. Ann. Math. Ser. A* **1984**, *5*, 559–568.
5. Babu, O.S.; Selvaraj, C.; Murugusundaramoorthy, G. Subclasses of bi-univalent functions based on Hohlov operator. *Int. J. Pure Appl. Math.* **2015**, *102*, 473–482. [CrossRef]
6. Frasin, B.A.; Aouf, M.K. New subclasses of bi-univalent functions. *Appl. Math. Lett.* **2011**, *24*, 1569–1573. [CrossRef]
7. Ali, R.M.; Lee, S.K.; Ravichandran, V.; Subramaniam, S. Coefficient estimates for bi-univalent Ma-Minda starlike and convex functions. *Appl. Math. Lett.* **2012**, *25*, 344–351. [CrossRef]
8. Altinkaya, S.; Yalcin, S. Faber polynomial coefficient bounds for a subclass of bi-univalent functions. *Comptes Rendus Math. Acad. Sci. Paris* **2015**, *353*, 1075–1080. [CrossRef]
9. Jahangiri, J.M.; Hamidi, S.G. Faber polynomial coefficient estimates for analytic bi-Bazilevič functions. *Mat. Vesnik* **2015**, *67*, 123–129.

10. Srivastava, H.M.; Mishra, A.K.; Gochhayat, P. Certain subclasses of analytic and bi-univalent functions. *Appl. Math. Lr.* **2010**, *23*, 1188–1192. [CrossRef]
11. Zireh, A.; Hajiparvaneh, S. Coefficient bounds for certain subclasses of analytic and bi-univalent functions. *Ann. Acad. Rom. Sci. Ser. Math. Appl.* **2016**, *8*, 133–144.
12. Xu, Q.H.; Gui, Y.-C.; Srivastava, H.M. Coefficient estimates for a certain subclass of analytic and bi-univalent functions. *Appl. Math. Lett.* **2012**, *25*, 990–994. [CrossRef]
13. Xu, Q.H.; Xiao, H.-G.; Srivastava, H.M. A certain general subclass of analytic and bi-univalent functions and associated coefficient estimate problems. *Appl. Math. Comput.* **2012**, *218*, 11461–11465. [CrossRef]
14. Aldawish, I.; Al-Hawary, T.; Frasin, B.A. Subclasses of Bi-Univalent Functions Defined by Frasin Differential Operator. *Mathematics* **2020**, *8*, 783. [CrossRef]
15. Srivastava, H.M.; Motamednezhad, A.; Adegan, E.A. Faber polynomial coefficient estimates for bi-univalent functions defined by using differential subordination and a certain fractional derivative operator. *Mathematics* **2020**, *8*, 172. [CrossRef]
16. El-Deeb, S.M.; Bulboacă, T.; El-Matary, B.M. Maclaurin Coefficient Estimates of Bi-Univalent Functions Connected with the q-Derivative. *Mathematics* **2020**, *8*, 418. [CrossRef]
17. Páll-Szabó, Á.O. On a class of univalent functions defined by Sălăgean integro-differential operator. *Miskolc Math. Notes* **2018**, *19*, 1095–1106. [CrossRef]
18. Sălăgean, G.S. Subclasses of univalent functions. *Lect. Notes Math.* **1983**, *1013*, 362–372.
19. Alexander, J.W. Functions which map the interior of the unit circle upon simple regions. *Ann. Math.* **1915**, *17*, 12–22. [CrossRef]
20. Fekete, M.; Szegő, G. Eine bemerkung über ungerade schlichte funktionen. *J. Lond. Math. Soc.* **1933**, *8*, 85–89. [CrossRef]
21. Bucur, R.; Andrei, L.; Breaz, D. Coefficient bounds and Fekete-Szegő problem for a class of analytic functions defined by using a new differential operator. *Appl. Math. Sci* **2015**, *9*, 1355–1368.
22. Dziok, J. A general solution of the Fekete-Szegő problem. *Bound Value Probl.* **2013**, *98*, 1–13. [CrossRef]
23. Kanas, S. An unified approach to the Fekete-Szegő problem. *Appl. Math. Comput.* **2012**, *218*, 8453–8461. [CrossRef]
24. Pommerenke, C. *Univalent Functions*; Vanderhoeck and Ruprecht: Göttingen, Germany, 1975.
25. Ma, W.; Minda, D. A unified treatment of some special classes of univalent functions. In Proceedings of the Conference on Complex Analysis 1992, Tianjin, China, 18–23 August 1992; pp. 157–169.
26. Shang, Y. A lie algebra approach to susceptible-infected-susceptible epidemics. *Electron. J. Differ. Equ.* **2012**, *2012*, 1–7.

© 2020 by the authors. Licensee MDPI, Basel, Switzerland. This article is an open access article distributed under the terms and conditions of the Creative Commons Attribution (CC BY) license (http://creativecommons.org/licenses/by/4.0/).

MDPI
St. Alban-Anlage 66
4052 Basel
Switzerland
Tel. +41 61 683 77 34
Fax +41 61 302 89 18
www.mdpi.com

Mathematics Editorial Office
E-mail: mathematics@mdpi.com
www.mdpi.com/journal/mathematics

www.ingramcontent.com/pod-product-compliance
Lightning Source LLC
LaVergne TN
LVHW070733100526
838202LV00013B/1223